Geographic Information Science and Systems

Geographic Information Science and Systems

Edited by **Marina De Lima**

CLANRYE INTERNATIONAL

New Jersey

Published by Clanrye International,
55 Van Reypen Street,
Jersey City, NJ 07306, USA
www.clanryeinternational.com

Geographic Information Science and Systems
Edited by Marina De Lima

International Standard Book Number: 978-1-63240-532-6 (Hardback)

Printed in China.

Contents

Preface

This book has been an outcome of determined endeavour from a group of educationists in the field. The primary objective was to involve a broad spectrum of professionals from diverse cultural background involved in the field for developing new researches. The book not only targets students but also scholars pursuing higher research for further enhancement of the theoretical and practical applications of the subject.

Geographic information science is a field of research that redefines geographic information by studying data and using different computational techniques. It gathers, interprets and analyses data to provide important and meaningful information. Geographic information systems on the other hand are the software used by this discipline. This book consists of various important and pivotal topics such as spatial analysis, data mining, data structures and algorithms, etc. This book delves into the techniques and technology employed in this field along with its latest applications. From theories to research to practical applications, case studies related to all contemporary topics of relevance to this discipline have been included in this book. It is a vital tool for all researching and studying geographic information science and systems.

It was an honour to edit such a profound book and also a challenging task to compile and examine all the relevant data for accuracy and originality. I wish to acknowledge the efforts of the contributors for submitting such brilliant and diverse chapters in the field and for endlessly working for the completion of the book. Last, but not the least; I thank my family for being a constant source of support in all my research endeavours.

Editor

1

Exploratory Testing of an Artificial Neural Network Classification for Enhancement of the Social Vulnerability Index

Ryan Hile [1,2,*] and Thomas J. Cova [1,2]

[1] Center for Natural and Technological Hazards, University of Utah, 260 S. Central Campus Dr. Rm. 270, Salt Lake City, UT 84112-9155, USA; E-Mail: cova@geog.utah.edu

[2] Department of Geography, University of Utah, 260 S. Central Campus Dr., Rm. 270, Salt Lake City, UT 84112-9155, USA

* Author to whom correspondence should be addressed; E-Mail: ryan.hile@geog.utah.edu

Academic Editors: Christoph Aubrecht and Wolfgang Kainz

Abstract: The Social Vulnerability Index (SoVI) has served the hazards community well for more than a decade. Using Utah as a test case, a state with a population exposed to a variety of hazards, this study sought to build upon the SoVI approach by augmenting it with a non-linear Artificial Neural Network (ANN). A SoVI was created for the state of Utah at the census block group level using five-year data (2008–2012) from the American Community Survey. The SoVI provided a dataset from which to train a neural network. The ANN was then used to classify a subset of the state to determine if it could provide a comparable classification of vulnerability. The ANN produced a vulnerability classification that was approximately 26% consistent with the SoVI created using the traditional approach. The differences in classifications were assessed using radar plots of block group variable averages to explore how the variables were handled in each classification. The results of this study warrant further investigation of the capabilities of an ANN-enhanced SoVI.

Keywords: artificial neural networks; social vulnerability; social vulnerability index; environmental hazards

1. Introduction

Contemporary research on the human dimensions of environmental hazards typically falls into three paradigms: (1) the exploration of social vulnerability that potentially contributes to enhanced disaster impact; (2) the exploration of disaster resilience, the capacity for communities and individuals to ameliorate and recover from the impact of a disaster; and (3) exploration of social perceptions that can enhance or mitigate the impact of a disaster [1–6]. While efforts in all three areas advance our understanding of the human component of environmental hazards and disasters, social vulnerability has the longest record of research in the environmental hazards discipline of the three paradigms [2].

Social vulnerability, as presented by Cutter [2,3], focuses on identification of vulnerability as it is constructed through social systems through exploration of demographic data to create a Social Vulnerability Index (SoVI). Exploration of this data to determine the factors that contribute to social vulnerability for an area, typically a U.S. Census enumeration unit, is conducted using Principal Components Analysis (PCA), which allows the data to drive factor selection for vulnerability. The loaded demographic variables in the PCA are identified to have specific representations of vulnerability based on our existing understanding of the role of social inequality on social vulnerability, and the PCA-identified factors are used to calculate the SoVI score for each enumeration unit. The SoVI has been demonstrated at a variety of scales, from the census block level to the county level across the United States [3,7] and at the county level in Norway [8], though the PCA approach has only been demonstrated at the county level [3,8].

While the SoVI appears to be the most ubiquitous social vulnerability assessment in the literature of the last decade and a half, it is not the only method by which vulnerability assessments can be made. Füssel [6] and Adger [9] have identified several approaches to vulnerability over the years, each with fundamental differences that have separated the approaches into niche-like areas of research based on initial assumptions, initial objectives, and end-point goals. Further, it is clear from Füssel [6] that there is a need for acceptance of a variety of valid frameworks for vulnerability assessment and appropriate use of the frameworks for given problems. The aim of this paper is not to limit advancement of vulnerability assessment to just one example of one approach at the expense of others, or to suggest the superiority of one method over another. The SoVI is of interest in this study as a linear assessment tool that offers the easiest test for the application of a non-linear method.

While the current method of SoVI construction has served analysts well for more than a decade, there are still areas for potential improvement. The PCA used in SoVI is a linear statistical method that can present problems for often-non-linear geographic human data. An approach to address this problem may lie in using an Artificial Neural Network (ANN). An ANN is a data-driven method similar to PCA, however it is a classification method, when used in a supervised capacity, which may eliminate complications from non-linear data [10,11].

This study compared the traditional SoVI approach with a modified SoVI approach with an ANN classification extension using a case study for the state of Utah. More specifically, the comparison between the traditional SoVI method and the ANN-classified SoVI identified how the two differed.

Through conducting this study we sought to answer the following questions:

1. How does an ANN SoVI result compare to the result of a traditional SoVI for a given region?
2. How and where do the results differ and why?

3. How can the ANN classification be interpreted relative to the traditional method regarding its strengths and weaknesses?

A point of concern in this approach is the use of the traditional SoVI result for a given region as a valid reference for the ANN SoVI. A common concern that arises in the discussion of the use of indices, in a broader sense, is that such indices are independent of their source data, that a given index has no direct relationship with what it is representing. The use of indices continues in spite of this challenge, however, as stronger alternatives often do not exist. Getting back to the concern of using one index to serve as the valid basis for a new index—basing the ANN SoVI on the traditional SoVI—it is important to note that the purpose of this study is not to make claims of index validity and appropriateness in a broader, theoretical sense. Such deliberation is beyond the scope of this project. As such, this study will proceed as though the traditional SoVI is a valid method, supported by more than ten years of use in the literature [2,3,7,8]. This is central to the purpose of assessing the viability of the use of an ANN to improve upon the traditional methodology.

2. Background

The concept of SoVI is a component from Susan Cutter's Hazards of Place (HoP) model, which seeks to integrate physical vulnerability and social vulnerability in an intuitive way [2,3]. The approach fuses two concepts in a novel way: vulnerability has to do with both proximity to hazards and political-economic factors. Cutter took this fusion a step further, however, and describes the concept of HoP as vulnerability that is a place-based characteristic of the vulnerable population where they live [2,7]. Vulnerability therefore varies at different scales, and takes into consideration generalization principles: the smaller the analysis scale, the more locally relevant the model will be for the population in that place. Place is implied to mean a human occupied space, an occupied location [2].

Cutter first introduced the HoP model with William Solecki in a 1989 paper on patterns of U.S. airborne toxic releases [12]. Cutter continued to work with the HoP concept through the 1990s, where the model expanded to include broader physical and social characteristics of hazard vulnerability [2]. The vulnerability model came to include hazard potential as a combination of risk and mitigation, which itself was broken into the related parts of social fabric, geographic context, biophysical vulnerability (akin to the proximity to hazard school of thought), and social vulnerability—similar to the social vulnerability aspects of political-economic concepts of social vulnerability) [2] (p. 78). Social fabric in this model represents the social and political background of the place and how that impacts social vulnerability, while geographic context accommodates more the physical characteristics of the place landscape as they operate together with biophysical hazards. The HoP model integrates these characteristics which results, when applied to a place, in a holistic assessment of its vulnerability. The joining of physical risk with a SoVI as the social vulnerability component has been done in several works by Cutter and others in the years following the creation of the HoP model to demonstrate its capabilities [2,3,7].

Further work involving the SoVI has gone on to explore how robust the SoVI is in terms of sensitivity and variance at different scales [3,13]. Schmidtlein *et al.* [13] determined that scale and minor changes in variable selection had little impact on the efficacy of the SoVI, a concern noted in Cutter's work. However, Tate [14] challenged that statistical bias, precision, and uncertainty are intrinsically part of the general hierarchical SoVI, thereby insisting that sensitivity analysis be included in the creation of a vulnerability index. Holand and Lujala [8] followed with an approach to adapt the SoVI to a new geographic context, in their study's case translating the United States-centric SoVI to apply to Norway. Holand and Lujala [8] altered the variable selection for the SoVI to better reflect the cultural, social, and political context of Norway. Taken together, this literature suggests that the SoVI is flexible in its construction and viable for results provided care is taken in construction of indices.

With the adaptive element identified by Holand and Lujala [8] coupled with index sensitivity analysis by Schmidtlein *et al.* and Tate [13,14], new possibilities are opened for advancing the power of the SoVI. One such opportunity becomes apparent when considering Stephen and Downing's [15] assessment of vulnerability to famine and food insecurity in Ethiopia in a comparison between three methods. The authors compared the commonly used Household Food Economy Approach and RiskMap (HFE) and Classification and Regression Tree (CART) methods, and then compared those two to a new approach utilizing an ANN.

An ANN is a machine-learning computation method capable of performing data exploration (in an unsupervised mode) and data classification (in a supervised mode), among other applications [16,17]. The method has a basic structure of three node layers: one layer of inputs with one node per variable input; one hidden layer of nodes that perform the analysis in the model, commonly one plus the number of input nodes; and one output layer, with the number of nodes equal to the number of class outputs desired [10,11,16]. The input nodes pass their values to each of the hidden nodes, which then perform the relationship analysis of the inputs, and then pass the classification out to the output nodes [10,11,16]. A common modification to this basic structure in modern studies using ANNs is the inclusion of back-propagation, whereby the result of the analysis in the hidden nodes is passed back to the network links between the input and hidden layers to apply an auto-adjusting weight to the network links to enhance the performance of the ANN [10,16]. The structure of the ANN and its relationship algorithm has been determined to implicitly capture non-linear relationships in data applied to an ANN [10,11,16]. The method does have a significant drawback with respect to social applications; training an ANN can require a significant amount of sample data, which, when considering a single-case model as SoVI has frequently been applied, could be problematic due to the reduction in dataset size for classification from training [17]. This problem could be alleviated with the use of a trained ANN for other case studies using the same input parameters, whereby data would only need to be reserved for model validation rather than for training. The convention put forth by Shahin *et al.* [17] recommends an optimal training set to be 70% of the data with 30% used for testing.

The direct application of ANNs in social problems has little presence in the literature. However, the use of ANNs in physical systems modeling is quite extensive. Examples in geography include, but are not limited to: rainfall estimation [18], landslide susceptibility [19], water quality [20], and solar energy potential [21]. These applications have utilized the classification potential for an ANN in developing predictive models for evaluating physical phenomena using input variables identified as key to understanding those phenomena. The SoVI is the product of attempting to parameterize social vulnerability with quantitative social data, which makes the SoVI not dissimilar to the physical phenomena in form. This similarity between SoVI and other analyses of physical phenomena, taken with the demonstration by Stephen and Downing [15] of an ANN in a vulnerability application, suggests that an ANN may be useful for assessing vulnerability.

Stephen and Downing [15] found that the ANN could produce comparable results to the commonly used HFE and CART methods, but more importantly performed some diagnostics on the functionality of the ANN in a vulnerability context. They found that their ANN was not sensitive to spatial scale and further determined that their ANN could accommodate data sets with different scales, which could suggest a possible way to address the Modifiable Areal Unit Problem in SoVI construction [15,22,23]. The ANN was also identified as having captured non-linear relationships in their data, even when the data were autocorrelated [15], consistent with statements about ANN performance by Fischer and Abrahart [16]. As noted by Stephen and Downing [15] and Fischer and Abrahart [16], topology between observations can also be maintained in some applications of the ANN. Ultimately, Stephen and Downing [15] opened up a new pathway for social vulnerability and the SoVI through integration of an ANN.

3. Experimental Section

To explore the modification of the SoVI method, we constructed an experiment to first assess vulnerability for Utah using the traditional method. From that point an ANN was constructed to expand the SoVI with a non-linear method to enhance the results. The following sections describe in greater detail the procedure used to explore this application of ANN to expanding SoVI. We chose the SoVI method to test an ANN as existing literature on SoVI [3,4,6,8,13–15] appears to provide a clear avenue for exploration.

3.1. Study Area

Utah is a state located in the Rocky Mountain West with a population in 2010 of 2,763,885 (Figure 1) [24]. The state has 29 counties, 18 of which have a population of 25,000 or fewer. Utah's capital is Salt Lake City, located in Salt Lake County, which is also the largest city in the state with a municipal population of 186,440 as of the 2010 decennial census [24]. The population distribution is non-uniform throughout the state, as a large portion of the population of the state is centered in the Ogden-Salt Lake-Provo corridor of the Wasatch Front, with the remainder of the state mostly rural.

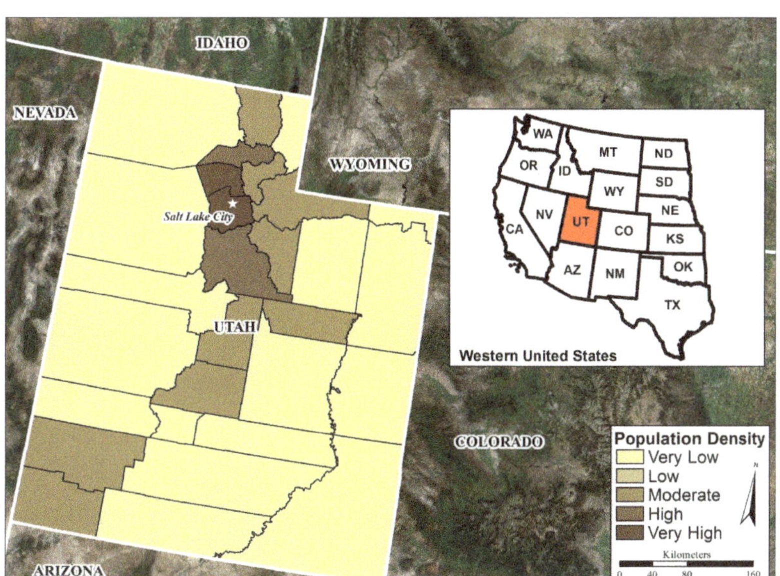

Figure 1. The state of Utah in the Western United States with county population by land area shown.

3.2. Population Data

To conduct our study, we took a subset of the United States Census Bureau's American Community Survey (ACS) data for the state of Utah for a five year period to account for as many social factors as possible from the traditional SoVI method [3]. The ACS surveys a small sample of the population each year to supplement and expand on the decennial census and to provide other statistics products useful for planners in a variety of contexts [25]. We selected the five year data for this study to reduce the effects of error, which is reduced through the combination of the samples from each of the years, thereby increasing the sample size. The period of our dataset is from 2008 to 2012.

The ACS data is aggregated at a variety of levels for different needs. We selected the smallest aggregation level that the ACS is published in for this study, the census block group level—the second smallest aggregation unit the Census Bureau uses in published data. This was done to capitalize on two key benefits: a relatively fine spatial scale to assess vulnerability and the largest possible number of aggregation units for the study area. The relative quality of the ACS is less important study than the ready availability of a large number of demographic variables no longer collected in the U.S. Census Long Form, particularly as the internal consistency of the data is the only validity concern for this study.

The five-year ACS data for Utah at the census block group level is composed of 1690 census block groups and contains a total of 2739 variables with respective margins of error. The subset of the variables used in this study totaled 60 variables from the ACS and 1 variable from the census block group geometry (Table 1). Some of the variables were combined where appropriate to produce more broadly descriptive variables, such as with public education attainment. The variables chosen for this study approximate a basic index covering the broadest themes of social vulnerability as presented by Cutter *et al.* [3]. These data provide a basic skeleton by which to test the ability of an ANN to perform as a classification method, rather than as a complete and accurate vulnerability assessment.

Table 1. Subset of the United States Census Bureau's American Community Survey (ACS) variables used in the study with descriptions and whether the variable was derived from a larger group of variables from the ACS data.

Variable Name and Number	Derived	Variable Description
MEDIAN_AGE (**1**)	No	Median age
PER_CAPITA_INCOME (**2**)	No	Per capita income
MED_VAL_OWN_OCC_HOUSING (**3**)	No	Median value of owner occupied housing
MED_RENT_RENT_OCC_HOUSING (**4**)	No	Median rent of renter occupied housing
NON_WHITE (**5**)	Yes	Proportion of population that is non-white
PC_POP_UNDER_5 (**6**)	Yes	Proportion of population under the age of 5
PC_POP_OVER_65 (**7**)	Yes	Proportion of the population over the age of 65
PC_CIVIL_LABOR_UNEMPLOYED (**8**)	Yes	Proportion of the civil labor force that is unemployed
AVG_PEOPLE_HOUSE (**9**)	No	Average number of housing occupants
PC_HOUSE_EARN_MORE_75K (**10**)	Yes	Proportion of population earning more than $75,000 per year
PC_POVERTY (**11**)	Yes	Proportion of population living below the poverty level
PC_RENT_OCC_HOUSING (**12**)	Yes	Proportion of housing occupied by renters
PC_MOBILE_HOME (**13**)	Yes	Proportion of occupied housing as mobile homes
PC_POP_OVER_25_NO_DIPLOMA (**14**)	Yes	Proportion of population over 25 with no high school diploma

Table 1. *Cont.*

Variable Name and Number	Derived	Variable Description
NUM_HOUSING_SQ_MI (**15**)	Yes	Housing density by square mile
PC_POP_IN_LABOR_FORCE (**16**)	Yes	Proportion of population in labor force
PC_EMP_EXTRACTIVE (**17**)	Yes	Proportion of labor force employed in extractive industry occupations
PC_EMP_TRANSCOMMUTIL (**18**)	Yes	Proportion of labor force employed in transportation, communications, and utilities occupations
PC_EMP_SERVICE (**19**)	Yes	Proportion of labor force employed in service occupations
PC_FEMALE (**20**)	Yes	Proportion of population that is female
PC_FEM_ONLY_HOUSE (**21**)	Yes	Proportion of households headed by a female with no spouse present
PC_HOUSE_SS_INCOME (**22**)	Yes	Proportion of households receiving social security income

These data were joined to the census block groups polygons in a Geographic Information System (GIS) and the final variables were calculated within the database. The joined data were exported from the GIS to the statistical software package R to perform further analysis.

3.3. Traditional SoVI Construction

The ACS variables attached to the census block groups for Utah were exported from the GIS into a shapefile format and read into R using the "maptools" package. The database table containing the variables for the state was translated into a data frame in R. Once the data were in the correct internal format in R, PCA was run on the data using the "prcomp" function. The data were scaled in the function to center all of the variables and to ensure the variance was not skewed by differences in variable magnitude. The resultant PCA analysis produced a total of 22 principal components of which 13 were selected as social factors with 87.8% of the variance explained (Table 2).

Table 2. Traditional Social Vulnerability Index (SoVI) factors with cardinality, factor name and proportion of variance explained (rounded), and dominant social variables with factor loading cardinality.

Factor 1 (−)	**Factor 2 (+)**	**Factor 3 (+)**	**Factor 4 (+)**	**Factor 5 (+)**	**Factor 6 (abs)**	**Factor 7 (−)**
Wealth (22)	Elderly (13.9)	Extractive (8.9)	Female (6.9)	Disadvantaged (5.5)	Employment (4.9)	Housing (4.4)
per capita (−)	med age (+)	renters (+)	female (+)	non-white (+)	unemployed (−)	median rent (−)
earn >75K (−)	over 65 (+)	house sqmi (−)	fem only home (+)	under 5 (+)	avg house size (−)	
	SSI receive (+)	extract emp (+)		female (+)	extract emp (+)	
					service emp (+)	
Factor 8 (−)	**Factor 9 (+)**	**Factor 10 (+)**	**Factor 11 (−)**	**Factor 12 (+)**	**Factor 13 (+)**	
Housing (4.1)	Race and Employment (4)	Disadvantaged (3.6)	Employment (3.5)	Disadvantaged (3.1)	Extractive Employment (3)	
median rent (−)	non-white (+)	mobile homes (+)	service emp (−)	under 5 (−)	mobile homes (+)	
	unemployed (−)	extract emp (+)		unemployed (−)	extract emp (−)	
		fem only home (+)		fem only home (+)		

From this selection of social factors, a function was created to additively combine the factors—using the variable loadings—into the final SoVI score. The function used to add the loadings is shown below:

$$-F1 + F2 + F3 + F4 + F5 + |F6| - F7 - F8 + F9 + F10 - F11 + F12 + F13 \quad (1)$$

The data were exported back to GIS to visualize social vulnerability for the state. The SoVI scores were symbolized using quantiles with five breaks. The breaks represent categorical vulnerability (Table 3).

Table 3. Vulnerability categories and their respective quantile break upper bound value.

Upper Bound of Category	Category
−1.59	Very Low
0.62	Low
2.44	Moderate
4.60	High
22.77	Very High

A new field was added to the database table to include the SoVI categories. The data were exported with the categories included into shapefile format to be read into R once more.

3.4. ANN SoVI Construction

The data were loaded into R using the "maptools" package once again. The database table was loaded into a data frame in R and unnecessary fields were stripped from the table (*i.e.*, the SoVI score table, GIS-generated fields, *etc.*). The data were then partitioned into two sets, one training set and one prediction set. The training set was 70% of the data, totaling 1183 census block groups, with the remaining 507 census block groups used for ANN classification. This training and testing set division was used following the convention established by Shahin *et al.* [17].

The ANN was created using the "nnet" package. The network had 22 input nodes, equal to the number of input variables. The hidden layer had a number of nodes equal to the input nodes plus one, or 23. The output layer had five nodes, equal to the number of social vulnerability categories (Figure 2).

The final ANN was run five times for categorical classification to find a common convergence value to determine that the model was running consistently and producing consistent results. A static seed value was also set in the process to ensure consistency in output. All of the classification runs converged at the same value, even when the model was run on other machines. The ANN was run multiple times with slight variations to the structure, however the structure presented here performed the best of the test cases.

3.5. Comparison of Traditional and ANN SoVIs

The comparison of the PCA-driven SoVI against the ANN-extended SoVI was conducted using change detection and radar plot comparisons of how each method handled the input social variables. For this study we focused primarily on the similarity between the PCA and ANN results, and we compared the performance of the methods based on the relative similarity of the results.

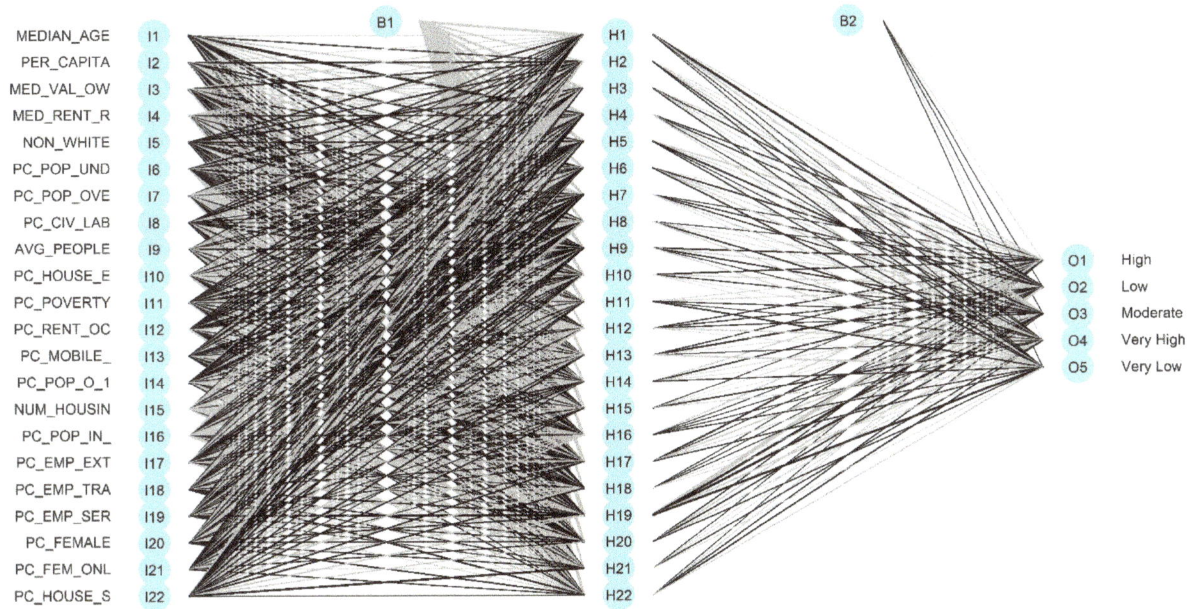

Figure 2. Structure of the neural network used to classify SoVI for Utah. The network contains 22 input nodes, 23 hidden nodes, and five output nodes.

The change detection employed identified a 'from–to' relationship between the traditional SoVI and the ANN-extended SoVI. A change detection matrix was created showing which blocks changed between methods and how the blocks were reclassified in the ANN-extended SoVI. The change detection matrix helped in determining the nature of classification between the two methods and visualized the differences between the methods.

The construction of radar plots allowed us to visualize the relative importance of each variable in each vulnerability class for both the traditional SoVI and the ANN-extended SoVI. Comparison of variable handling, the importance of each variable in each vulnerability class, allowed us to assess how both methods classified vulnerability and to determine if significant deviations in variable handling were present.

4. Results and Discussion

4.1. Traditional SoVI Discussion

The PCA scores from the traditional SoVI were categorized into classes of vulnerability based on quantiles (Figure 3). The represented social vulnerability for the state of Utah reveals a largely random pattern of vulnerability. Some clustering, identified using Local Moran's I statistic, adjacent to cities in the state, but not exclusive to the cities, was found randomly around the state (Figure 4). This pattern is to be expected given the strong rural-urban population concentration for the state of Utah as seen in Figure 1. High and very high vulnerability block groups are generally found in the south and eastern parts of the state, while low and very low vulnerability block groups are scattered throughout the state, however those classes are more common in the central and northwestern parts of Utah.

It is interesting to note that the traditional method produced a few instances of potential misclassification. The method classified census block groups with zero population as having high vulnerability. These block groups were used to test if the ANN would reclassify the block groups more appropriately, with the PCA results passed to the ANN in different trials, corrected (the misclassified block groups were reclassified) and uncorrected (the misclassified block groups were unchanged). Figure 3 displays the uncorrected traditional SoVI result to compare with the ANN SoVI result with the uncorrected block groups shown below.

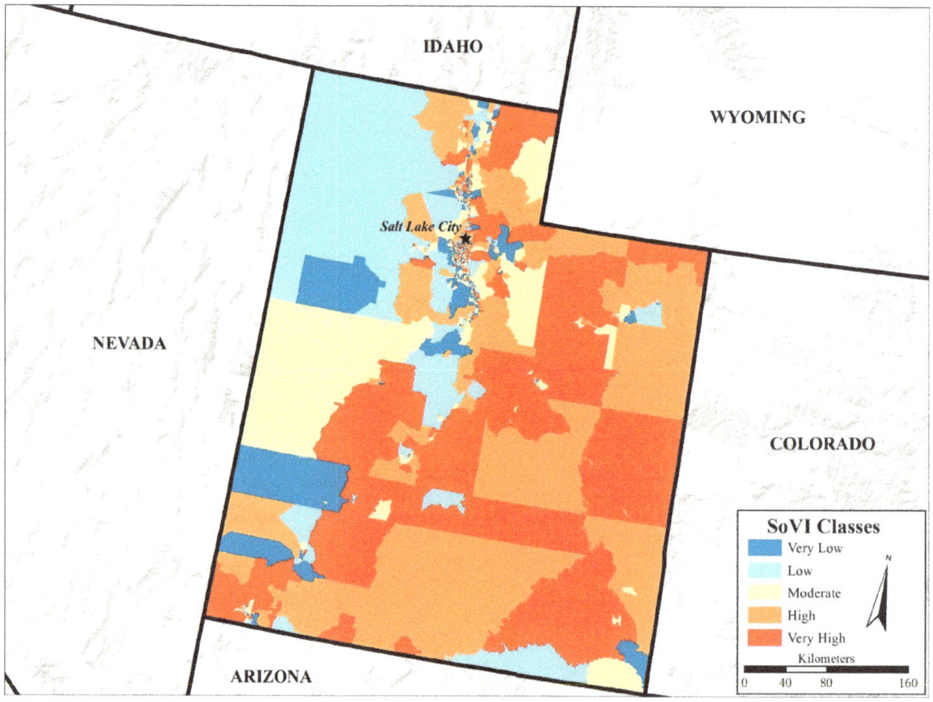

Figure 3. The SoVI constructed for the state of Utah using the traditional approach.

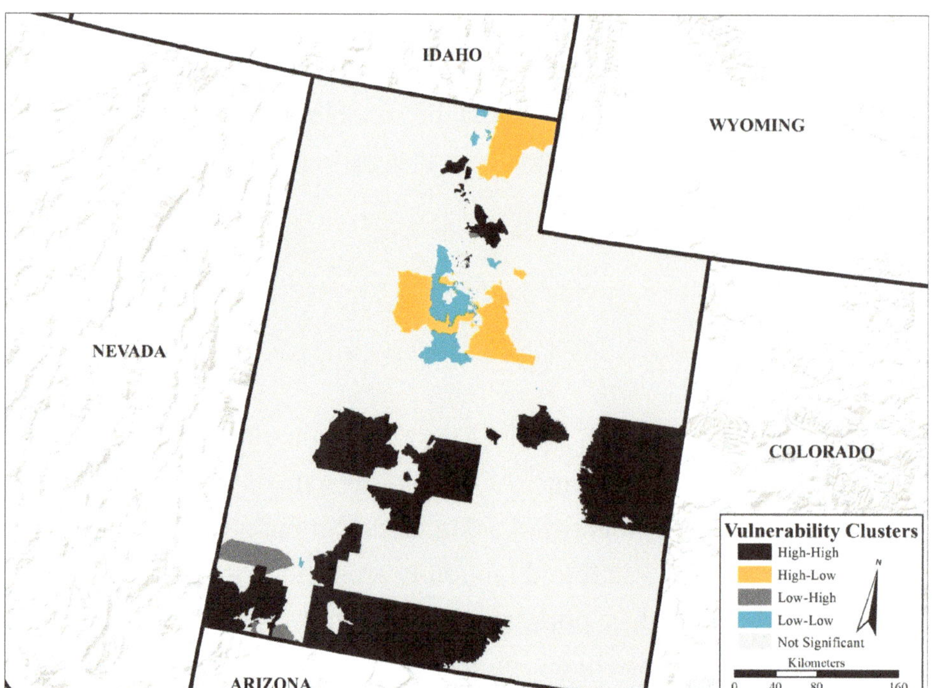

Figure 4. Clustering of vulnerability categories for Utah based on the traditional SoVI using the Local Moran's I statistic. The pattern demonstrates some areas of clustering near populated cities.

The SoVI result from the traditional approach is largely consistent with understanding of vulnerability in Utah based on qualitative assessment by the authors. Further, the pattern is similar to a smaller scale study using different datasets for Salt Lake County in Utah [26]. The authors accept the traditional SoVI used in this study, with the limitations noted in the previous paragraph, as a valid result.

4.2. ANN SoVI Discussion

The ANN was trained using 70% of the data (demonstrated Stephen and Downing [15] and supported Fischer and Abrahart [16]), totaling 1183 census block groups. The remaining 507 block groups were used to test the classification capability of the trained ANN (Figure 5). The classification by the ANN produced a roughly evenly distributed classification. Each of the classes was of a size within one standard deviation of the mean number of block groups in each vulnerability class; however, the very low class had 129 block groups. It may appear that the very low class is slightly biased. Comparison between the traditional SoVI and the ANN-extended SoVI revealed some of the nature of the ANN classification, as discussed in the following section. It is also worth noting that the ANN classification produced consistently the same results with a specified seed value in R over multiple model runs, indicating stability in the classification.

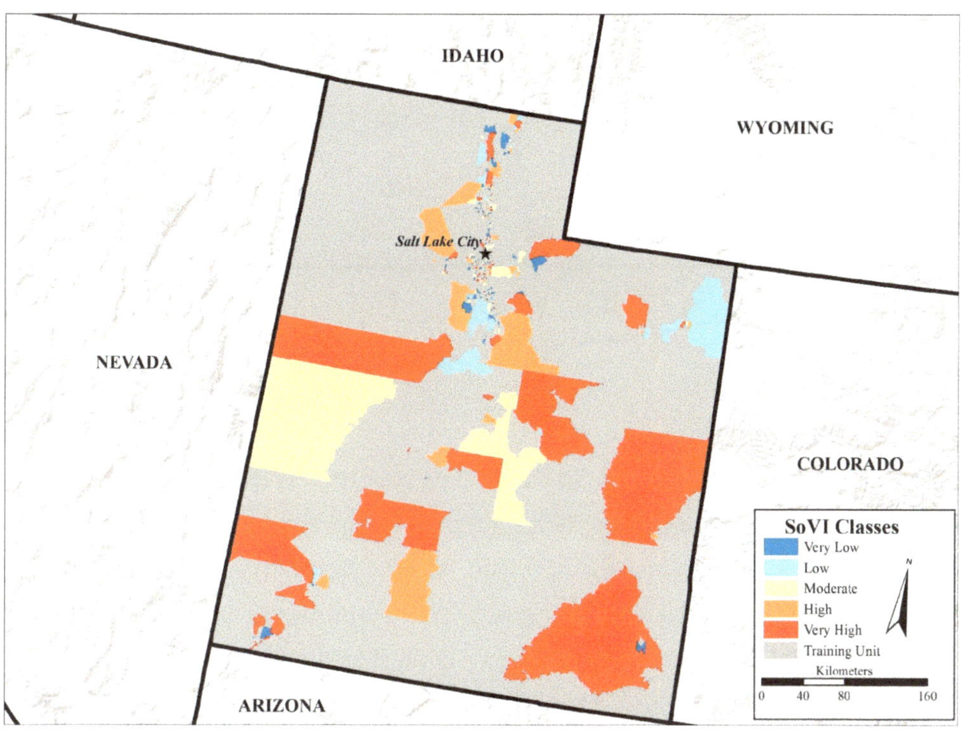

Figure 5. The SoVI for the state of Utah constructed using the Artificial Neural Network (ANN)-extended method.

4.3. Traditional SoVI and ANN SoVI Comparison

Comparing the two results necessitated the extraction of the PCA loadings and the ANN network weights. These data are not directly comparable, so two comparison methods were devised to assess both methods: change detection between the results of both methods and radar plot assessment of the variable handling in each method. To aid in the comparison, we used the cardinality and internally relative magnitudes of the PCA loadings and the ANN network weights.

The first part of the comparison was conducted through the analysis of a change detection plot between the traditional SoVI and the ANN SoVI with an accompanying change matrix (Figure 6). The first thing of note is the agreement between the two methods. The ANN-extended SoVI produced a classification that was consistent with the traditional SoVI over 26% of the classified block groups. While this may seem low, it is important to note that the two methods should produce different results; it is therefore encouraging to find some (but not total) agreement between the methods. Where the methods differ there are opportunities for further investigation into how the ANN differed from the PCA.

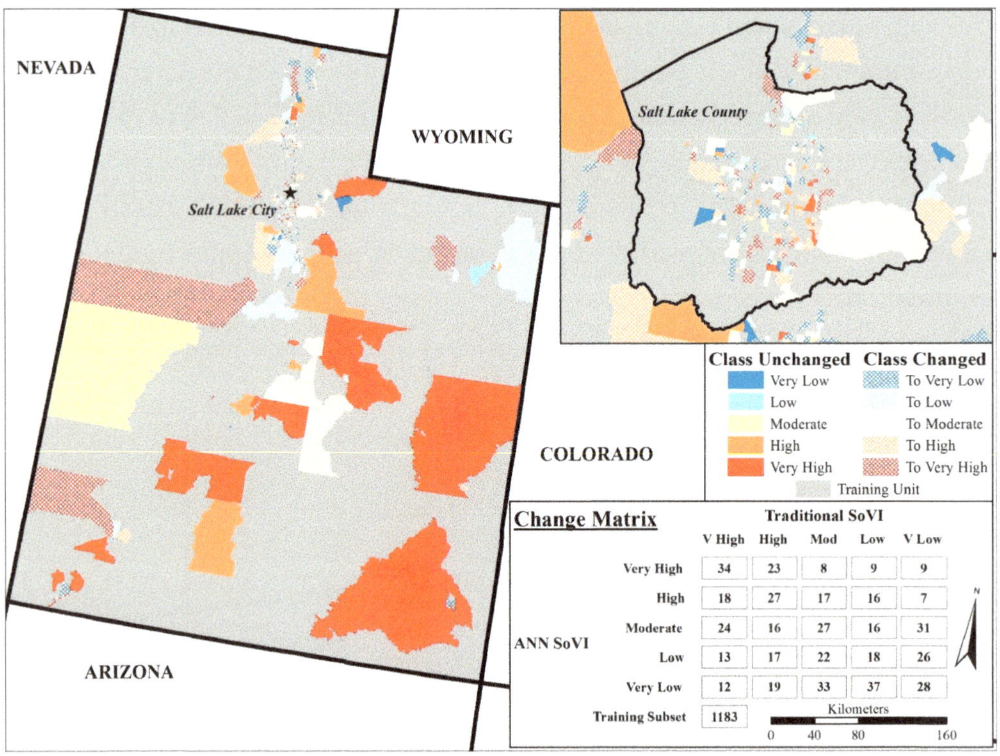

Figure 6. Comparison of the traditional SoVI classes and the ANN-extended SoVI classes for the state of Utah. The legend includes a change matrix showing how the traditional SoVI classified the census block groups against how the ANN-extended SoVI classified the same census block groups for the classification subset. The symbolized classes indicate census blocks that remained unchanged between methods as well as a generalized representation of which census block changed to each vulnerability class. An inset map of Salt Lake County is included to show the most densely populated portion of the study area.

Referring back to the previous section and the possible concern of classification bias in the very low vulnerability class, we can see from the change matrix that the two methods had the second highest agreement in the very low class. Further, the bulk of the reclassification came from block groups classed as moderate and low vulnerability in the traditional SoVI. Based on this information it appears that there may be a bias to the very low vulnerability class. Many of the block groups changed from their traditional SoVI classification through the ANN classification, and it is noteworthy to point out that this occurred from each class to each other class. This may be evidence that the ANN was able to capture non-linear relationships in the data and express that in its classification.

Using the misclassified block groups from the traditional approach as an extra diagnostic, we found that the ANN classified the (potentially) misclassified block groups into the same category as the traditional approach. This result is encouraging as the ANN may have been forced to adjust to accommodate the abnormal block groups. This additional point of consistency between the two classification methods suggests that the ANN is capturing some of the relationships that the PCA also found. The training set included examples of the high category that were not invalid, as well, and was consistent in some of those classifications with the traditional SoVI, but it also classified some of the high block groups into other vulnerability classes. This may indicate that the ANN found a non-linear relationship between the social variables that caused it to class those block groups differently.

It could further be suggested, from a visual appraisal of Figure 6, that the rural areas of Utah dominate the training set for the ANN. A closer look at Salt Lake County—the center of the most densely populated region in Utah—reveals that the block groups within this densely populated region are well represented in the training data as well as in the classification set.

The second part of the comparison was conducted by qualitatively assessing the social variables of the block groups and the classification results from both methods. By constructing radar plots with each variable—the variables were standardized and averaged for each vulnerability class—represented on its own X-axis, we were able to determine the tendency of variable importance in the classifications (Figure 7). The radar plots reveal relative internal consistency for how both the traditional approach and the ANN classified the block groups. From Figure 7 (axis numbers correspond to variables in Table 1) we can see that the tendency for very high and moderate lined up well between both the traditional classification and the ANN classification (Figure 7a,c). Clear variation in variable handling between both methods can be seen in the very low vulnerability class (Figure 7c). This difference between the two classification methods may be a case where the ANN is handling the variable relationships in a non-linear way.

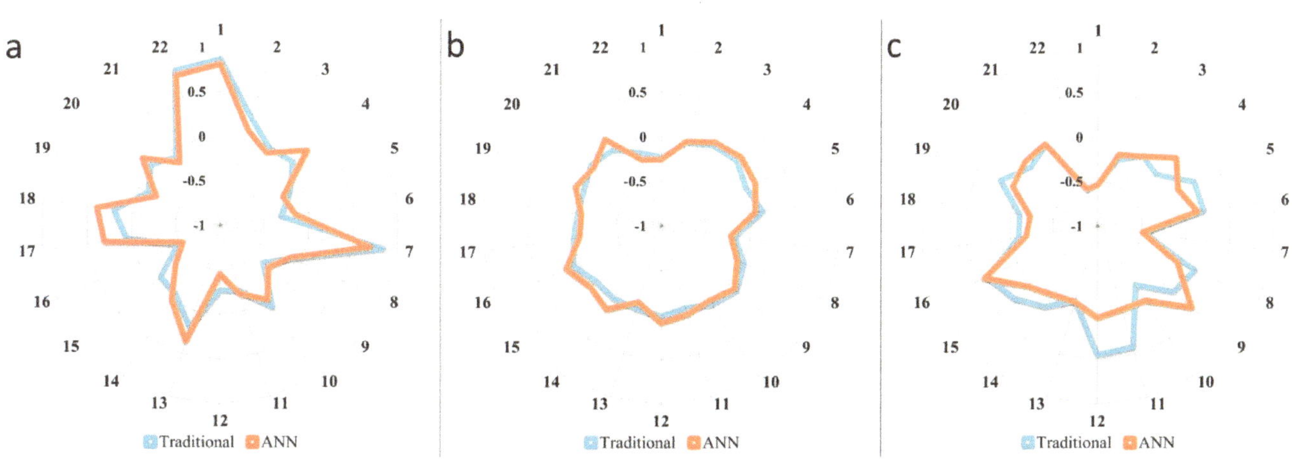

Figure 7. A comparison of the averages of the social variables within the very high (**a**); moderate (**b**); and very low (**c**) vulnerability classes between the traditional SoVI and the ANN-extended SoVI.

Through analysis of the differences between the traditional SoVI and the ANN-extended SoVI classifications, we were able to determine that the results from both methods were reasonable and that the two methods could be readily compared. The comparison revealed situations where the non-linear nature of the ANN's classification may have been beneficial, resulting in the difference seen in Figure 6. The stability of the ANN classification as the network reached convergence helps to strengthen these findings, in addition to the inter-method consistency checked using invalid block groups.

Due to the exploratory nature of this study—to determine if the ANN can enhance the traditional SoVI method—the margin of error term for the ACS variables was not used with the study data. While we did not feel this was necessary for this study, other researchers employing the methodology described in this paper are encouraged to ensure that the margin of error is included to provide the most accurate results.

5. Conclusions

This study compared the traditional SoVI method of vulnerability classification with a modified form utilizing an ANN. Both the traditional approach and the ANN approach produced reasonable vulnerability classifications for the state of Utah with respect to the basic parameterization used for this study. The methods were in agreement for 26% of the common block groups. This partial agreement shows that both methods are capable of identifying some of the same relationships between the input variables; the difference in how relationships were handled demonstrates that the ANN captured non-linear relationships that the PCA could not. Further, agreement between the methods for the invalid block groups, classed as high vulnerability in both approaches, demonstrates cross-method consistency—the ANN was forced to classify the invalid block groups the same as the traditional approach, likely due to the out-of-ordinary nature of variable values for those block groups.

The ANN-extended SoVI demonstrated some bias in classification in the high and very low vulnerability classes, with noticeable differences in how the traditional classification and the ANN classification handled variables for those classes. The radar plots of the social variables for each vulnerability class reveal that the ANN handled the variables for the other three classes similarly to the traditional approach. This internal data-handling consistency between the methods strengthens the finding that the ANN captured non-linearity between the variables, consistent with statements in the literature [10,11,16], as can be seen below (Figure 8). Consistency between model runs of the ANN further strengthen the classification result, as there was no variation in the results between runs and the convergence was identical in each test run.

Using the methodology outlined in this paper, we can summarize our findings with respect to our study questions. Firstly, the traditional SoVI and the ANN-extended SoVI both produced reasonable results for the state of Utah—the classifications of vulnerability from both methods were reasonable for the population when compared to existing work in the region [26,27]. Secondly, the results between the methods were different, but partially consistent. The ANN classified 26% of the classification subset consistently with the traditional SoVI, indicating that both methods were able to capture similar relationships between input variables. However, the ANN classified the remaining subset differently, in some cases to one vulnerability class away from the original (e.g., from very high to high). Analysis of the variable tendencies in the vulnerability classes for both methods (using the radar plots) shows that both methods handled the variables in largely the same way, with some notable differences. The divergence between the two methods indicates that the ANN was able to capture non-linear relationships in the variables which were not captured with linear PCA. Finally, because the ANN did capture non-linearity in the variables for its classification, the ANN should be a viable pathway to enhancing the way SoVI is constructed in the future, and could be useful in other quantitative vulnerability assessment methodologies. The consistency between the methods shows that the ANN was able to capture the linear relationships identified with PCA, which means that the strengths of the PCA method can be preserved when using an ANN. Invalid data, however, presents problems for both methods, despite the strength of consistency found using invalid data; analysts will need to be vigilant to ensure quality data is used when using the ANN, as is the case with the traditional approach.

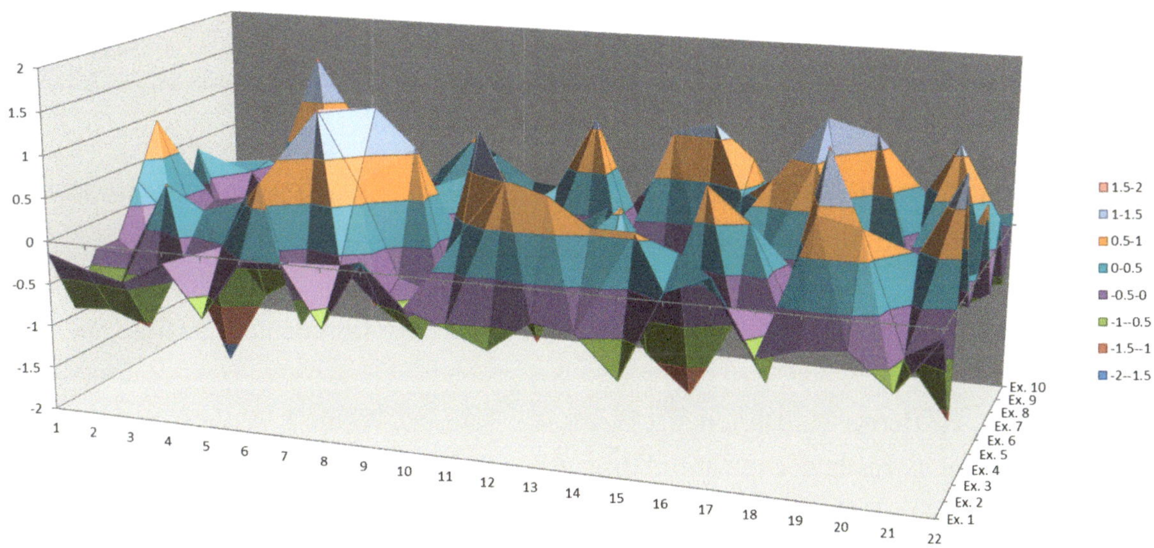

Figure 8. Evidence of non-linear relationships between the input variables. Ten examples were selected: two of each class from the traditional SoVI that changed to the two most distant classes in the ANN-extended SoVI (e.g., one high to low, one high to very low, one moderate to very high, *etc.*).

The methods to explore the use of ANN to enhance the SoVI method outlined in this study demonstrate that this approach can produce a viable alternative to the well-established approach to creating a SoVI for a region. Indeed, the ANN retains the strengths of the existing method with few of its weaknesses, in addition to its own strengths, particularly the handling of non-linear data-relationships. Further exploration of this method will demonstrate how capable the ANN can be for SoVI analysis, as well as how it may best be implemented for creating a SoVI. Future exploration of implementing an ANN in other vulnerability assessment methodologies has the potential to progress the field of vulnerability assessment as a whole.

Acknowledgments

The authors would like to extend their gratitude to the United States Census Bureau for collecting and distributing the demographic data in this study and to the R Project for Statistical Computing for all its members' contributions to the open-source statistical software used in this study. They would also like to thank the Center for Natural and Technological Hazards and the Department of Geography at the University of Utah for facilitating this study. Thanks are also due to Richard Medina, Neng Wan, and Simon Brewer for their advice and support of this study.

Author Contributions

Ryan Hile and Thomas J. Cova conceived and designed the experiment; Ryan Hile performed the experiment; Ryan Hile analyzed the data; Thomas J. Cova contributed analysis commentary and support; Ryan Hile wrote the paper.

Conflicts of Interest

The authors declare no conflict of interest.

References

1. Blaikie, P.; Cannon, T.; Davis, I.; Wisner, B. *At Risk: Natural Hazards, People's Vulnerability and Disasters*; Routledge: London, UK, 1994.
2. Cutter, S.L. Vulnerability to environmental hazards. *Prog. Hum. Geogr.* **1996**, *20*, 529–539.
3. Cutter, S.L.; Boruff, B.J.; Shirley, W.L. Social vulnerability to environmental hazards. *Soc. Sci. Q.* **2003**, *84*, 242–261.
4. Cutter, S.L.; Barnes, L.; Berry, M.; Burton, C.; Evans, E.; Tate, E.; Webb, J. A place-based model for understanding community resilience to natural disasters. *Glob. Environ. Chang.* **2008**, *18*, 598–606.
5. Wachinger, G.; Renn, O.; Begg, C.; Kuhlicke, C. The risk perception paradox—Implications for governance and communication of natural hazards. *Risk Anal.* **2013**, *33*, 1049–1065.
6. Füssel, H. Vulnerability: A generally applicable conceptual framework for climate change research. *Glob. Environ. Chang.* **2007**, *17*, 155–167.
7. Cutter, S.L.; Mitchell, J.T.; Scott, M.S. Revealing the vulnerability of people and places: A case study of Georgetown County, South Carolina. *Ann. Assoc. Am. Geogr.* **2000**, *90*, 713–737.
8. Holand, I.S.; Lujala, P. Replicating and adapting an index of social vulnerability to a new context: A comparison study for Norway. *Prof. Geogr.* **2013**, *65*, 312–328.
9. Adger, W.N. Vulnerability. *Glob. Environ. Chang.* **2006**, *16*, 268–281.
10. Somers, M.J.; Casal, J.C. Using artificial neural networks to model nonlinearity: The case of the job satisfaction-job performance relationship. *Organ. Res. Methods* **2009**, *12*, 403–417.
11. Mao, J.; Jain, A.K. Artificial neural networks for feature extraction and multivariate data projection. *IEEE Trans. Neural Netw.* **1995**, *6*, 296–317.
12. Cutter, S.L.; Solecki, W.D. The national pattern of airborne toxic releases. *Prof. Geogr.* **1989**, *41*, 149–161.
13. Schmidtlein, M.C.; Deutsch, R.C.; Piegorsch, W.W.; Cutter, S.L. A sensitivity analysis of the social vulnerability index. *Risk Anal.* **2008**, *28*, 1099–1114.
14. Tate, E. Uncertainty analysis for a social vulnerability index. *Ann. Assoc. Am. Geogr.* **2013**, *103*, 526–543.

15. Stephen, L.; Downing, T.E. Getting the scale right: A comparison of analytical methods for vulnerability assessment and household-level targeting. *Disasters* **2001**, *25*, 113–135.
16. Fischer, M.M.; Abrahart, R.J. Neurocomputing—Tools for geographers. In *GeoComputation*, 1st ed.; Openshaw, S., Abrahart, R.J., Eds.; Taylor and Francis: London, UK, 2000; pp. 187–217.
17. Shahin, M.; Maier, H.; Jaksa, M. Data division for developing neural networks applied to geotechnical engineering. *J. Comput. Civ. Eng.* **2004**, *18*, 105–114.
18. Bellerby, T.; Todd, M.; Kniveton, D.; Kidd, C. Rainfall estimation from a combination of TRMM precipitation radar and GOES multispectral satellite imagery through the use of an artificial neural network. *J. Appl. Meteorol. Climatol.* **2000**, *39*, 2115–2128.
19. Biswajeet, P.; Saro, L. Utilization of optical remote sensing data and GIS tools for regional landslide hazard analysis using an artificial neural network model. *Earth Sci. Front.* **2007**, *14*, 143–151.
20. Singh, K.P.; Basant, A.; Malik, A.; Jain, G. artificial neural network modeling of the river water quality—A case study. *Ecol. Model.* **2009**, *220*, 888–895.
21. Mellit, A.; Pavan, A.M. A 24-hr forecast of solar irradiance using artificial neural network: Application for performance prediction of a grid-connected pv plant at trieste, Italy. *Sol. Energy* **2010**, *84*, 807–821.
22. Openshaw, S. *Concepts and Techniques in Modern Geography No. 38: The Modifiable Areal Unit Problem*; Geo Books: Norwich, UK, 1984.
23. Wong, D. The Modifiable Areal Unit Problem (MAUP). In *The SAGE Handbook of Spatial Analysis*, 1st ed.; Fotheringham, A.S., Rogerson, P.A., Eds.; SAGE: London, UK, 2009; pp. 105–124.
24. American FactFinder Community Facts. Available online: http://factfinder.census.gov/faces/nav/jsf/pages/community_facts.xhtml (accessed on 26 March 2015).
25. American Community Survey. Available online: http://www.census.gov/acs/www/ (accessed on 26 March 2015).
26. Hile, R.; Chaney, P.L. Mapping social vulnerability for earthquake hazards in Salt Lake County, Utah. *Auburn Univ. J. Undergrad. Scholarsh.* **2013**, *2*, 48–53.
27. SoVI 2006–10—30 Variable | State and Regional Maps. Available online: http://webra.cas.sc.edu/hvri/products/sovi2010_maps.aspx (accessed on 26 March 2015).

2

A GIS-Based Web Approach for Serving Land Price Information

Yongjun Yang [1], Yaqin Sun [1,*], Songnian Li [1,2], Shaoliang Zhang [3], Kuoyin Wang [1], Huping Hou [1] and Shishuo Xu [1]

[1] School of Environment Science and Spatial Informatics, China University of Mining and Technology, Daxue Road No.1, Xuzhou 221116, China;
E-Mails: yangyongjuncumt@163.com (Y.Y.); snli@ryerson.ca (S.L.); 3409719@163.com (K.W.); hphou@163.com (H.H.); cumtxushishuo@126.com (S.X.)

[2] Department of Civil Engineering, Ryerson University, 350 Victoria St., Toronto, ON M5B 2K3, Canada

[3] Low Carbon Energy Institute, China University of Mining and Technology, Daxue Road No.1, Xuzhou 221116, China; E-Mail: slzhang@cumt.edu.cn

* Author to whom correspondence should be addressed; E-Mail: cumtsyq@163.com

Academic Editor: Wolfgang Kainz

Abstract: Participants in the land market are usually hampered to browse and analyze the land price information due to the lack of information sources and available analysis tools. A service-oriented GIS-based web system was developed to provide a practical solution, its essential data sources contain basic geographic elements and benchmark land price (BLP)-related information. Core models for land price analysis were implemented, including land price index, spatial distribution, and parcel appraisal. The system was developed based on a four-level Browse Server (B/S) architecture using GIS and web service technologies, which enables the publishing, browsing, and analysis of the land price information via the Internet. With effective functionalities, the system has been employed in a project for updating BLP in a case study city located in China. The main advantage of the GIS-based web approach lies in its integration of spatial-temporal analysis models and web GIS technology, which allows more investors and administrators with limited domain knowledge to obtain further understanding on the change pattern and spatial distribution of land price by an online means. The experience in the case study city demonstrates that the approach has strong practicality for land price information services.

Keywords: GIS; system development; benchmark land price; analysis model; web technology

1. Introduction

Land and property transactions play significant roles in governments' revenues and financial institutions' banking policies, which is particularly the case for the developing countries with rapid urbanization and heavy reliance on land-based finance [1–4]. Land is a special kind of commodity, which cannot be moved from place to place. Before making judicious decisions to purchase or rent, investors and property brokers should carry out an analysis of the land market. On-the-spot gathering of the information about the basic condition of parcels, such as their price and location, is often costly [5,6]. Such transaction costs are indeed information costs and a fundamental impediment to land mobility [7,8]. Hence, there are great demands for land price information supply and deep processing. However, it is not an easy task to satisfy these demands by traditional information presentation means because land price is a kind of special geographical information which varies both spatially and temporally [9–12].

Providing information sources is one of the crucial aspects for reducing transaction costs. Many municipalities and governments gather and store information related to the use, the ownership, and the value of the land using cadaster and land registry [13–15]. Traditionally, the public can access such information by an inefficient means, delving into archives. Some more effective information technologies have been applied to improve this situation. Since 1986, databases have been used to store all land properties. London and Toronto began e-registration in 1999, which has since been adopted by many developing countries with the development of information communications technology [16–20]. Although many information systems that aim at land monitoring and management have been built in recent years, only a few of them are specialized in providing urban land price information to the public. For example, the real estate information systems employed by local governments in many countries have been developed mainly to enhance information gathering and management [21–23]. Scholars in Germany have developed a GIS-based system that offers brokers information to support hotel room pricing [24]. A web page with GIS tools in Spain serves as an information platform for leasing rural land plots [25]. However, these research works indicate the importance of available information about land related transactions.

On the other hand, analysis tools are another aspect, which is of particular importance in land transactions. Considering that the value of property is primarily determined by its location [26], people realized the importance of geographical information system (GIS) techniques [27–29]. GIS provides not only a geo-database to store relevant data with spatial relationships, but also a visual representation of information. Land price research has long recognized GIS as an ideal tool for appraisal processes and spatial analysis [30–33];for instance, creating land price maps by using spatial distribution models [31,32]. Although researchers can gain further understanding of the characteristics of urban land prices by these professional tools or systems installed on isolated computers, for general participants of the land market it is difficult to carry out land price analysis if they do not have access to such professional analysis tools.

Regarding land price services, both information sources and analysis tools are essential, given the geographical characteristics of land price. This paper presents a novel, GIS-based web approach that combines land price analysis models and web GIS technologies. Firstly, the functional requirements and core models of land price services are described based on the review of benchmark land price (BLP) and data sources. The architecture, development, and use cases of the system are then explained, followed by an application in a project entitled "Updating of BLP in Feng County (Jiangsu Province, China)". The interactive web-based graphical user interfaces enables users to acquire land price information efficiently on the internet.

2. Backgrounds

The urban land market in China has been put into operation in recent years, in which land-use rights (LUR) of urban land (for each type of land use, generally including commercial, residential, industrial) can be transferred between land-users. Participants in the land market mainly include administrator, seller (state or existing user), buyer (new land-user), and brokers [34,35]. Land price (LUR's price) may depend on private treaty, tender, auction or bargain under the evaluation of location and other conditions of land parcels. BLP refers to the average LUR's price established within a specific time period in a particular area for a particular land use. It provides an important reference for the government macro-control premium, land use tax collection, setting base prices for the LUR of state-owned land transferred, appraisal of the land parcel, negotiation of land price between buyer and seller, and reasonable adjustment of land use structure [36,37].

Local land administrations regularly monitor land prices and update BLP so as to manage land pricing and ensure macro-economic regulation for land development activities and LUR transactions [38,39]. The information generated from these works is usually published by means of the Internet or public documents, which gives the participants opportunities to analyze and appraise the land price. Such work is recognized as an information base to support efficient land market operations and an important task of Chinese Land and Resources Information 12th Five-Year Plan aimed at improving the informationization level of the land market. Currently, there are two official websites, China Urban Land Price Dynamic Monitor (http://www.landvalue.com.cn/) [40] and Jiangsu Land Market Website (http://www.landjs.com/) [41], where people can browse land price information and some transaction cases in different cities via the Internet but without the support of GIS-based analysis tools.

Researchers concerning the establishment of the land market in China emphasized the importance of making information available to the public on a fee-for-service basis and the role of GIS, considering that the young land market with lacking transaction cases needs to be improved in a market mechanism [36–39]. Without effective information sources and available analysis tools, participants of the land market will be hampered to browse and analyze the land price, such as its change and distribution in different urban regions. Therefore, a service-oriented, land-price information system will be more helpful if it provides some analysis functions that perform online analysis of effectively published land price information. Literature study and the use requirements analysis of the actual project indicate some very basic needs for such a system. They are summarized as follows: (1) regular publishing of land price information with online map, especially the information generated from land price monitoring (LPM) and BLP updating, and (2) analysis of land price, including the land price index, spatial distribution modeling of land price, and appraisal of land parcels based on LPM and BLP.

3. Data Sources

According to China National Standards "Regulations for valuation on urban land (GB/T18508-2014)" and "Regulations for Gradation and Classification on Urban Land (GB/T18507-2001)", after dividing a city into several homogenous regions, land price of these regions are appraised and graded to establish a land price system, called the benchmark land price (BLP). BLP is updated every three years. Meanwhile, in order to monitor the dynamics of the urban land price, a number of standard parcels are selected as monitoring points, and their prices are appraised quarterly. The information about BLP and those monitoring points, such as standard condition, correction coefficient, location, and other data are opened to the public. These data, in the form of digital or paper map and documents, can be collected from local land administrations. To organize these data, two geographic data sets have been established, benchmark land price and land price monitoring. Apart from these, basic geographic data and land use planning data are also taken into consideration for the purposes of online map services and land price analysis. Table 1 lists the data sets and their contents. The processing of map data includes digitizing paper maps and documents, unifying coordinate reference systems, and building a topological structure using GIS. Attribute data of each map features are stored in a relational database and linked to each graphic element. The established geo-database serves as the data source of land-price information service.

Table 1. Data sets for serving land price information.

Name of Data Set	Feature Layers	Attribute Data
Basic geographic data	Road, river, topography, building, *etc.*	Name, category, code, grade, *etc.*
Land use planning	Land use situation, overall plan for land utilization, city planning.	Land use type, city function zones, land use conditions, degree of development, *etc.*
Benchmark land price	Benchmark land price of commercial, residential, industrial land.	Land price, grade, average condition, table of adjustment factors, appraisal time, *etc.*
Land price monitoring	Monitoring points of commercial, residential, industrial land.	Land use type, land price, location, tenure, ownership, appraisal time, *etc.*

4. Analysis Models

This section describes the analysis models that are used in building the proposed GIS-based, web land price information system.

4.1. Land Price Index

As mentioned above, the price of standard parcels are appraised quarterly so as to monitor the urban land price. This means that there is a long series of price figures for each monitoring point. The land price index (LPI) is employed for the purpose of detecting and comparing the change of land price of each point. As a relative indicator of the trend of land price, LPI is actually a ratio between two land prices at different appraisal times. Depending on the base appraisal time, there are two types of LPI, chain (LPI_c), and fixed base (LPI_f), can be determined by the formula in Equations (1) and (2) and the land price monitoring data sets, where P_0, P_t, and P_{t-1} are the land price of monitoring point at base, t, and t-1quarter, respectively.

$$LPI_c = \frac{P_t}{P_{t-1}} \times 100 \quad (1)$$

$$LPI_f = \frac{P_t}{P_0} \times 100 \quad (2)$$

4.2. Spatial Distribution Model

In urban areas, it is not appropriate to use land price index to reflect the distribution of land price because land price index is in the form of point data instead of regional data. For different land use types (e.g., commercial, residential, and industrial), characterizing the spatial distribution is essential for understanding the structure of land price and predicting values for those areas without monitoring points [30,31]. Kriging is an interpolation technique of which the surrounding measured values are weighted to obtain a predicted value for an un-sampled location. It has been found that this technique has great advantages to analyze complex spatial land price distribution [42–44]. Equation (3) illustrates that the predicted value is a function of known values and their weights. Based on regionalized variable theory, the kriging technique assumes that the spatial variation in the data being modeled is homogeneous across the surface. That is, the same pattern of variation can be observed at all locations on the surface. In Equation (3), $Z^*(x_0)$ and $Z(x_i)$ represent the predicted and the measured values (land price monitoring points) of land price at point x_0 and x_i, respectively. λ_i are the weights of the surrounding measured values, which are based on the distance between the measured points and the predicted points, and the overall spatial arrangement among the measured points. To obtain unbiased and optimal estimation of predicted values ($Z^*(x_0)$), there are two constrained conditions in Equation (4) that should be satisfied. The first one is that the sum of λ_1, λ_2, …, and λ_i should be equal to 1, and the second is that the variance between $Z(x_i)$ and $Z^*(x_0)$ is minimum.

$$Z^*(x_0) = \sum_{i=1}^{n} \lambda_i Z(x_i) \quad (3)$$

$$\begin{cases} \sum_{i=1}^{n} \lambda_i = 1 \\ Var[Z^*(x_0) - Z(x_i)] \rightarrow min \end{cases} \quad (4)$$

The solution of Equation (4) relies on a construction of a Lagrange function (interested readers can find more details from Appendix A in reference [44]). Equation (5) is called the equation set of kriging by which λ_i can be solved, where $\gamma(x_i, x_0)$ is the semi-variance between predicted and measured value at points x_0 and x_i, $\gamma(x_i, x_j)$ is the semi-variance between two measured values at point x_i and x_j. μ is the Lagrange multiplier used for calculating of minimum variance. By using the data provided by the land price monitoring dataset and interpolation technique, a predicted surface of the land price of an entire urban area can be obtained.

$$\begin{cases} \sum_{i=1}^{n} \lambda_i \gamma(x_i, x_j) + \mu = \gamma(x_i, x_0) \\ \gamma(x_i, x_j) = \frac{1}{2} Var[Z(x_i) - Z(x_j)] \\ \gamma(x_i, x_0) = \frac{1}{2} Var[Z(x_i) - Z(x_0)] \end{cases} \tag{5}$$

4.3. Parcel Appraisal Model

Common methods of parcel appraisal, such as the cost approach, sales comparison approach, and income capitalization approach, are not applicable due to the lack of historical transactions or income benchmarks. In this situation, the land datum value method based on BLP is recommended to be another option according to "Regulations for valuation on urban land (GB/T18508-2014)". This method is especially suitable for those areas whose regional transactions are inactive or the land market is immature, and is widely used for fast evaluation of a number of parcels when there is a large area of housing demolition or land development. BLP is an average price of a region with average land conditions at a specific time point. Obviously, this average price cannot serve as the price of each parcel within the region because the heterogeneity of land parcels will be overlooked. Practically, each parcel holds different land conditions, such as their regional (business density, accessibility, *etc.*) and individual conditions (size, shape, *etc.*), plot ratio, degree of development, *etc.* In applying the land datum value method, the land conditions of the subject parcel are compared with the average land conditions stored in the dataset of the benchmark land price, and the difference is the basis to determine the adjustment coefficients from the table of adjustment factors accordingly. Mathematically, the calculation of the land price of a subject parcel (P) includes multiplying the regional average land price (P_b) by adjustment coefficients and adding adjustment value. Hence, P can be determined by Equation (6), where P_b is the regional average land price, K_i is the ith adjustment coefficient of the regional or individual condition, K_j is the jth adjustment coefficient of the appraisal time, plot ratio, and tenure, and D is the adjustment value of the degree of development.

$$P = P_b \times \left(1 \pm \sum K_i\right) \times K_j + D \tag{6}$$

5. System Design and Prototyping

A GIS-based web land-price information system was designed and developed following a prototyping methodology to satisfy the requirements discussed in previous sections. A four-level B/S architecture was used for the system development, where analysis models are part of GIS/application server components.

5.1. System Design

5.1.1. System Architecture

The purpose of designing and developing this system is to provide a platform to support online publishing, browsing, and analyzing of land price information. The architecture of the system is illustrated in Figure 1. The system adopts a four-level Browser/Server architecture that includes an interface tier, web services tier, business logic tier, and data tier. The interface tier provides client visualization of land price information. In addition, in this tier, the browser can send requests and receive analysis results related to land price. With published map objects and geo-processing services, the web services tier acts as a gate between the client and the server, and connects the interface tier and the business logic tire. The business logic tier is the core of the system, where land price analysis models are carried out when there is a request from the web browser. Finally, at the fourth level, the data tier, since there is a variety of spatial data and land price attribute information, a spatial and non-spatial database can be used to store and handle these data.

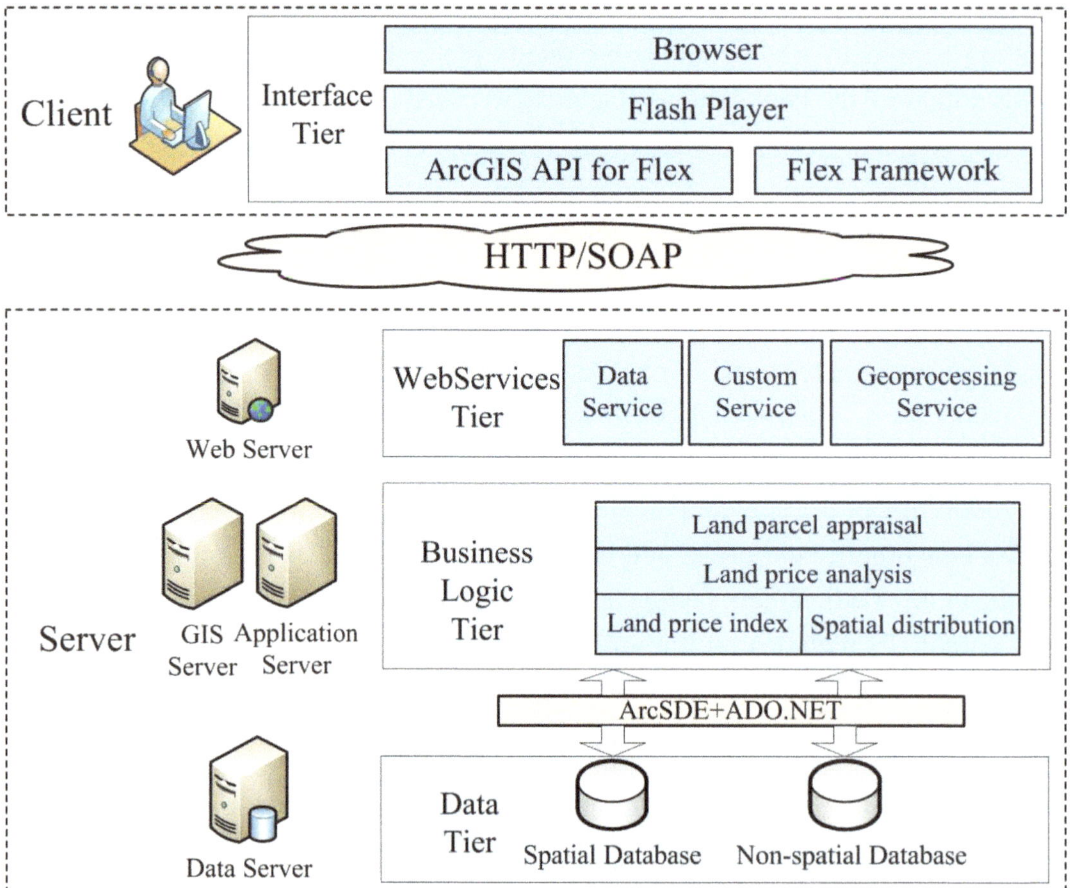

Figure 1. System architecture.

5.1.2. Users and User Requirements

Users of the GIS-based web information system consist of participants of land market such as sellers, buyers, and brokers. The user requirements presented in Figure 2, as a use case diagram of the Unified Modeling Language (UML), describe the functional requirements of the system. The system can be abstracted to some functional modules, such as basic map operation, information publishing, browsing, querying and analysis of land price, parcel appraisal, management and maintenance of data, *etc.* The users can be divided into three types of roles: *administrator,* from local land administration, *investor*, and *general user*. Different roles have different access privileges in the system. Administrators have the highest right to use almost all functions of the system, but mainly for publishing land price information by updating the database regularly. As for investors, in-depth analysis of land price, including the land price index, spatial distribution modeling, and appraisal of interesting parcels, are essential procedures before they make a wise decision of investing. General users with limited knowledge and authority can utilize the functional modules, for example, basic map operation, browsing, and querying, to access the land price information published by land administrators.

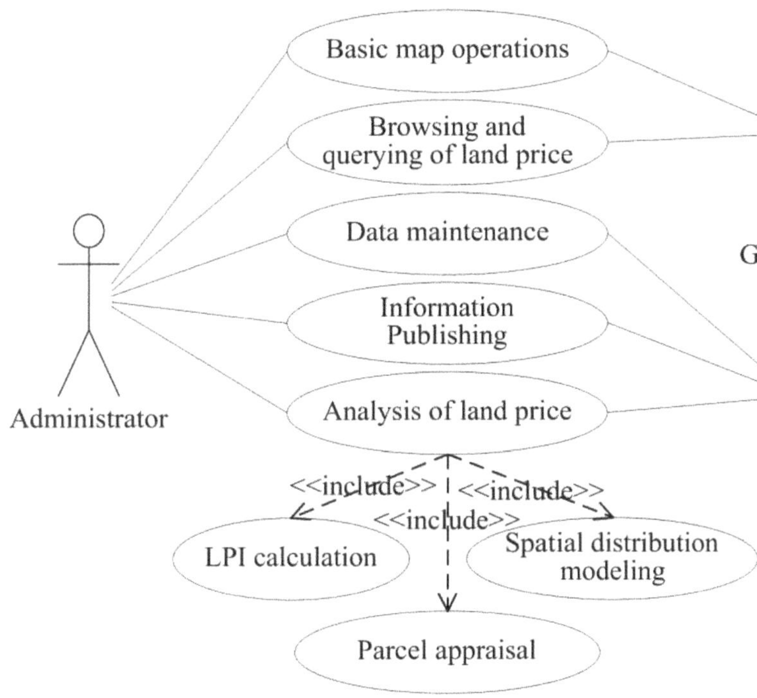

Figure 2. Use case.

5.2. Prototype Development

To facilitate the development of the proposed approach, Feng County was selected as a case study city. It was selected because it is a typical county-level city in China, whose land market has gradually become active but without sufficient information support.

The county is located in Jiangsu Province of Eastern China with a total urban area of 50.89 km^2. As a city experiencing rapid economic development and large expansion of urban areas in recent years, the Feng County's urban population rose from 83,900 to 397,700 and Gross Domestic Product (GDP) increased from 2.68 billion yuan to 32.00 billion yuan during the period of 1994–2014. Before the Chinese reform and opening-up, the urban area of the Feng County was restricted to the old town district situated in the Midwest. The land use right of a large amount of state-owned land has been transferred to the investors for the purpose of land development in the process of expansion to the south and east. As a result, a new city center equipped with better infrastructure has been formed in the central urban area. In the past three years, the LUR of more than 6.86 km^2 of land has been transferred and the land price (LUR's price) has held a growth rate around 1% quarterly. Although Feng County is one of the 2717 county-level cities in mainland China that have established BLP [45], the publication of BLP information still relies on traditional ways such as paper-based maps and documents. Therefore, adopting the GIS-based web information system has considerably practical significance and demonstrable effect.

5.2.1. Data Preparation

As part of the project "Updating of BLP in Feng County (Jiangsu Province, China)", the land resources bureau decided to shift the traditional way to publish land price information. We had collected the latest city maps, reports of BLP, LPM database, and land use planning of the urban area in Feng County. ArcGIS was used for data processing and establishment of land price datasets. Under the authority of administrator, the datasets were uploaded to the server of the system after validity checking.

5.2.2. System Implementation

Based on the system architecture designed above, a land-price information system with web map services via the Internet had been developed using such technologies as ArcGIS Server 10.0, ArcSDE, Flex Builder 4.6, and SQL Server 2008. As shown in Figure 1, due to its advantage of being easy to integrate with other software, SQL Server was selected as a database management system in this development. ArcSDE and ADO.NET technology were adopted as communication tools between the business logic tier and the data tier to facilitate management of spatial and attribute data in a relational database. ArcGIS Server was employed to ensure the display, query, and analysis of land price information. In this system, in order to provide effective land price services via the Internet, the Flex technology was used for the interface tier. Flex is a kind of Rich Internet Application (RIA) technology which lets users not only create nice look-and-feel and highly interactive interfaces, but also share some work of the servers, e.g., map layer control, map query, *etc.* Flex client uses Simple Object Access Protocol (SOAP) to call the data and function services provided by the server. By using ArcGIS API for Flex, map and other application tasks are integrated into the system. Additionally, two open source technologies, Open Flash Chart and Flash Away3D (version 4.07), were used to display spatial distributions of land prices in the forms of charts and 3D views.

5.2.3. Main Functions

Figure 3 shows the main interface of the prototype system, whose elements are organized around the county's map. The implementation of the main functions relies on several components shown in the user interface: (1) a map frame for viewing of spatial land price information (labeled as 1 in Figure 3), (2) universal GIS tools for manipulating maps, such as "zoom in", "zoom out", "identify", "print map", *etc.* (labeled as 2), (3) menu items for controlling the interesting map layers of land price data sets to be displayed (labeled as 3), and (4) an information board for browsing land price-related news and downloading documents (labeled as 4). Button-driven handling and scrolling ensures that the users can easily control the displayed map layers and browse the interested land price information. In addition, the requested results can be presented in the map frame and can be downloaded as images and documents. Compared with the traditional way, the proposed approach makes it a convenient and public-oriented way to publish land price information.

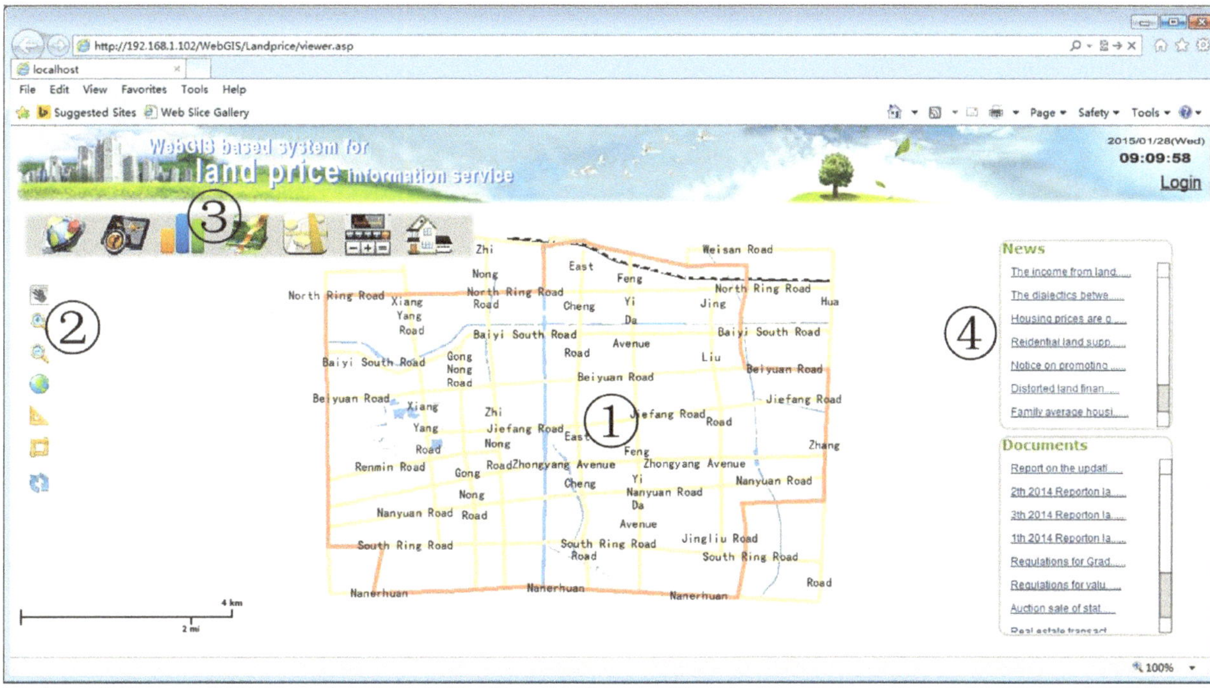

Figure 3. Main interface of the prototype system.

(1) Information publishing

After the system receives the request from the user via the user interface to browse and query the land price information, it retrieves the datasets stored in the data tier and then presents the inquiry results in the map frame. Apart from this, the publishing of land price information also includes the management and maintenance of data sources. Administrators of the land market have the right to upload new datasets to replace the old regularly, especially the datasets of benchmark land prices (every three years) and land price monitoring (every quarter). The automated process includes validating and updating the datasets layer by layer. In 2014, BLP and LPM were first opened to the public by this GIS-based web approach instead of using traditional methods.

(2) LPI calculation

LPI calculation is designed to know the trend of land price. To run this function (see an example in Figure 4), users need to select the objective land price monitoring points using either the map or the attribute-based query, and input time range, labeled as 1 and 2, respectively, in Figure 4. The system is designed to record these two parameters and send the instructions to the server to start the function. The server-side program performs the required computation based on the equations in Section 4.1 and the data retrieved from the data sets of land price monitoring. Line graphs are used to display the results within web browsers. To acquire a glimpse into the tendency of land price in the Feng County, LPI_f of four randomly selected land price monitoring points were determined by using the function of LPI calculation (the result is labeled as 3). The line chart illustrates that all of them maintain the same increasing trend with a maximum growth rate of more than 7% (land price monitoring point No. 003J12) during the time range of the four quarters in 2014.

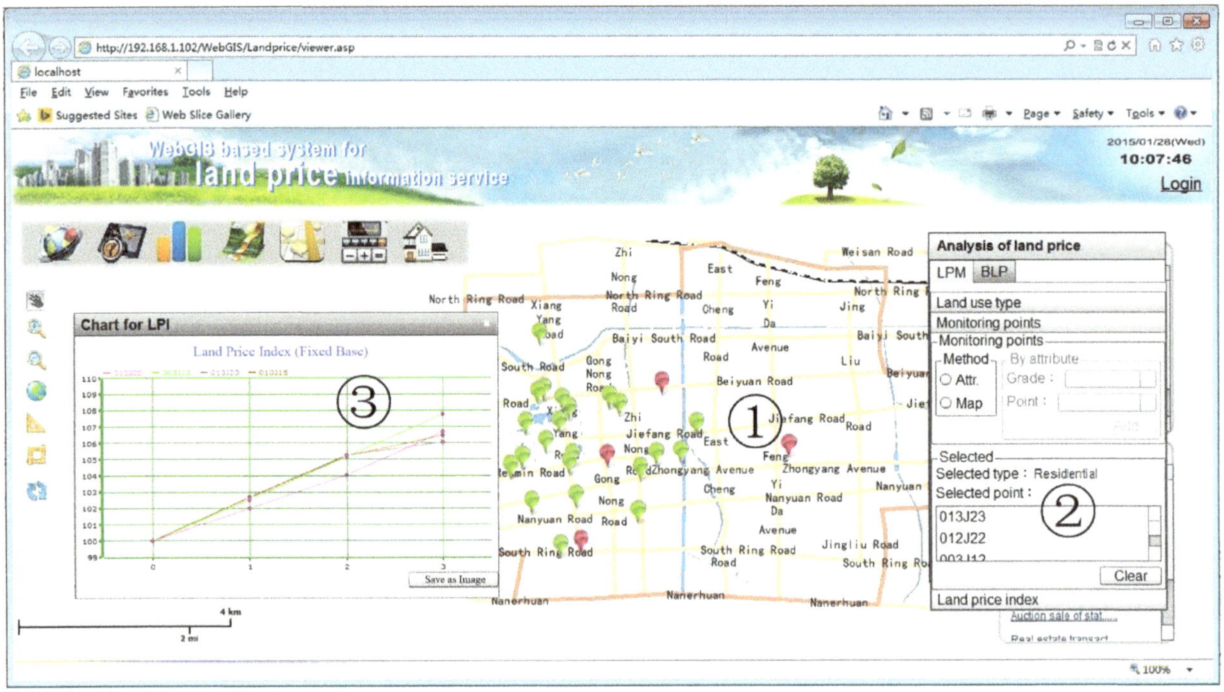

Figure 4. Land price calculation of selected monitoring points.

Figure 5. Spatial distribution modeling of land price.

(3) Spatial distribution modeling

Based on the model presented in Section 4.2, the system allows investors or other users to explore the spatial structure of land price for a particular land use type. The two required parameters, land use type and appraisal time, are listed in a drop-down menu for users' selection. There are several steps to model land price distributions after the server receives the submitted parameters from the user interface, including data retrieve and check, modeling the semi-variogram, cross-validation of results, and prediction of the land price surface. All of these processes are handled by server-side components, and the results are displayed on the map viewing frame of the user interface. A three-dimensional tool is provided to visualize the surface of the land price from different angles. Figure 5 gives an example of modeling the distribution of the land price with an industrial land use type and an appraisal time in the first quarter in 2014. The result is displayed as a grid surface in the map viewing frame (labeled as 1) and a three-dimensional visualization (labeled as 2). The section contour indicates that the Feng County's industrial land price has distribution patterns characterized by "high in central area, low in fringe area (range from red to green)" (labeled as 3), which is mainly in accordance with the quality of infrastructure in this case city.

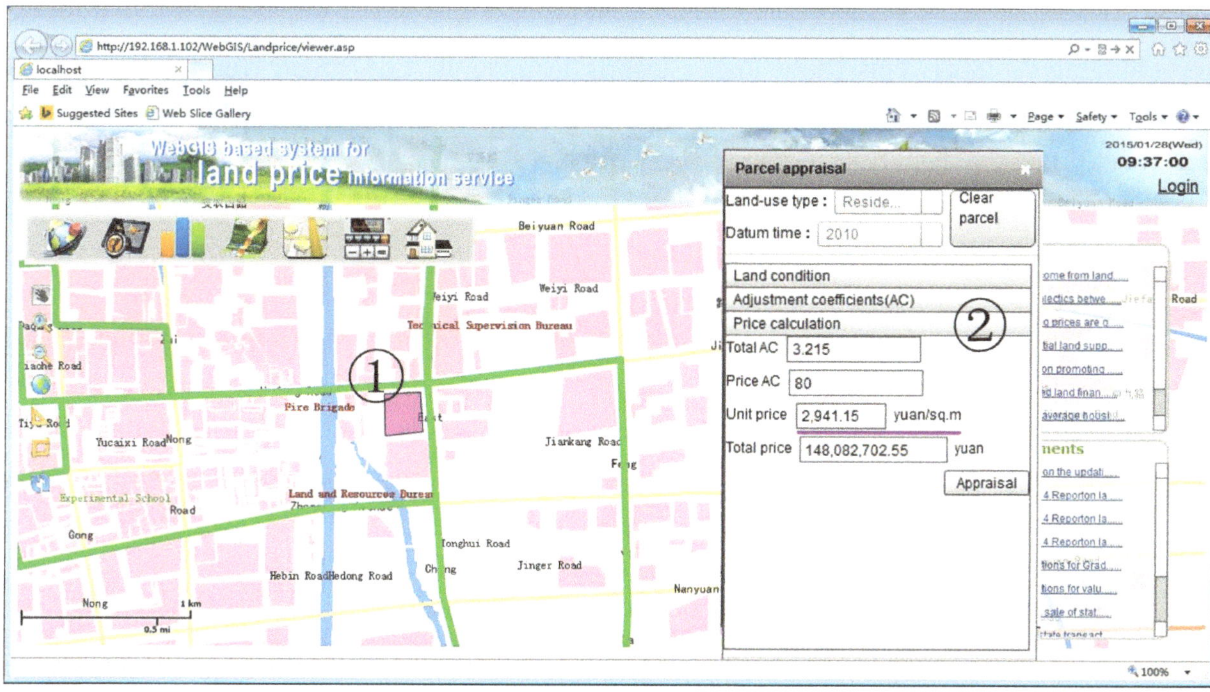

Figure 6. An example of using pracel appraisal function.

(4) Parcel appraisal

This function is intended for appraising the parcels of interest. Here, the term "parcel" is employed to describe a plot within an ownership boundary. The operation of this function starts from drawing or picking up a parcel (a polygon) on the map frame and specifying a few parameters of land conditions (*i.e.*, land use type, regional and individual conditions, plot ratio, and degree of development) on the pop-up form (labeled as 1 and 2, respectively, in Figure 6). The server first figures out whether the land conditions of the parcel conform to the city's land use planning. If negative, for example, the specified plot ratio is higher than the control standards of the district where the parcel is located, users will be prompted to change the parameters. Then, the server determines the adjustment coefficients of land conditions by making a comparison of the parcel and BLP region. The final step is to calculate the land price using the land datum value method and send back the results to the user interface. Based on the land datum value method and web-based approach, investors or administrators can achieve a quick and batch appraisal without physically appearing in the Feng County and collecting massive materials. The example in Figure 6 shown the price of the interested parcel is appraised as 2941.15 yuan per square meter (underlined in Figure 6). Since Feng County is undergoing urban expansion and renewal, more than 35 parcels (LUR) had been transferred during 2014. Their appraised values computed by the system have proved to be practical in providing a reference for the determination of floor price before parcel transaction. These results generated from the web system have given people further understanding and application of land price rather than tedious information.

6. Conclusion

Operation of urban land markets calls for an information base to support efficient data sharing and analysis. Great progress has been made in both GIS and land price theory areas. However, land price information service remains unsatisfied due to the lack of information sources and available spatial-temporal analysis tools. This paper makes efforts to integrate GIS-based web technology and land price analysis models to provide a novel approach for serving land price information. After reviewing the background and analysis models of land price, a system with a four-level B/S architecture was designed and implemented to meet the requirements of participants of land market.

The application in the case study city demonstrates that the novel approach helps connect land information providers and users, *i.e.*, land administrators are able to publish BLP-related information, thus reducing paper works, and other participants of the land market can query such information simply using web browsers through handy functions. Another advantage is that the system has implemented the spatial-temporal analytic techniques that consider the geographical features of land price, which allows investors and other users to conduct further analyses to understand the change in pattern and spatial distribution of land prices. With practical functionality and effective architecture, the combination of land price-related models and GIS techniques shows its potential to apply the approach to some other cities and to be extended as a model to the development of similar information systems.

However, the analytic methods and information sources of land price remain limited in the developed system. In the near future, the promotion of the system should include providing more analysis models, such as sales comparison approaches for parcel appraisal. Apart from the BLP-related data provided by land administrators, the volunteered geographic information is another promising data source; for this, the system can provide a platform for those system users to upload their data, such as the comparable cases of land transaction.

Acknowledgments

This work reported in this paper was supported by the National Natural Science Foundation of China (Grant No.41301433), and the Priority Academic Program Development of Jiangsu Higher Education Institutions. The authors would like to thank anonymous reviewers for their constructive comments.

Author Contributions

Yongjun Yang, Yaqin Sun and Songnian Li conceived and designed the study. Yongjun Yang, Kuoyin Wang and Shishuo Xu developed the system. Huping Hou collected and processed the data. Yongjun Yang and Yaqin Sun wrote the paper. Songnian Li and Shaoliang Zhang contributed in the preparation of the manuscript. All authors read and approved the manuscript.

Conflicts of Interest

The authors declare no conflict of interest.

References

1. Kettani, O.; Oral, M.; Siskos, Y. A multiple criteria analysis model for real estate evaluation. *J. Glob. Optim.* **1998**, *12*, 197–214.
2. Cao, G.; Feng, C.; Tao, R. Local "land finance" in China's urban expansion: Challenges and solutions. *China World Econ.* **2008**, *16*, 19–30.
3. Tao, R.; Su, F.; Liu, M.; Cao, G. Land leasing and local public finance in China's regional development: Evidence from prefecture-level cities. *Urban Stud.* **2010**, *47*, 2217–2236.
4. Lin, G.; Yi, F. Urbanization of capital or capitalization on urban land? Land development and local public finance in urbanizing China. *Urban Geogr.* **2011**, *32*, 50–79.
5. Buitelaar, E. A transaction-cost analysis of the land development process. *Urban Stud.* **2004**, *41*, 2539–2553.
6. Deininger, K.; Savastano, S.; Carletto, C. Land fragmentation, cropland abandonment, and land market operation in Albania. *World Dev.* **2012**, *40*, 2108–2122.
7. Dahlman, C.J. The problem of externality. *J. Law Econ.* **1979**, *22*, 141–162.
8. Zevenbergen, J.; Bogaerts, T. Alternative approaches for successful cadastral systems. In Proceedings of the 22nd Urban and Regional Data Management Symposium, Delft, The Netherlands, 11–15 September 2000; pp. 67–76.
9. Paz, P. Determinants of housing prices in Spanish cities. *J. Prop. Invest. Financ.* **2003**, *21*, 109–135.
10. Cichociński, P.; Dąbrowski, J. Spatio-temporal analysis of the real estate market using geographic information systems. *Real Estate Manag. Valuat.* **2013**, *21*, 72–82.
11. Nanda, A.; Yeh, J. Spatio-temporal diffusion of residential land prices across Taipei regions. *SpringerPlus* **2014**, *3*, 505–520.
12. Saefuddin, A.; Widyaningsih, Y.; Ginting, A.; Mamat, M. Land price model considering spatial factors. *Asian J. Math. Stat.* **2012**, *5*, 132–141.
13. Williamson, I.; Ting, L. Land administration and cadastral trends—A framework for re-engineering. *Comput. Environ. Urban Syst.* **2011**, *25*, 339–366.
14. Bennett, R.; Wallace, J.; Williamson, I. Organising land information for sustainable land administration. *Land Use Policy* **2008**, *25*, 126–138.
15. Mađer, M. Land information management-Cadastre. *Cartogr. Geoinf.* **2012**, *11*, 186–186.
16. Bordyuzha, A. Development of land cadastral system in Ukraine. *Balanc. Nat. Using* **2012**, *5*, 65–69.
17. Divithure, H.; Tang, C. Evolution of land registration and cadastral survey systems in Sri Lanka. *Surv. Rev.* **2013**, *45*, 126–135.
18. Kurwakumire, E. Digital cadastres facilitating land information management. *South Afr. J. Geomat.* **2014**, *3*, 64–77.

19. Hanstad, T. Designing land registration systems for developing countries. *Am. Univ. Int. Law Rev.* **1997**, *13*, 647–703.
20. Azad, B.; Faraj, S. E-Government institutionalizing practices of a land registration mapping system. *Gov. Inf. Q.* **2009**, *26*, 5–14.
21. Nasralla, Z.H.; Aleqabie, H.J.; Thamer, Z. Design and implementation of Karbala real estate information system. *J. Kerbala Univ.* **2013**, *1*, 240–248.
22. De Vries, P.; Faber, R. Towards a real estate monitoring information system in Romania. *Romanian Econ. J.* **2009**, *32*, 187–214.
23. Stanfield, D.; Raço, M. Land markets, information, and a property registration system in Albania. *Comput. Environ. Urban Syst.* **1994**, *18*, 133–141.
24. Kisilevich, S.; Keim, D.; Rokach, L. A GIS-based decision support system for hotel room rate estimation and temporal price prediction: The hotel brokers' context. *Decis. Support Syst.* **2013**, *54*, 1119–1133.
25. Porta, J.; Parapar, J.; García, P.; Fernández, G.; Touriño, J.; Doallo, R.; Óncga, F.; Santé, I.; Díaz, P.; Miranda, D.; *et al.* Web-gis tool for the management of rural land markets application to the land bank of Galicia (NW Spain). *Earth Sci. Inf.* **2013**, *6*, 209–226.
26. Pagourtzi, E.; Assimakopoulos, V.; Hatzichristos, T.; French, N. Real estate appraisal: A review of valuation methods. *J. Prop. Invest. Financ.* **2003**, *21*, 383–401.
27. Zeng, T.Q.; Zhou, Q.M. Optimal spatial decision making using GIS: A prototype of a real estate geographical information system (REGIS). *Int. J. Geogr. Inf. Sci.* **2001**, *15*, 307–321.
28. Rodriguez, M.; Sirmans, C.F.; Marks, A.P. Using geographic information systems to improve real estate analysis. *J. Real Estate Res.* **1995**, *10*, 163–173.
29. Li, H.; Yu, L.; Cheng, E.W. A GIS-based site selection system for real estate projects. *Constr. Innov.* **2005**, *5*, 231–241.
30. Pagourtzi, E.; Nikolopoulos, K.; Assimakopoulos, V. Architecture for a real estate analysis information system using GIS techniques integrated with fuzzy theory. *J. Prop. Invest. Financ.* **2006**, *24*, 68–78.
31. Tsutsumi, M.; Shimada, A.; Murakami, D. Land price maps of Tokyo metropolitan area. *Procedia Soi. Behav. Sci.* **2011**, *21*, 193–202.
32. Hu, S.; Cheng, Q.; Wang, L.; Xu, D. Modeling land price distribution using multifractal IDW interpolation and fractal filtering method. *Landsc. Urban Plann.* **2013**, *110*, 25–35.
33. Tsutsumi, M.; Seya, H. Measuring the impact of large-scale transportation projects on land price using spatial statistical models. *Pap. Reg. Sci.* **2008**, *87*, 385–401.
34. Xie, Q.; Parsa, A.R.G.; Redding, B. The emergence of the urban land market in China evolution, structure constraints and perspectives. *Urban Stud.* **2002**, *39*, 1375–1398.
35. Liu, Y.; Zheng, B.; Huang, L.; Tang, X. Urban residential land value analysis: Case Danyang, China. *Geo-Spat. Inf. Sci.* **2007**, *10*, 228–234.
36. Li, L.H.; Walker, A. Benchmark pricing behaviour of land in China's reforms. *J. Prop. Res.* **1996**, *13*, 183–196.
37. Xu, Z.; Li, Q. Integrating the empirical models of benchmark land price and GIS technology for sustainability analysis of urban residential development. *Habitat Int.* **2014**, *44*, 79–92.

38. Wang, R. The structure of Chinese urban land prices: Estimates from benchmark land price data. *J. Real Estate Financ. Econ.* **2009**, *39*, 24–38.
39. Ding, C. The benchmark land price system and urban land use efficiency in China. *Chin. Geogr. Sci.* **2001**, *11*, 306–314.
40. China Land Surveying and Planning Institute. China Urban Land Price Dynamic Monitor. Available online: http://www.landvalue.com.cn/ (accessed on 3 July 2015).
41. Jiangsu Land and Resources Department. Jiangsu Land Market Website. Available online: http://www.landjs.com/ (accessed on 3 July 2015).
42. Martinez, M.G.; Lorenzo, J.M.M.; Rubio, N.G. Kriging methodology for regional economic analysis: Estimating the housing price in Albacete. *Int. Adv. Econ. Res.* **2000**, *6*, 438–450
43. José-Ma, M.; Beatriz, L. Estimating housing price: Kriging the mean. *Int. Adv. Econ. Res.* **2006**, *12*, 419–419.
44. Johnston, K.; Ver Hoef, J.M.; Krivoruchko, K.; Lucas, N. *Using ArcGIS Geostatistical Analyst*; ESRI: Redlands, CA, USA, 2001; Volume 300.
45. Dong, W. Benchmark Land Price in China: Present State and Outlook. Available online: http://www.fig.net/pub/fig2013/ppt/ts06e/TS06E_dong_6561_ppt.pdf (accessed on 3 July 2015).

3

Simulating and Communicating Outcomes in Disaster Management Situations

Michal Lichter *, Asher Yair Grinberger and Daniel Felsenstein

Department of Geography, Hebrew University of Jerusalem, Mount Scopus 91900, Israel;
E-Mails: asherya.grinberger@mail.huji.ac.il (A.Y.G.); msdfels@mscc.huji.ac.il (D.F.)

* Author to whom correspondence should be addressed; E-Mail: mlichter@gmail.com

Academic Editors: Christoph Aubrecht and Wolfgang Kainz

Abstract: An important, but overlooked component of disaster managment is raising the awareness and preparedness of potential stakeholders. We show how recent advances in agent-based modeling and geo-information analytics can be combined to this effect. Using a dynamic simulation model, we estimate the long run outcomes of two very different urban disasters with severe consequences: an earthquake and a missile attack. These differ in terms of duration, intensity, permanence, and focal points. These hypothetical shocks are simulated for the downtown area of Jerusalem. Outcomes are compared in terms of their potential for disaster mitigation. The spatial and temporal dynamics of the simulation yield rich outputs. Web-based mapping is used to visualize these results and communicate risk to policy makers, planners, and the informed public. The components and design of this application are described. Implications for participatory disaster management and planning are discussed.

Keywords: agent based modeling; Web GIS; spatial computing; hazard management

---

1. Introduction

Communicating the likely outcomes of catastophic events to potential stakeholders is an integral part of disaster management. Building community resilience is as much about improving accessibility to information and arousing awareness of hazards, as it is about engaging engineering solutions.

However, this aspect of post-disaster management has been less investigated as it deals with longer term, less tangible outcomes. Short-term protective fixes for hazards are always easier to quantify, visualize, and promote [1]. In the case of coastal flooding, for example, defensive mechanisms can be easily illustrated and animated while the disequilibrating effects on population and land use are harder to convey.

The situation is further confounded by the fact that technology for potentially conveying these outcomes is moving ahead faster than the ability to generate applications. This means that much potential information that could be made available lies dormant for lack of suitable means of communication. In addition, the frequency and intensity of natural disaster events is becoming ever more extreme and less predictable. This serves to underscore the importance of a long-term perspective over a short-term hazard response. Aside from the immdediate needs for flood evacuation, disease prevention, building reconstruction, and the like, disaster management needs to also be concerned with raising awareness and preparedness through communicating plausible outcomes.

This paper illustrates how some of the analytic tools of geo-informatics can be harnessed for both generating and conveying outcomes in disaster management to a broader audience. For generating disaster outcomes, we use agent-based simulation. We utilize dynamic web mapping as the vehicle for communicating these outcomes. The next section discusses some of the uses of web GIS for disaster management. This is followed by a concise description of the analytic modeling framework in Section 3. The simulation results are presented in Section 4 where emphasis is placed on the very different outcomes implied by each disaster. Some of the technical features of the design of the platform including data formats and ancillary functionalities are described in Section 5. In Section 6 we present the web platform for communicating these outcomes and discuss its features. Finally, we conclude with some implications for public participation in disaster management arising from the increasing transparency of methods and outcomes.

2. Web Mapping for Disaster Management

Web GIS has been heralded as a key component in hazard management and vulnerability assessment [2]. It extends desktop GIS capabilities to an internet environment and thus encourages the development of applications that are accessible, dynamic and interactive. In this respect, it releases disaster managers from the tasks of data collection and map generation and allows them to focus on visualization and analysis. As maps are an intuitive and user-friendly medium for communicating risk [3], it is not suprising that the development of web-based mapping has been conceived as a central axis in incorporating public participation in disaster management. Little *et al.* [4] for example, show how web-based geovisualization tools can both encourage stakeholder involvement and public input into emergency management. The result is a framework that goes beyond improving disaster response and can contribute to the wider organizational realms of training, awareness enhancement and team building.

Many examples exist in which web mapping has been utilized in emergency management. For example, Hagemeier-Klose and Wagner [5] evaluate the use of web mapping services in communicating flood risk, and Kwan and Lee [6] analyze the potential use of real-time 3D GIS in the case of terror attacks. In fact, web mapping has raised its public profile through a series of natural and man-made disasters starting with the WTC attack on New York in 2001 and progressing through the ravages of Hurricane Katrina in New Orleans (2005), the Haiti and Christchurch earthquakes in 2010 and 2011, the Tokohu earthquake and tsunami (2011), and Superstorm Sandy (2012).

It should be noted that web-mapping these catastrophic events encourages collective and participatory activity and affords access to large scale data hitherto unavailable. Crowdsourcing, volunteered information, map mashups, the development of synthetic big data and the like, all challenge the conventional information chain [7]. While web mapping is proclaimed as the ultimate democratizer that delivers information, empowers the public, and reduces the digital divide, caution needs to be taken to avoid methodological pitfalls. Web-based information allows for immediate change in spatial resolution. This can encourage misuse with respect to scale and issues of ecological fallacy. In addition, the ease in which data overlays can be performed can lead to suggestive, but spurious, correlations. Thus, while web-based platforms for disaster management are here to stay, care needs to be invested in their execution and design. They need to deliver outcomes in a seamless and non-technical fashion and these outcomes needs to have been generated in a plausible manner. It is to these two issues that we now turn.

3. Disaster Simulation Framework

We illustrate how geo-information can be generated by using an agent-based (AB) framework to simulate the long-term consequences of a disaster. While other modeling frameworks exist for disaster analysis (see for example, the suite of applications using multi-regional input-output modeling in Richardson *et al* [8]), the value of AB models lies in their ability to create high-resolution representations of the urban environment. The long-term indirect effects of an event are reflected in the behavioral responses of the agents. "Shocks" generated by a disaster are mediated through the aggregate behavior of "agents" (households, workers, land developers, firms, city authorities, and intervention agencies). These agents create complex network patterns of change. The patterns are not predicatable through simply aggregating individual agent behavior. However, the micro-scale interactions between individual agents can be modeled within a computable system grounded in the basic tenets of micro-economic behavior. This gives a rich set of opportunities for understanding the reactions of affected populations under varying conditions, times frames, levels of aggregation and spatial scales. Additionally, the outcomes of the disaster can be tied to local, place-specific circumstances. This frees disaster management from the constraints imposed by coarse administrative borders. Additionally AB models can represent dynamics at high levels of temporal resolution. Therefore, the AB framework has been readily applied in the context of natural disaster scenarios such as flooding, fires, and earthquakes [9–13].

AB models are based upon three elements: the environment, the agents, and a set of rules guiding agent-agent and agent-environment interactions [14]. The latter may be defined within the model based on social, economic, and spatial decision rules. The first two however are exogenous starting conditions of the simulation. As spatial socio-economic and urban data is usually available at aggregates such as census tracts, agent-level data must be generated synthetically. Here we use data disaggregation techniques to create "big" spatial data in which census tract-level socio-economic values are synthetically distributed over buildings, households, and individuals (see Figure 1). The allocation algorithm preserves aggregate census tract-level values. The resulting database is used to represent the starting conditions upon which supply and demand dynamics following a disaster are simulated. These procedures are used to evaluate the long term effects of two scenarios: an earthquake and a missile attack both simulated on the same area in downtown Jerusalem, Israel.

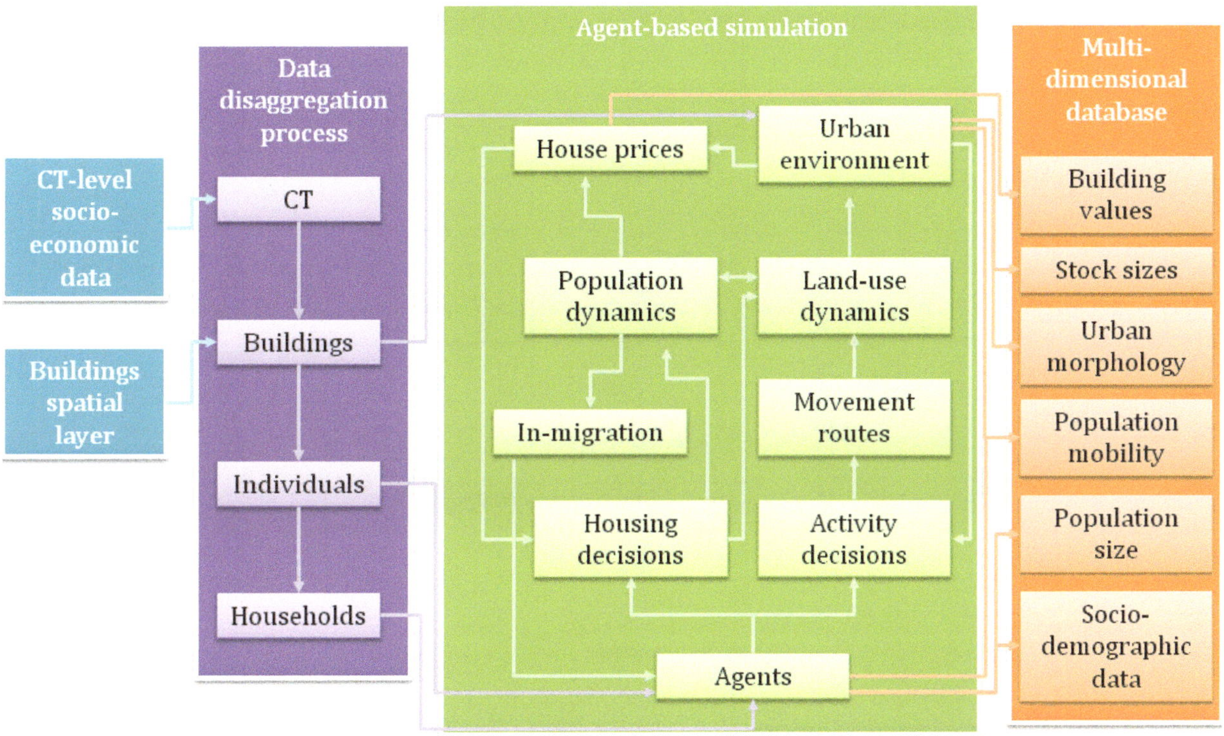

Figure 1. The Analytic Framework.

3.1. Spatial Context

The simulation is applied to the Jerusalem city center, a mixed-use area including two major commercial centers (the Mahaneh Yehuda enclosed market and the CBD), a number of other commercial and public-use venues and many low-rise residential building (see Figure 2). Three major traffic arteries traverse the area: Agripas St. and Jaffa St. (light-railway route) running north-west to south-east and King George St. running north-south. The area comprises 22,243 residents, 717 residential buildings (243Th sqm), 119 commercial buildings (505Th sqm) and 179 governmental/public buildings (420Th sqm). The occurrence of both scenarios in the area is probable. It is situated 30 km north-west of the active Dead Sea fault and while the area is relatively seismically stable, most of the buildings in this part of the city were constructed prior to the introduction of earthquake-related building codes. This makes them prone to damage in such a scenario [15]. In addition, the area was a focal point for various terror attacks over the years, including missile attacks in the summer of 2014.

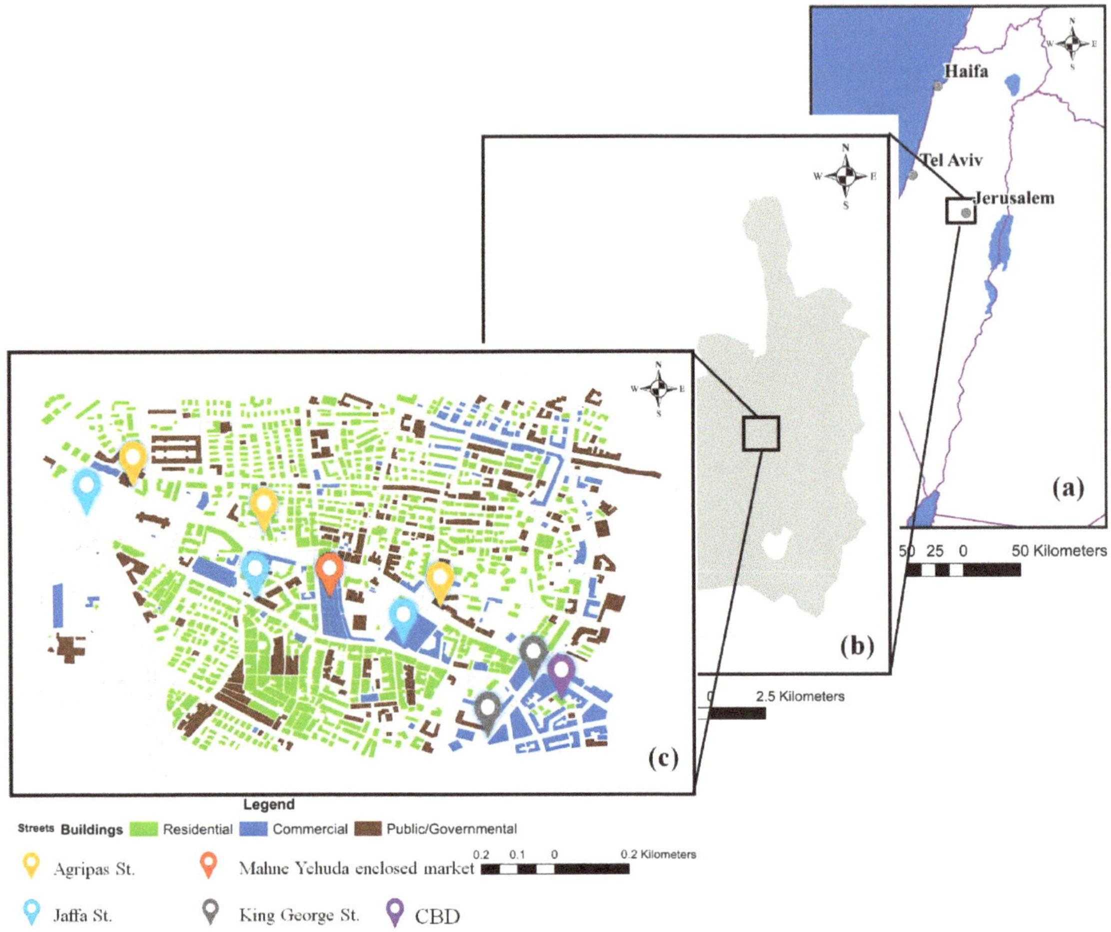

Figure 2. The study area (**a**) Israel; (**b**) Jerusalem municipal boundaries; and (**c**) study area.

3.2. Data Disaggregation

The literature offers a number of data disaggregation techniques such as population gridding [16], areal interpolation [17], dasymetric representation [18,19] proportional iterative fitting [20,21], and dynamic population modeling [22,23]. The method we use here is more in line with the technique used by Harper and Mayhew [24,25]. We combine administrative data available at a coarse spatial level and a detailed buildings GIS layer in order to create spatial representations of individuals and households within buildings and allocate synthetic socio-economic values to them. This involves two stages of down-scaling and one stage of clustering. We use an allocation algorithm to disaggregate from census tracts (CTs) to buildings and from buildings to individuals. We then cluster individuals into households (see Figure 1). At each stage, the dataset is populated with the following values:

- Buildings: land-use, floor-space, number of floors, building values, number of households.
- Households: number of members, earnings, car ownership
- Individuals: household membership, disability, participation in the work force, employment sector, age group, workplace location.

The sources of the data used in this process are the 2008 Israeli Census (for households, individuals and earnings, disability, age, labor force participation and employment by sector), a GPS-survey (for workplace location), National Tax Authority data (residential property value per meter), and capital stock estimates [26].

Socio-economic values are allocated to buildings in the first stage in proportion to their floor-space. See Lichter and Felsenstein [27] for a full articulation of the allocation method. As Equation (1) illustrates for population size, individual buildings values are calculated by multiplying CT-level densities with buildings-level floor-space:

$$Pop_b = fs_b * \frac{Pop_c}{fs_c} \tag{1}$$

where *Pop* is population size, *fs* is floor-space, *b* is individual building and *c* is census tract.

Floor-space is calculated according to aerial footprint and height in meters, assuming a floor-height of 5 m for residential buildings and 7 m for non-residential buildings. These figures were derived by comparing the calculated sum total of floor-space over all buildings by use with total national built floor-space. This proportional allocation process necessarily entails a loss of data due to the division of integers (e.g., population) by fractions (e.g., floor-space). The SQL-based allocation algorithm compensates for this by adjusting the floating point figures rounding threshold for each variable separately. In this manner, the algorithm verifies that CT control totals are met.

At the second stage, each of the individuals is given a unique id that is tied to a specific building and is located at a random location within the building. Next, demographic values (e.g., age, disability, workforce participation) are allocated to individuals so that the entire set of residents within a building represents the distribution of socio-economic variables within it. This distribution corresponds to the CT distribution from the previous stage. Under this allocation system, the socio-demographic structure of households in multi-unit buildings is homogenous while for single household units it is variable. Current research uses a more refined down-scaling method whereby disaggregation is based on representative distributions grounded in the national census rather than on floor-space area. Finally, individuals within a building are clustered into heterogeneous households. These represent a "traditional household" including both adults and children when possible. The clustering algorithm iterates through individuals and aims to create new household entities which are not identical but are closely similar in terms of age representation. Each household has a unique id, is assigned to a building, and individuals are assigned to it. This process results in a high-resolution spatial database at the national scale that includes accurate synthetic representations of 7,354,200 individuals allocated to 771,226 residential buildings (out of 1,075,904 buildings).

3.3. Agent-Based Simulation Dynamics

The synthetic big database is used to characterize the starting conditions of the urban simulation in terms of both environment and agents. Each synthetic representation of an individual and a household is transformed into an agent. A socio-economic profile, which consists of a set of variables unique to each agent type, is defined for each agent-age group, disability, employment status, employment sector, workplace location, household membership for individuals and monthly income, car ownership, and members for households. Values are allocated to these variables according to the characteristics of the relevant synthetic entity. Place of employment for agents working within the study area is the only variable whose values are generated within the model. Agent profiles are translated into behavior expressed through the operation of basic logical decision rules and constraints. Agents are rational, utility-seeking entities whose preferences reflect a mix of behavioral assumptions—satisficing behavior [28], residential segregation [29], and risk evasiveness.

In accordance with agent-based modeling, the dynamics of the city are characterized "bottom-up". The individual actions of atomic units accumulate as aggregate changes. Households and their members are defined as agents. Their movements through buildings over the road network (representing the environment) over both diurnal and long-term temporal scales leads to environmental change (see Figure 1). The model is implemented within the Repast Simphony 2.0 modeling platform [30]. Each model iteration reflects one day in the urban setting in which agent behavior affects land-use, urban morphology, capital stock values, and population dynamics. The model is fully described in Grinberger, Lichter and Felsenstein [11].

Agent behavior revolves around two location decisions-residential location choice made at the household level and individual-level activity decisions. These decisions are based on a combination of constraints (such as budget constraint) and preferences (such as segregative residential tendencies). Equation 2 shows an example of such a decision process in the case of residential location:

$$h_{hh} = b_j \Rightarrow \left[\frac{I_{hh}}{3} > HP_j\right] * \left[k_{hh} > \frac{\Phi\left(\frac{\bar{I}_b - \bar{I}_h}{I_{\sigma_h}}\right) + \Phi\left(\frac{\bar{A}_b - \bar{A}_h}{A_{\sigma_h}}\right)}{2}\right] = 1 \qquad (2)$$

where: h is home location for household hh, b is a building considered as new residence place, $\{x\}$ is a binary expression where 1 if x is true and 0 otherwise, HP is monthly cost of living in building j, k is a randomly drawn preference value for the household, $\Phi(x)$ is the cumulative normal distribution value for x, $\bar{A}$,$\bar{I}$ are building-level average age and income respectively and A_σ, I_σ are standard deviation values for age and income in a building.

In the above equation, the constraint is a budget constraint. The randomly drawn preferences reflect segregative behavior by representing limited tolerance to changes in the demographic nature of the residential neighborhood. Similarly, activity decisions are based on land-use suitability and attractiveness in relation to random preferences. Attractiveness is related to the nature of the environment of the location, distance to current location (which is weighted by the mobility profile of the individual) and floor-space volume. In both decision processes, the first location to satisfy both constraints and preferences is selected in Accordance with satisficing behavior. Number of activities per day is set for each agent according to its mobility profile (age, disability, household car ownership) and employment profile (work force participation, workplace location). Residential mobility is motivated either by exogenous migration probabilities (intra- and inter-urban) or by changes to the environment (land-use change or destruction by disaster). In-migration is also considered where the number of potential new households in each iteration is dependent upon an in-migration/out-migration ratio and proportional to the number of vacant dwelling spaces.

The bottom-up dynamics described above capture demand-side interactions. A comprehensive urban model must also relate to supply side activities. Such interactions are simulated here by conceptualizing different spatial units-CTs, buildings, and residential units—as environmentally sensitive entities within a top-down environmental influence procedure (see Figure 1). This process is expressed by changes in land-use patterns and residential units' monthly (rental) prices. Land-use changes are based on the ratio of floor-space volume to local average traffic loads. We assume that higher volume traffic reflects more revenue from visits (represented by traffic) and greater commercial activity. Low ratio values (high traffic loads in relation to floor-space volume) encourage commercial activity at the expense of residential supply, while high values make the success of commercial activity less probable. The logistic probability of land-use change is based on the standardized cumulative exponential distribution values of floor-space and traffic loads in order to avoid inflation of small commercial uses and deflation of large land uses. Land-use changes affect house prices, along with changes in supply (number of residential buildings in a CT) and demand (the population size of the CT). Changes to these elements lead to changes in average house prices per meter in the CT. Rising service levels (number of non-residential units) drive up prices and in line with standard economic theory, prices will drop with increase in supply or decrease in demand. CT-level changes trickle down to the level of the individual building proportionate to floor-space volume. In addition, building values are adjusted according to their accessibility to non-residential (*i.e.*, service) functions. Greater accessibility makes a building more valuable. Finally, the value of each residential unit within a building is calculated, assuming uniform values within a building. Values are transformed into monthly housing costs in accordance with the population's willingness-to-pay (represented by the budget constraint in Equation (2)).

3.4. Simulated Scenarios

We simulate two very different scenarios (Table 1). The first relates to multiple missile attacks. These represent long term continuous, low level attrition of the urban system. The attacks are random in space, time and quantity (magnitude) and there is no single focus of the event. The second scenario is an earthquake. This delivers a catastrophic one-time shock to the urban system and has a defined focus (aftershocks not withstanding). Disaster management in both cases relates to the effects on the urban system in terms of speed to recovery, population and land use change and shifts in urban morphology.

Table 1. Characteristics of Scenarios Simulated.

Scenario	Scenario Duration	Effects	Event Dynamics	Scale	Number of Event Foci
Multiple Missile Attack	Long term	Temporary	Diffuse	Citywide	Many
Earthquake	Long term	Permanent	Focused	Citywide	One

Differences between the scenarios are formulated in relation to their temporal extent, intensities, direct and indirect effects. While the earthquake may affect an entire urban area, its impact is determined in relation to one focal location. Missile attacks, on the other hand, have more localized impacts that diffuse through the urban system via multiple foci. Moreover, effects of an earthquake tend to be irreversible (total destruction of buildings) while, in the missile attack scenario, such impact is rare and the temporal-behavioral effect is more dominant. Both the first missile attack and the earthquake occur at day 50 (*i.e.*, the 50th simulation iteration) in order to afford a "run-in" period for the urban system to stabilize. While the earthquake shock occurs only once, the missile-attacks continue to appear until the simulation ends. The focal center of the shock is determined randomly to avoid place-based bias.

The earthquake is simulated as spreading outwards from the epicenter with declining intensity. The direct physical effect of this shock is manifested through the collapse of buildings. This effect is probabilistic in nature. The chance of a building being damaged is proportional to its distance from the epicenter and its height. The road nearest to a collapsed building is blocked and remains so until the building is restored. The restoration period is proportional to floor-space area. Upon physical restoration, the building may or may not also restore its functionality. This formalization of the hazard is necessarily simplistic. Given our aim of assessing socio-economic rather than physical resilience, we feel this approach is justifiable. The high variance in the distribution of physical damage across multiple simulations allows for estimating resilience levels that are not dependent on a specific event.

The missile attack is simulated as a series of multiple local shocks appearing every day and varying in number from 0–10. If a missile hits a road, it remains blocked for that iteration. If a building is hit, there is a 5% probability (depending on a randomly drawn value) of serious damage which results in the same physical effect as building collapse in the earthquake scenario. In addition to this effect, missile hits impact decision making processes. The accumulation of hits in a single place makes it less attractive as a residential location (reflecting risk-evasiveness), inducing out-migration and reducing in-migration. This is formalized in the behavior of residents. If their close neighborhood (50 m radius) is hit more than three times over the preceding 30 iterations, they are given a 20% probability of relocating. Such buildings have a zero chance of attracting new residents. Activity location decisions are also affected as agents avoid choosing destinations that were hit over the course of the current day. Exogenous shocks therefore directly affect urban dynamics via physical changes (destruction of buildings and disruption to movement) and psychological effects (risk evasiveness). Due to the model dynamics, these direct effects lead to a second round of indirect effects.

4. Simulation Outcomes

The results of the simulations include time-series data at high-level spatial resolution. They relate to changes in capital stock and population size, building values, urban morphology, and functionality. The richness of the outcomes facilitates analysis of urban development at multiple scales as well as across different scenarios. This section presents examples for macro and micro analysis and comparisons that unveil the main processes of change. These results relate to 25 simulations per scenario where each simulation comprises of 1010 simulated days. The results reported are the simulation averages.

4.1. Macro Trends

Table 2 presents a macro comparison of event effects in the two scenarios. We observe the tendency of key variables to return to pre-shock values and converge over time (over different values). Variables are considered convergent (*i.e.*, reach equilibrium) if they show no significant change over the last 50 iterations or more. The values in Table 2 indicate the day at which on average the last significant change was registered until the end of the simulation.

Table 2. Recovery of the Urban System by Simulated Scenario.

Parameter	Variable	Earthquake			Missile Attack		
		Average Final Change (% of Pre-Shock Value)	**Frequency of Equilibrium (out of 25 Simulations)**	**Average Duration to Achieve Equilibrium (Days)**	**Average Final Change (% of Pre-Shock Value)**	**Frequency of Equilibrium (Out of 25 Simulations)**	**Average Duration to Achieve Equilibrium (Days)**
Population	Population	67.85	24	397	70.39	3	944
	Average Income	50.53	11	950	50.62	6	954
Residential stock	Residential Stock Size (# buildings)	88.34	25	332	93.30	15	948
	Average Residential Value	96.12	22	677	90.93	7	951
Non-residential stock	Non-Residential Stock Size (# buildings)	142.43	23	670	73.55	4	948
	Average Non-Residential Value	78.61	25	385	89.45	12	926

Population and residential stock variables show similar behavior in the two scenarios in regard to regaining pre-event values. It seems that both types of shocks induce an out-flow of migration which includes more wealthy households. This leads to a slightly reduced and cheaper housing stock. Other variables point to striking differences between the scenarios. First, the immense growth of non-residential stock in the earthquake scenario is replaced by a reduction in size in the missile attack. Moreover, convergence over time is rarely achieved in the latter, while it frequently occurs in the former. These results may be attributed to the nature of the shocks. The missile attacks call for continuous adaptation in face of an on-going event. This erodes reorganizational ability that may be more attainable in the case of a one-time shock such as an earthquake.

4.2. Micro Scale Change

Figure 3 presents micro-scale results relating to the different population and land-use dynamics resulting from the simulated disasters. In this figure, buildings are characterized at each point in time in terms of the most frequent land-use and the average Social Vulnerability Index (SVI) of resident households over the simulations. Following Felsenstein and Lichter [31] we use four key social variables obtained from the national census (2008) to define social vulnerability scores of households. The SVI relates to demographic characteristics which may enhance or constrain a household's adaptation capabilities. The index is constructed as follows:

$$SVI_{hh} = 0.5I_{hh} - 0.2A_{hh} + 0.1Car_{hh} - 0.2Dis_{hh} \quad (3)$$

where *hh* is a household, *I* is monthly income, *A* is the share of dependent members, *Car* is car ownership, and *Dis* is the share of household members with disability. All values are standardized.

The variables comprising the SVI have been chosen with discretion. "Household income" represents a direct measure of socio-economic status. "Disabilities" represent the share of disabled persons in a household. It combines the share of persons who are unable or have difficulty walking, hearing, seeing, have memory problems or unable to dress and shower independently. "Age" portrays the size of the dependent population in the household and combines the percentage of persons over 65 and under 18. "Car" depicts households with no car, one car or more. While this can be construed as a measure of wealth it is also a measure of a capacity of evacuate in the case of an extreme event. We regard this indicator with caution since in large cities for example, it is not always an efficient indicator of wealth and some hazards such as sea level rise, do not require rapid evacuation. Therefore, this indicator is assigned a weight of 0.1 in the overall index. We divide the national distribution of each variables quintiles (1: least vulnerable, 5: most vulnerable) and assign each household area to its respective class.

The strong direct impact of the earthquake leads to immediate consequences. The population is forced to relocate and the initial northeast-southwest divide between more and less vulnerable populations (Figure 3, $t = 50$) is fractured (Figure 3, $t = 100$). In contrast, the missile attacks do not create such an accentuated result. The same process of dispersal and re-concentration only appears after damage has accumulated over time (Figure 3, $t = 1000$). This is because the effects on accessibility caused by physical destruction lead to indirect effects on the land-use system. As traffic disperses due to the inaccessibility of some roads, new commercial functions may develop in areas which attract more traffic at the expense of existing commercial venues located near previously busy roads. As traffic is constantly diverted by the continual flow of shocks in the missile scenario, new commercial functions struggle to survive. Accordingly, commercial activity goes into slow and protracted decline, as evidenced by the succession of commercial venues becoming vacant along King George St. and in the north-east corner of the study area (Figure 3, t = 1000).

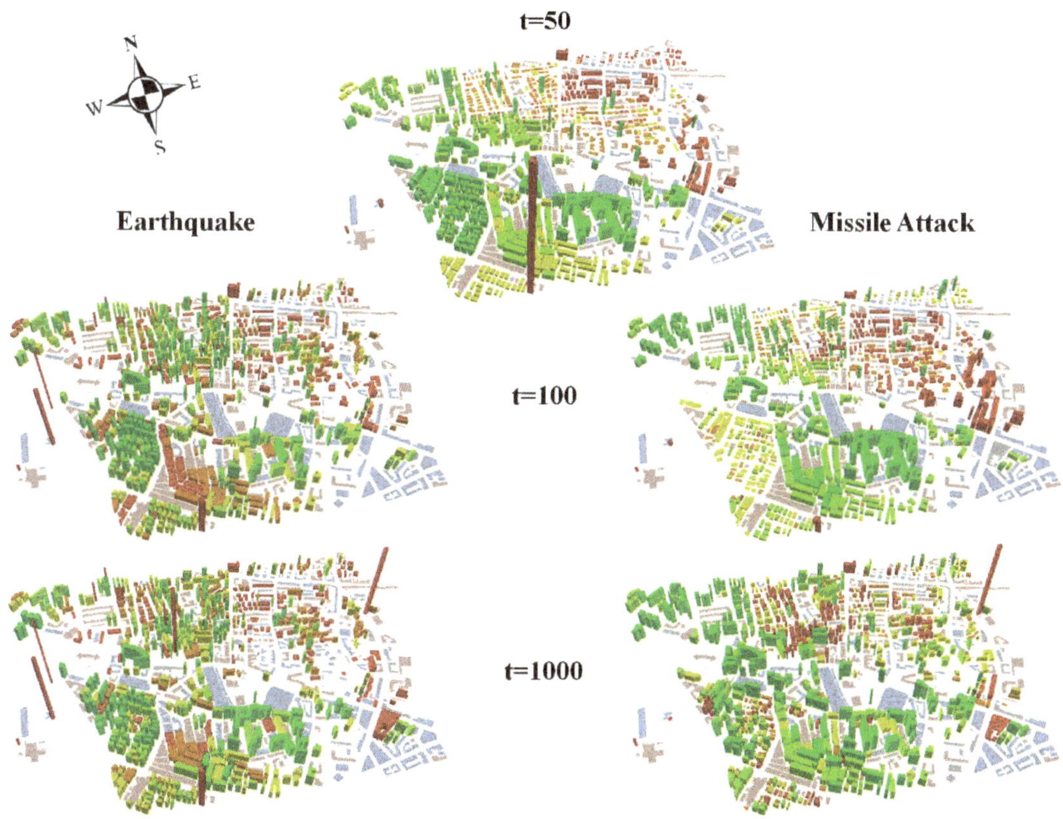

Figure 3. Changes in land-use and population demographics by building at discrete time points. Flat (2D) buildings are non-residential: grey for vacant, blue for commercial, and pink for public/governmental. 3D buildings are residential, where both height and color represent average SVI scores. Building height represents absolute values, *i.e.*, buildings with negative values have "positive" heights. Color categories are ordered by quantiles: red signifies lower scores (high vulnerability) and green higher scores (low vulnerability).

On the other hand, in the earthquake scenario decline is accelerated. This enables a stabilization of traffic patterns, around new commercial centers that have succeeded to reorganize. As new functions attract more traffic, an agglomerative process occurs. The new clusters of commercial activity appearing in the areas to the southwest and northeast of the market (Figure 3, t = 1000) indicate such a process. Once commercial activity rejuvenates and land-uses patterns become fixed, fluctuations in house prices decrease and the population is able to re-organize. This process is visibly less prominent but can be detected as some of the less vulnerable clusters seem to grow in strength, such as the one south of the market (Figure 3, t = 1000). The continuous fluctuations in traffic patterns and in the land-use system under the missile scenario require agents to constantly adapt and disturb any attempt to achieve stability. This attrition effect is associated with constant low-grade shocks.

These contradicting micro-spatial patterns of reorganization *versus* decay explain the propensity of the system to reach stability in each of the scenarios, as reported in Table 2. Thus, while low-resolution spatial analysis may be useful in identifying patterns of aggregate change, utilizing the spatio-temporal richness of the data allows for insights regarding the causes of such patterns. As can be seen from the level of complexity in Figure 3, it is not easy to communicate these results, especially if both time and space are visualized at high-resolutions.

5. Design Steps

The multi-dimensional simulation outputs result in a voluminous database of spatial and temporal urban dynamics. The first challenge is to take these vast amounts of data and turn them into meaningful information. The second challenge is to design a visualization platform that will enable a comprehensible display of a large set of meaningful multi-dimensional results. We design a dedicated web-based application that allows interactive visualization and querying of the outputs in an intuitive and user-friendly fashion. (see http://ccg.huji.ac.il/AgentBasedUrbanDisaster/index.html). Using a web browser, the user can generate time animation visualizations in the form of maps and graphs without previous experience in GIS or spatial data handling. This requires a dedicated database design and construction of selected outputs to allow efficient and rapid application response and data extraction. We do this by using DB design that does not always follow strict DB design standards, but rather contains some flat tables to enable lateral data charting, displayed in pop-ups, graphs, and charts.

5.1. Visualization Platforms

Our results are visualized using three main platforms (Figure 4):

(i) Google Maps API is used as a 2D web-mapping platform.
(ii) Google Earth API is used as a 3D display platform.
(iii) Google Charts API is used as a non-spatial graphic visualization platform for aggregate results.

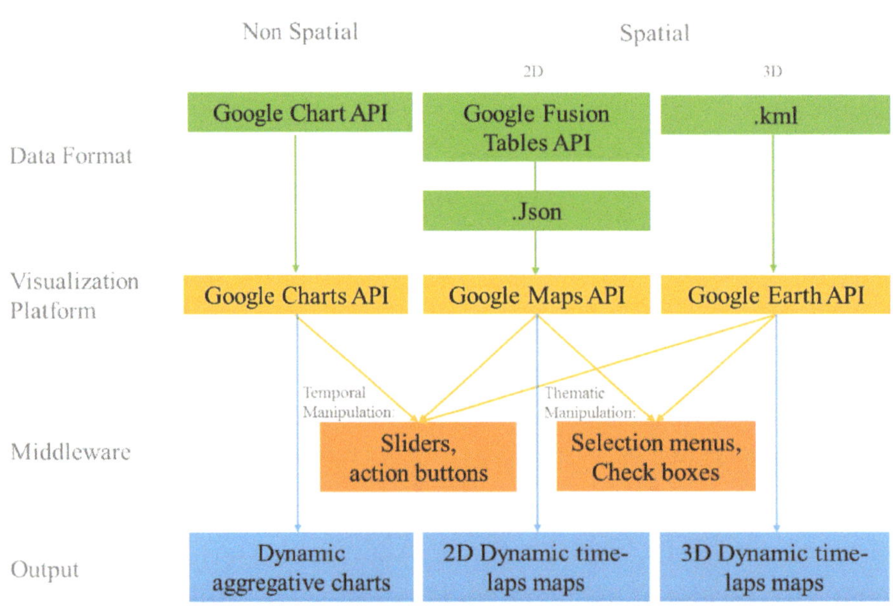

Figure 4. Web visualization design roadmap.

5.2. Data Formats

Each of the above mentioned platforms require different formats of input data, and sometimes accepts more than one input format interchangeably (Figure 4). The input data has to be designed so that it can be animated over both temporal and thematic dimensions.

(i) Google Fusion Tables API-We use Google Fusion Tables API to feed spatial vector 2D layers into Google maps. Each feature in the layers contains time series of values for each variable. These are displayed using SQL queries generated in the background when an action button is clicked, when a slider button is dragged or when a feature is clicked upon.
(ii) Json data files—are used to create heat map visualizations using the 2D Google Maps API. Each time an action button is turned on or the slider is dragged back or forth, the time increment is changed, variable values are changed and the heat map draped over the study site changes accordingly.
(iii) Graphs displayed using Google Charts API are based on a specific data format that feeds into this API. A JavaScript loop is written in order to generate the time lapse visualization initiated with a click of an action button.
(iv) 3D kml files are generated in ArcScene to create the color and height symbology. Each file contains features representing the values of a variable over a different time interval. An action button click or a drag of a slider changes the time interval of choice and calls for the appropriate kml layer to be displayed and the previous layer to be turned off.

5.3. Middleware and Ancillary Functionalities

We use various JavaScript libraries and APIs in order to add middleware functionalities to the application. These middleware interact with the web-mapping platform to provide ancillary capabilities [32] such as time laps animation, selection of variable to show, *etc.*

(i) Action buttons—initiate an automatic looping time lapse animation of a chosen variable in a map or graph in the different visualization platforms.
(ii) Sliders—allow the user to manually slide through the visualization of a certain variable in the different visualization platforms.
(iii) Selection menus—allow the user to browse through and choose a variable for display. A change in the choice of a variable also initiates a change in the displayed legends to fit the relevant variable.

6. Communicating Outcomes

Traditionally, research outputs are communicated through scientific publications and reports. These are limited in the amount of textual and visual information they contain. These constraints are compounded as the sophistication and volume of outputs increases. Furthermore, public participation in planning and decision-making is gaining increased currency [33,34]. The new consumers of information invariably do not have access to traditional sources of scientific information generating a need for communicating spatial information to professionals and the public alike in a comprehensible and intuitive manner.

Communication of complex and information intensive research results to end-users from wide-ranging backgrounds is a challenging task. The increase in sophistication, volume, and complexity of modeling urban dynamics and the exponential growth of computing power and big data, compound this challenge. Web-based cartographic spatial and temporal visualization technologies can function as a bridge between the research environment in which outputs are generated and the user. We develop a web-based application that serves as a means to communicate outputs generated using an agent-based simulation model to potential end-users, such as urban engineers and evacuation planners. Communicating spatial information in this way helps to increase transparency and opens the door for public awareness and participation in planning processes post disaster. Our application allows the user to browse through four types of spatial and non-spatial visualization techniques.

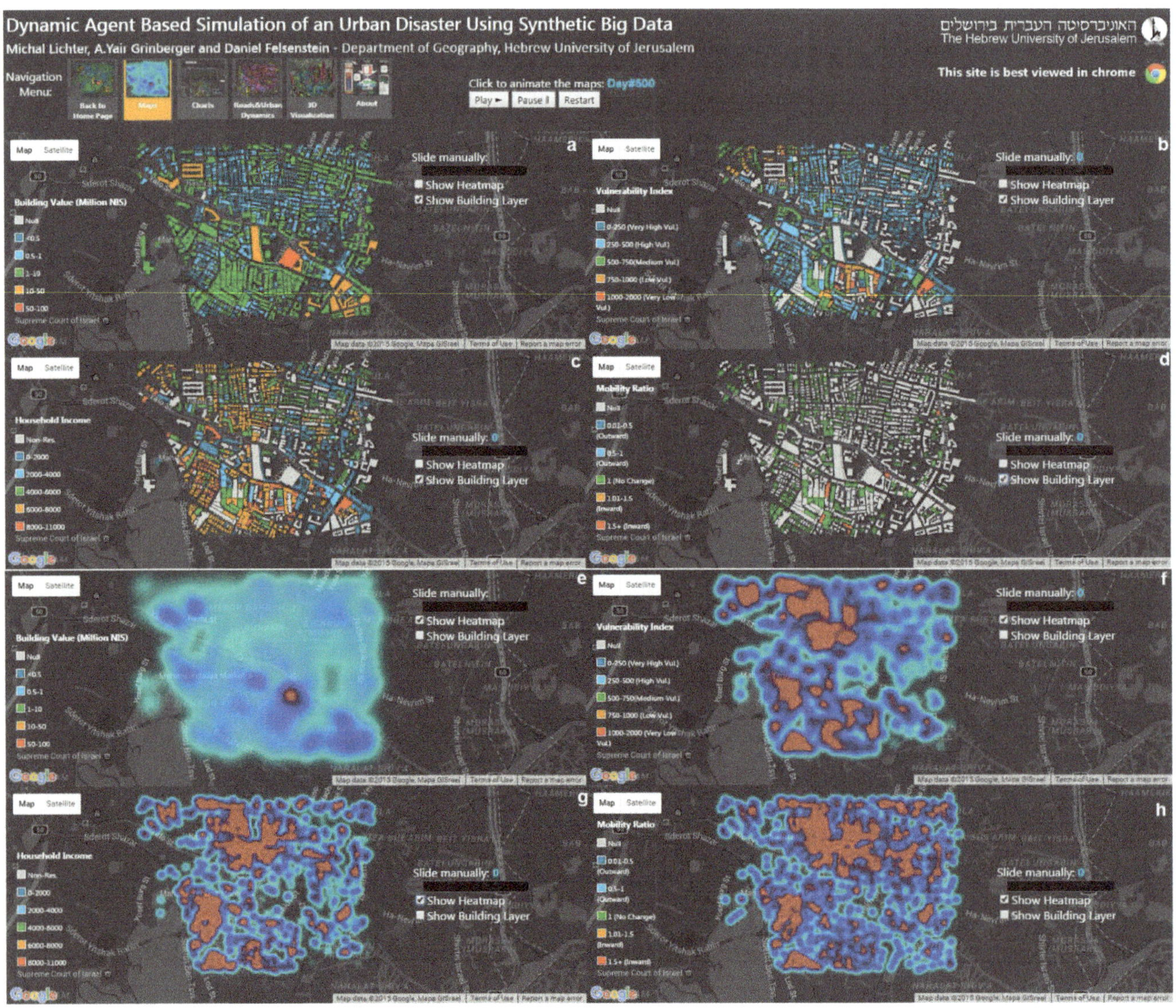

Figure 5. The map comparison panels displaying (**a–d**) vector based discrete buildings and (**e–h**) continuous heatmap surface for building value (a,e), vulnerability index (b,f), household income (c,g) and mobility ratio (d,h).

6.1. Map Comparison Panel

In this visualization the user can view four different maps of four different dynamically changing variables in the aftermath of a disaster (missile attack or earthquake): building value, vulnerability index, household income and mobility ratio. The maps are not static but rather can be animated to display a sequence of time lapse portrails of each variable. Each map contains "shots" in space and time of the variables from before the event ($t + 0$) to the time of the event ($t + 50$) to three years after the event ($t + 1000$) in time steps of 50 days. The animated maps of each variable are presented in two formats from which the user can choose. The first is a discrete vector building layer format which displays the change in variable values over time through a change in building color. With a mouse-click on each building, the user can trigger a popup window with all the properties attached to the building (Figure 5). The second format is a heatmap—this is a continuous surface draped over the study site which portrays high and low concentrations of a phenomenon using hot and cold colors. Areas where a certain variable displays high values, such as building values or mobility ratios, will be displayed in red. Areas with low values are displayed in blue. This enables the user to easily identify spatial trends and configurations, dynamically adjusting with the zoom level of the map. It does not however, allow value extraction by a click of the mouse. Both formats allow the user to animate the maps by clicking on the "play" button to automatically animate the maps over time or use a slider to manually change the a maps time steps.

6.2. Dynamic Graphs Panel

In this panel, change in parameters is charted over dynamic-queryable graphs (Figure 6). For example, variables relating to population dynamics, such as the number of inhabitants in the area and their monthly earnings at each point in time are displayed over the earnings of new in-migrants. This shows the increase in total poulation accompanied by a drop in the total earnings in the study area over the entire three year period, post-earthquake. Specifically, the high variance in household income of in-migrants can be noted. Other graphs display the change in value of residential and non-residential buildings out of total number of buildings in each category. Due to the model dynamics, this can constantly change as buildings are either destroyed or become uninhabitable after an event, as they become rehabilitated or as they change use from residential to non-residential and *vice versa*. While this visualization is not spatial, it enables the display of aggregated results (macro analysis) related to the entire study area over time and lets the user query the graphs with a mouse click.

6.3. Roads and Urban Dynamics

This is a vector-based time lapse visualization which allows the animation of five different variables over time relating to change in the number of passengers along the road network in the study site (Figure 7). This change in traffic volume ultimately drives dynamic processes in the model through changing accessibility. Here too, the time lapse visualization is based on a 50 day interval over three years. The user can choose one variable to display with the road network: land use, building value, vulnerability index, household income, and mobility ratio. Since this is a vector based visualization, the user can query elements such as buildings and road segments in the map, using a mouse, and extract the properties of each element at each point in time. This visualization can be animated using the relevant buttons and can be manually manipulated using a time lapse slider.

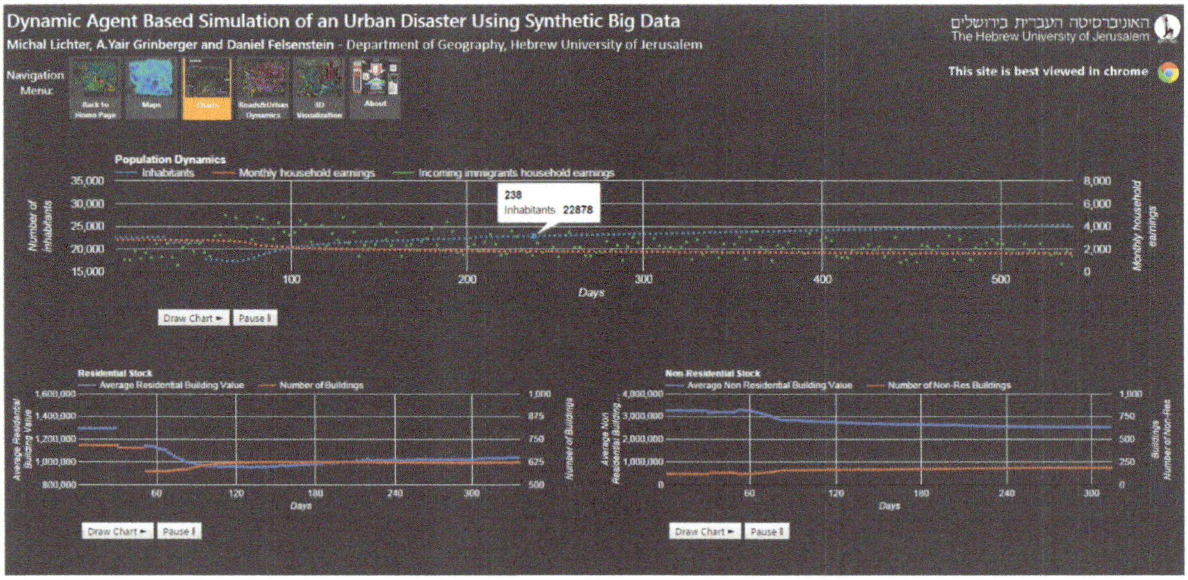

Figure 6. Querying the dynamic charts by clicking on points in the graph representing the value of a variable at a certain point in time.

Figure 7. Vector based 2D visualization of various variables over the road network.

6.4. 3D visualization

This visualization uses 3D visualization techniques to display the change in variables over time. We use the color of buildings to display change in each building land use and we use building height to display the change in building value, vulnerability index, household income and mobility ratio. Due to data volume constraints we limit the visualization to one year post the event with a 50-day time interval (Figure 8). This visualization also enables the user to familiarize the study site using 3D with Google

Earth as the platform. Consequently, the user can rotate the scene and change the angle of presentation.

Figure 8. 3D visualization of various variables. Color represents the main land use of a building and height represents value of the variable.

7. Conclusions

In terms of simulating outcomes, this paper has shown that diverse disaster situations result in very different outcomes. Over the long term, the city tends to recover from the earthquake and reaches equilibrium for key indicators over a period of 400–600 days. With the missile attack, things are rather different. The random, low-grade shocks erode re-organization capacity and the city never really recovers. This is also reflected at the micro level. The earthquake induces a process of dispersal and re-concentration of population and commercial activity. In contrast, the missile attacks cause residential and commercial clusters to decay as their capability to re-group is never allowed to materialize.

Just as these outcomes are very different, the methods of communicating them need to be delivered accordingly. The earthquake impacts need to be imparted to a population engaged in moving and re-adjusting post-disaster. In contrast, the missile outcomes need to be communicated to a population that stays put but gradually suffers from attrition. This implies that disaster management needs to move beyond providing engineering fixes and relate to wider process that differentiate across population groups affected by the disaster.

Improving accessibility to information is one route towards enahncing resilience to shocks. Along with the explosion of available information through enhanced computer power and techological progress, broader societal change also demands democratization of crisis management and increased citizen empowerment in the recovery process. The centralized, linear, top-down model of disaster management is slowly being augmented by a networked community-based approach [35]. This mode of management is grounded in data pooling and public input through crowdsourcing and is lubricated by resources such as OpenStreetMap and GeoCommons. In this respect, the Internet acts as the great facilitator. It encourages open standards and simplified interfaces and generally makes information generation more transparent and democratic.

The web-based delivery of likely disaster outcomes not only encourages public participation in rebuilding and rejuvenation but also differentiates across the types of response required to withstand the shock. In our simulated cases, mitigating the effects of the earthquake point to the need for assisting recovery in new locations, encouraging personal mobility, and removing regulatory contraints to the physical recovery of communities. For the missile attack case, a very different suite of interventions may be relevant. These relate to community preservation and stabilization, for example stemming the tide of out-migration and bolstering local social services.

Author Contributions

Database preparation, visualization and web design were done by Michal Lichter; Articulation and execution of the modeling framework was done by Yair Grinberger and Daniel Felsenstein; Writing was undertaken as a joint effort.

Conflicts of Interest

The authors declare no conflict of interest.

References

1. Koch, J.V. Costs of defending against rising sea levels and flooding in Mid-Atlantic metropolitan coastal areas: The basic issues. *J. Reg. Anal. Policy* **2010**, *40*, 53–60.
2. Tate, E.; Burton, C.G.; Berry, M.; Emrich, C.T.; Cutter, S.L. Integrated hazards mapping tool. *Trans. GIS* **2011**, *15*, 689–706.
3. Dransch, D.; Rotzoll, H.; Poser, K. The contribution of maps to the challenges of risk communication to the public. *Int. J. Digit. Earth* **2010**, *3*, 292–311.
4. Little, R.G.; Loggins, R.A.; Wallace, W.A. Building the right tool for the job: Value of stakeholder involvement when developing decision-support technologies for emergency management. *Nat. Hazards Rev.* **2015**, doi:10.1061/(ASCE)NH.
5. Hagemeier-Klose, M.; Wagner, K. Evaluation of flood hazard maps in print and web mapping services as information tools in flood risk communication. *Nat. Hazards Earth Syst. Sci.* **2009**, *9*, 563–574.
6. Kwan, M.P.; Lee, J. Emergency response after 9/11: The potential of real-time 3D GIS for quick emergency response in micro-spatial environments. *Comput. Environ. Urban Syst.* **2005**, *29*, 93–113.
7. Roche, S.; Propeck-Zimmermann, E.; Mericskay, B. GeoWeb and crisis management: Issues and perspectives of volunteered geographic information. *GeoJournal* **2011**, *78*, 21–40.
8. Richardson, H.W.; Park, J.; Moore, J.E.; Pan, Q. (Eds.) *National Economic Impact Analysis of Terrorist Attacks and Natural Disasters*; Edward Elgar: Cheltenham, UK, 2014.

9. Chen, X.; Zhan, F.B. Agent-based modelling and simulation of urban evacuation: Relative effectiveness of simultaneous and staged evacuation strategies. *J. Oper. Res. Soc.* **2006**, *59*, 25–33.
10. Crooks, A.T.; Wise, S. GIS and agent-based models for humanitarian assistance. *Comput. Environ. Urban Syst.* **2013**, *41*, 100–111.
11. Dawson, R.J.; Peppe, R.; Wang, M. An agent-based model for risk-based flood incident management. *Nat. Hazards* **2011**, *59*, 167–189.
12. Grinberger, A.Y.; Felsenstein, D. Bouncing back or bouncing forward? Simulating urban resilience. *Urban Des. Plan.* **2014**, *167*, 115–124.
13. Grinberger, A.Y.; Felsenstein, D.; Lichter, M. Simulating urban resilience: Disasters, dynamics and (synthetic) data. In *Planning Support Systems and Smart Cities*; Geertman, S., Stillwell, J., Ferreira, J., Goodspeed, R., Eds.; Springer: , Cham, Switzerland, 2015; pp. 99–119.
14. Macal, C.M.; North, M.J. Tutorial on agent-based modeling and simulation. In Proceedings of the 37th Conference on Winter Simulation, Orlando, FL, USA, 4–7 December 2005; pp. 2–15.
15. Salamon, A.; Katz, O.; Crouvi, O. Zones of required investigation for earthquake-related hazards in Jerusalem. *Nat. Hazards* **2009**, *53*, 375–406.
16. Linard, C.; Gilbert, M.; Tatem, A.J. Assessing the use of global land cover data for guiding large area population distribution modelling. *GeoJournal* **2010**, *76*, 525–538.
17. Reibel, M.; Bufalino, M.E. Street-weighted interpolation techniques for demographic count estimation in incompatible zone systems. *Environ. Plan. A* **2005**, *37*, 127–139.
18. Eicher, C.L.; Brewer, C.A. Dasymetric mapping and areal interpolation: Implementation and evaluation. *Cartogr. Geogr. Inf. Sci.* **2001**, *28*, 125–138.
19. Mennis, J. Generating surface models of population using dasymetric mapping. *Prof. Geogr.* **2003**, *55*, 31–42.
20. Beckman, R.J.; Baggerly, K.A.; McKay, M.D. Creating synthetic baseline populations. *Transp. Res. Part Policy Pract.* **1996**, *30*, 415–429.
21. Pritchard, D.R.; Miller, E.J. Advances in population synthesis: Fitting many attributes per agent and fitting to household and person margins simultaneously. *Transportation* **2011**, *39*, 685–704.
22. Bhaduri, B.; Bright, E.; Coleman, P.; Urban, M.L. LandScan USA: A high-resolution geospatial and temporal modeling approach for population distribution and dynamics. *GeoJournal* **2007**, *69*, 103–117.
23. Martin, D.; Cockings, S.; Leung, S. Developing a flexible framework for spatiotemporal population modeling. *Ann. Assoc. Am. Geogr.* **2015**, *105*, 754–772.
24. Harper, G.; Mayhew, L. Using administrative data to count local populations. *Appl. Spat. Anal. Policy* **2012**, *5*, 97–122.
25. Harper, G.; Mayhew, L. Applications of population counts based on administrative data at local level. *Appl. Spat. Anal. Policy* **2012**, *5*, 183–209.
26. Beenstock, M.; Felsenstein, D.; Ben Zeev, N. Capital deepening and regional inequality: An empirical analysis. *Ann. Reg. Sci.* **2011**, 599–617.
27. Lichter, M.; Felsenstein, D. Assessing the costs of sea-level rise and extreme flooding at the local level: A GIS-based approach. *Ocean Coast. Manag.* **2012**, *59*, 47–62.
28. Simon, H.A. A behavioral model of rational choice. *Q. J. Econ.* **1955**, *69*, 99–118.

29. Schelling, T.C. Dynamic models of segregation. *J. Math. Sociol.* **1971**, *1*, 143–186.
30. Crooks, A.T.; Castle, C.J.E. The integration of agent-based modelling and geographical information for geospatial simulation. In *Agent-Based Models of Geographical Systems*; Heppenstall, A.J., Crooks, A.T., See, L.M., Batty, M., Eds.; Springer: Houten, The Netherlands, 2012; pp. 219–251.
31. Felsenstein, D.; Lichter, M. Social and economic vulnerability of coastal communities to sea level rise and extreme flooding. *Nat. Hazards* **2014**, *71*, 463–491.
32. Batty, M.; Hudson-Smith, A.; Milton, R.; Crooks, A. Map mashups, Web 2.0 and the GIS revolution. *Ann. GIS* **2010**, *16*, 1–13.
33. Dunn, C.E. Participatory GIS—A people's GIS? *Prog. Hum. Geogr.* **2007**, *31*, 616–637.
34. Elwood, S. Grassroots groups as stakeholders in spatial data infrastructures: Challenges and opportunities for local data development and sharing. *Int. J. Geogr. Inf. Sci.* **2008**, *22*, 71–90.
35. Kawasaki, A.; Berman, M.L.; Guan, W. The growing role of web-based geospatial technology in disaster response and support. *Disasters* **2013**, *37*, 201–221.

4

Communicating Thematic Data Quality with Web Map Services

Jon D. Blower [1,*], Joan Masó [2], Daniel Díaz [2], Charles J. Roberts [1], Guy H. Griffiths [1], Jane P. Lewis [1], Xiaoyu Yang [3] and Xavier Pons [4]

[1] Department of Meteorology, Harry Pitt Building, University of Reading, Reading RG6 6BB, UK; E-Mails: c.j.roberts@reading.ac.uk (C.R.); guy.griffiths@reading.ac.uk (G.G.); j.p.lewis@reading.ac.uk (J.L.)

[2] Grumets Research Group, CREAF, Edifici C, Universitat Autònoma de Barcelona, 08193 Bellaterra, Catalonia, Spain; E-Mails: Joan.Maso@uab.cat (J.M.); d.diaz@creaf.uab.cat (D.D.)

[3] Computer Network Information Center, Chinese Academy of Sciences, Beijing 100190, China; E-Mail: kxy@cnic.cn

[4] Grumets Research Group, Dep Geografia, Edifici B, Universitat Autònoma de Barcelona, 08193 Bellaterra, Catalonia, Spain; E-Mail: Xavier.Pons@uab.cat

* Author to whom correspondence should be addressed; E-Mail: j.d.blower@reading.ac.uk

Academic Editor: Wolfgang Kainz

Abstract: Geospatial information of many kinds, from topographic maps to scientific data, is increasingly being made available through web mapping services. These allow georeferenced map images to be served from data stores and displayed in websites and geographic information systems, where they can be integrated with other geographic information. The Open Geospatial Consortium's Web Map Service (WMS) standard has been widely adopted in diverse communities for sharing data in this way. However, current services typically provide little or no information about the quality or accuracy of the data they serve. In this paper we will describe the design and implementation of a new "quality-enabled" profile of WMS, which we call "WMS-Q". This describes how information about data quality can be transmitted to the user through WMS. Such information can exist at many levels, from entire datasets to individual measurements, and includes the many different ways in which data uncertainty can be expressed. We also describe proposed extensions to the Symbology Encoding specification, which include provision for visualizing uncertainty in raster data in a number of different ways, including contours, shading and bivariate colour maps. We shall also describe new open-source implementations of the new specifications, which include both clients and servers.

Keywords: GIS; data quality; uncertainty; WMS; OGC; INSPIRE; GEOSS; SDI

1. Introduction

The Web Map Service standard (WMS, [1]), published by the Open Geospatial Consortium and standardized by ISO as ISO19128:2005 [2], has been one of the most successful standards enabling the interoperability of geographic information. It allows clients to request georeferenced map images representing a wide variety of data sources, together with accompanying metadata. The standard is implemented in a wide range of open-source and commercial software, including GeoServer, MapServer, ArcGIS Server, Cadcorp SIS and ncWMS [3] (see http://www.opengeospatial.org/resource/products/compliant for a list of certified compliant software). An overview of the standard can be found in Section 5.1 below. WMS is used as the visualization standard in many spatial data infrastructures such as INSPIRE [4] and the Global Earth Observation System of Systems (GEOSS) [5].

A common concern with Spatial Data Infrastructures (SDIs) in general, is that the user is frequently not provided with information regarding the quality of the underlying data. This is particularly important in the field of scientific data, in which the user often wishes to know about the uncertainty of the data. Although a number of methods are in use for visualizing uncertain data (such as those suggested by the INTAMAP project for visualizing uncertainty of interpolated maps [6], some examples for mapping uncertainty of weather ensemble data [7,8] and several others described in [9]) these are mostly not supported by WMS implementations, or are not supported in a standardized, consistent fashion; furthermore the relevant symbology standards (e.g., Symbology Encoding (SE), [10]) do not provide the required flexibility to accommodate these more complex visualization types, particularly for raster data. Some solutions for using WMS and SE in uncertainty visualization for vector data are given in [11], which focuses on use cases in spatial planning.

In this paper we describe a new profile of WMS for data quality (WMS-Q), which provides a number of mechanisms for using WMS to communicate data quality at a number of levels, from the dataset level to the level of individual samples (pixels). WMS-Q is entirely compatible with the current version of the WMS standard (1.3.0) and considers a number of underlying data types (including continuous and categorical data). Section 2 introduces the concept of "data quality", Section 3 clarifies the terminology we will use in the paper, Section 4 explores current means of representing data quality, Section 5 describes the design of WMS-Q, Section 6 briefly describes some current implementations of the profile, Section 7 describes how WMS-Q has been integrated into a "quality-aware" SDI, and finally Section 8 identifies areas of future work.

2. What is "Data Quality"?

Data quality is a concept that is difficult to define precisely. It means different things to different communities and it is sometimes confused with the amount of information. It is defined by the ISO

8402:1994 [12] standard as the "totality of characteristics of a product that bear on its ability to satisfy stated or implied needs". In the geospatial domain, it includes many measures including spatial accuracy, temporal accuracy, consistency, completeness, scope and attribute accuracy. There is a general consensus that quality is a *subjective* concept and each user defines a dataset as having good quality when it fulfils his expectations, that is, it *fits his purpose*. Unfortunately, data producers cannot anticipate the expectations of all future users. Thus, in practice they try to quantify some aspects by comparing the dataset with other reference sources that are accepted to be correct and provide a numeric quantification of the discrepancies in terms of uncertainty metrics and error statistics. Alternatively, they create a specification for their product, and for each dataset, they provide a confirmation of the product specifications conformity [13]. When these approaches are applied to the data, a quality indicator [14] is generated and attached to the metadata.

Data quality can be applied to different components of the data: spatial (e.g., positional accuracy), temporal (temporal accuracy), thematic (e.g., completeness, thematic accuracy, classification correctness, *etc.*) [14–16]. Some aspects covered by this paper apply to all of the above but the focus is on thematic accuracy.

3. Terminology Used in This Paper

Different communities use different terms to represent similar concepts, or the same term for different concepts. The WMS-Q profile can be applied in many different communities, including Earth Observation, climate science, meteorology, oceanography, terrestrial studies and more. Therefore we must carefully define the terms we use in this paper, which may be different from common usage in particular communities:

Dataset: The term *dataset* is used in its most general sense of "a logical collection of data". The information in a dataset may be recorded physically in a single file or a set of files and may represent a single snapshot in time, or time-evolving information. The choice of how to group data into datasets is highly community-specific; typically, data providers will decide how to group information into logical units that make sense to them and their users. Discovery metadata [17,18] is typically generated at the dataset level.

Variable: A *variable* is a measured quantity, such as temperature, velocity or a land use classification. It can be observed or calculated. In this paper we use "variable" to represent the concept; the values and their uncertainty are recorded in *fields* (see below).

Sample: A *sample* is an individual measurement or observation of a single variable. A pixel in an Earth Observation image, or a grid cell in the results of a numerical simulation, may contain several samples, one for each variable that is recorded in that pixel or cell.

Component: If the values of a variable within a dataset are uncertain, then each sample of this variable comprises more than one piece of information (for example, components could be the "most likely" value of the variable and the upper and lower confidence range representing its uncertainty). Each piece of information is known as a *component*. Different ways of expressing uncertainty are described in Section 4.3 below.

Field: A *field* is a collection of values of a single component. Typically this will be recorded in a data file as a multidimensional array of scalar (numeric) values.

We illustrate these concepts through two examples.

Example 1: The Sea Surface Temperature *dataset* from the European Space Agency's Climate Change Initiative (CCI-SST, [19]) provides measurements of two *variables*: sea surface temperature and sea ice area fraction, recorded on a global grid at 1/20 degree (~5 km) resolution. The *sample* measurements of both variables are uncertain and therefore the values of each variable are expressed as two *components*, one representing the mean value and one representing its variance. The dataset therefore contains four scalar *fields* (two for each variable).

Example 2: The Landcover Classification of Landsat Barcelona-Girona scenes [20] *dataset* provides information on the land cover of that region of Catalonia, derived from satellite imagery. Each *sample* (*i.e.*, pixel) within the dataset contains the best estimate of the land cover type (expressed as a classification code), some alternative classification values plus several measures of confidence that the classification is correct. The dataset therefore contains one *variable* (land cover type) expressed as a number of scalar *fields*: some categorical (classifications) and some continuous (confidence measures).

4. Encodings of Data Quality

4.1. The ISO Suite of Standards

In current SDIs, data quality information can be included within structured metadata documents that describe geospatial datasets [21]. There are some metadata standards produced by International Standards Organization Technical Committee 211 (ISO-TC/211), the Federal Geographic Data Committee (FGDC), European Committee for Standardization Technical Committee 287 (CEN/TC-287), the International Cartographic Association (ICA) and others that include some or all of the quality components. The ISO19115:2003 [22] model (and its counterpart XML encoding, ISO19139:2007 [23]) is perhaps the most commonly-adopted standard for describing geospatial metadata. ISO19113:2002 [24] defined the quality principles that were included in the later standard ISO19157:2013 [14], which also provides details about quality measures and the metadata elements for quality documentation that can be incorporated to the general ISO19115 geospatial metadata model. The European SDI that results from the INSPIRE directive is one example of a multinational data catalogue that relies on these standards.

In addition to quality indications, lineage or provenance information is also considered a component that helps users to assess fitness-for-purpose. Provenance information includes information about the producer, which can influence the user trust, and also formal documentation about the processes, algorithms and data sources involved in the elaboration of a dataset [25]. ISO19115 incorporates provenance information through the LI_Lineage structures.

4.2. Building on the ISO Model

The GeoViQua project (http://www.geoviqua.org) built on these standards to provide new mechanisms to exchange quality information from both the data producer's and the data user's point of view. The GeoViQua Producer Quality Model (PQM) builds on ISO19157 and includes new support for discovered issues, workarounds, usage, citations, goodness-of-fit statistics and validation. The PQM is complemented by the User Quality Model (UQM), which captures user feedback on datasets, including numerical ratings with text justification; user comments and reports of usage and problems identified;

citation of publications; tags to facilitate data discovery; supplementary quality reports; and information on any spatial, temporal or thematic focus of the feedback. The schemas defining both the PQM and UQM can be found at http://schemas.geoviqua.org/GVQ/4.0/.

XML (e.g., ISO19139 documents) is a machine-readable format that is not well suited for direct presentation to the user. Tools are required to both present metadata and edit metadata [26]. These tools are able to produce human readable reports, e.g., in the form of HTML pages [27] using different strategies, XSLT being a popular choice for turning XML metadata document sources into human-readable HTML. Even though a human readable report is easily interpreted by an operator that is familiar with it, the work load to compare tens or hundreds of documents is impractical. More sophisticated visualization tools can be elaborated to represent the content of a metadata in a simplified and compact image that represents the existence of some information in the metadata by colouring facets of a GEO label [28] or by representing side by side some metadata values in a normalized scale such as a parallel coordinate plot [29].

4.3. Vocabularies for Describing Data Quality

The ISO standards provide a framework for structuring data quality information, but need to be used in combination with comprehensive controlled vocabularies that enable the precise description of quality information. As we shall describe in Section 5, these vocabularies can be used to convey the precise semantic meaning of WMS Layers.

UncertML (http://www.uncertml.org/) is a vocabulary of terms that can be used to describe the dispersion of values of uncertain variables and the components that comprise them. In UncertML, uncertainty can be represented in three main ways:

1. As individual **realisations** of an uncertain variable. For example, the uncertainty of a temperature field could be expressed by running a simulation ten times under different conditions, and recording the results of each simulation in a data file as a separate field.
2. As **summary statistics**. Instead of recording each realisation individually, the data provider may choose to calculate statistics describing the spread of values at each measurement location. For example, the spread may be expressed as a mean field and a variance field.
3. As **probability distribution functions** (PDFs). This is the most powerful representation of uncertainty and gives a functional form for the probability of a variable having a given value at a given location. In a data file, a PDF may be represented by specifying the functional form of the PDF as metadata (e.g., Gaussian or log-normal) and giving the values of each parameter of the distribution as a separate field.

In this way, a set of scalar fields can be suitably "tagged" with UncertML terms to allow the user to understand that they represent the components of an uncertain variable and, when taken together, provide a measure of the variable's uncertainty.

The QualityML vocabulary [30] builds on UncertML [31] and ISO19157 to provide a more comprehensive suite of terms that describe a broader range of quality measures. These terms include statistics that summarise the uncertainty of an entire field (e.g., the root-mean-square error) and terms that describe uncertainties in categorical data (e.g., the commission error or the Kappa coefficient).

The NetCDF-U specification (currently under discussion within the Open Geospatial Consortium, [32]) provides a mechanism for using the richness of the UncertML vocabulary within NetCDF files. QualityML could also be used in NetCDF for the same purpose.

5. Design of WMS-Q

This section describes the main features of the WMS-Q profile. The current version of the full WMS-Q profile document can be found at http://www.geoviqua.org/Docs/WMS-Q_v2.d_final.pdf.

5.1. Overview of WMS

We briefly describe here the main relevant features of the WMS standard (version 1.3.0), in order to aid understanding of the WMS-Q extensions we describe in the sections below.

The data provided by a WMS server are divided into *Layers*, which represent the basic units of information. The server provides metadata about these Layers through a machine-readable service metadata document (also called the "Capabilities" document) that can be requested by clients through the *GetCapabilities* operation. All Layers have a human-readable Title field. Further metadata can be provided through external documents (both human-readable and machine-readable), which are linked to Layers in the service metadata document through the *MetadataURL* tag.

Georeferenced images (maps) of the Layers are requested through the *GetMap* operation. There are two ways of controlling the appearance of a map image. The simplest method is for the server to advertise fixed *Styles*, each of which is identified by a name that can be requested in the *GetMap* request. A more sophisticated method can be provided using a WMS server that is enabled with the *Styled Layer Descriptor (SLD)* [33] extension. In this case, the client has much closer control over the appearance of the requested map image by sending the server an XML document encoded using the *Symbology Encoding (SE)* standard. Such a server is often referred to as an *SLD/SE-enabled* WMS server.

The WMS standard permits Layers to be arbitrarily nested within a service metadata document. Layers that have a <Name> property are *displayable* (which means that they can be requested in a GetMap operation). WMS permits any Layer to be displayable, even if that Layer contains child Layers. Layers that are not displayable do not have a <Name> and are usually used as a grouping mechanism for child Layers. A WMS service metadata document therefore contains a hierarchy of Layers. The WMS standard specifies rules (Table 7 in [1]) that govern the inheritance of properties of parent Layers by child Layers; these rules are important to the design of WMS-Q: see Section 5.6.

The *GetFeatureInfo* operation allows a WMS client to retrieve more information about a specified pixel in a map image. The WMS standard does not specify a particular format for the information that is returned by the server and so implementations can differ widely in their behaviour.

5.2. Design Goals for WMS-Q

The design goals of the WMS-Q profile were:

1. To maintain compatibility with version 1.3.0 of the WMS standard. In other words, WMS-Q aims to be a *profile* of WMS, not an *extension*. Therefore, instead of adding new functionalities, it provides a set of rules on how to use the existing mechanisms permitted by the WMS standard

to convey dataset, sample and variable level quality. In this way, standard WMS clients will be able to read information from a WMS-Q.

2. To re-use existing general methods for expressing data quality, where appropriate (e.g., UncertML, see Section 4.3 above), but to avoid methods that are highly specific to particular communities.
3. To be independent of any particular format or convention for data or metadata storage.
4. To focus on conveying the *thematic accuracy* of both categorical and continuous raster data (including its uncertainty), but allow techniques to be more widely applicable in future (e.g., to vector data).

5.3. Identification of Conformance to WMS-Q

In WMS 1.3 and previous versions, there is no standard mechanism at present for advertising to clients that a WMS service instance conforms to a particular profile. In WMS-Q we use a Keyword at the top level of the service metadata document, with the value "WMS-Q" in the vocabulary "http://www.geoviqua.org/def/doc/conventions/vocabulary". Sample Capabilities documents illustrating this are referenced in Section 5.6.

5.4. Dataset-Level Quality

In WMS-Q, datasets (see "Terminology", above) are represented by Layers that are not displayable, but act as containers for child layers representing variables. These Layers can be arbitrarily nested within other non-displayable ones, as chosen by the data provider, to impose a logical structure. (In this case the non-displayable Layers act as "folders" and typically appear as such in the legend or the layer list in WMS client software.)

Data quality is expressed at the Dataset level by associating the relevant Layer with a MetadataURL that points to a suitable, standardized metadata document (e.g., an ISO19115 dataset descriptor). However, we recommend that data quality information be associated more precisely at the Variable level where possible.

5.5. Variable-Level Quality

In WMS-Q, Variables are represented by Layers that are immediate children of Layers that represent Datasets. Quality information at this level is provided, as with Dataset-level information, by external documents linked through MetadataURL tags. The linked ISO19115 data metadata documents will have one or more DQ_DataQuality elements reporting overall quality indicators represented by estimators or statistical summaries (e.g. root-mean-square errors or confusion matrices). Since a ISO19115 document can be used to convey several DQ_DataQuality elements, it is important to keep a link of each DQ_DataQuality to the original WMS-Q layer. This can be achieved using the sub-elements of the MD_Scope, and document level = "layer" and levelDescription = *layer name*. Then a linkage of a MD_Distribution can be used to describe the link to the WMS service as a whole.

If the Variable does not contain sample-level quality information, then the Layer representing the Variable will be displayable and therefore the client can request map images of the Variable. If the Variable does contain sample-level quality information, its corresponding Layer must be non-displayable, acting only as a container for child Layers that represent the *components* of the samples, as will explained in the next section. This Layer contains the Keyword "qualityCollection" from the "http://qualityml.geoviqua.org/1.0/" vocabulary. It may also contain Keywords from other vocabularies—for example, if the uncertainty of the Variable at the sample level is expressed using a set of summary statistics (see Section 4.3 above), the Layer may be tagged with the Keyword "statisticsCollection" from the UncertML vocabulary.

5.6. Sample-Level Quality

If the Variable contains sample-level quality information all sample components of a Variable are represented as direct children of the Variable's Layer (at the bottom level of the hierarchy, *i.e.*, each component is a leaf node in the tree of Layers). These child Layers are displayable and are given Keyword tags that describe the semantics of the component, using terms from vocabularies such as QualityML or UncertML where possible.

Although sometimes it is convenient to visualize these components (child Layers) individually, it is often highly desirable for a WMS client to be able to display a single map image that represents the variable as whole, perhaps by overlaying contours of some uncertainty metric (e.g., variance) over a raster image representing the mean field (see Figure 1). In some cases, this effect can be achieved by requesting separate images from the server (one for each component) and composing them on the client. However, this method does not work for some methods of visualizing data quality (for example, bivariate colour maps, see below). For this reason, WMS-Q servers provide an extra displayable Layer as a sibling of the other child Layers that represent components. This "special" child Layer represents the Variable as a whole and is given the Keyword "qualityComposition". It is the first Layer to be listed as a direct child of the Layer representing the Variable (in this case it can also have a metadata document describing its quality attached in the MetadataURL as above). It is intended that clients regard this "special" child Layer as a sensible default portrayal of the uncertain variable.

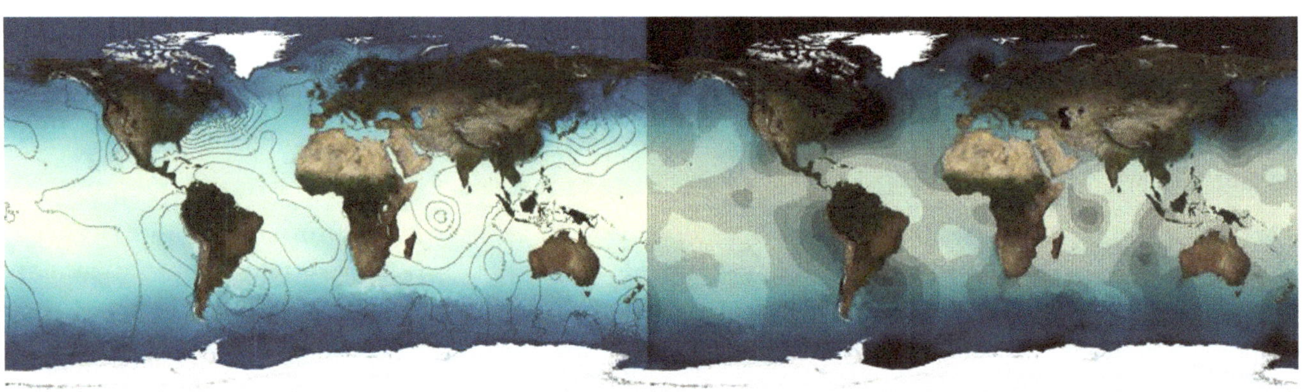

(a) **(b)**

Figure 1. *Cont.*

(c) (d)

Figure 1. Sample visualisations of sea surface temperature and its uncertainty from the Sea Surface Temperature dataset from the European Space Agency's Climate Change Initiative (CCI-SST) dataset [19], generated by the ncWMS-Q software (see Section 6.1) using the proposed extensions to the Symbology Encoding (SE) specification (see Section 5.8). From top left: (**a**) temperature encoded as lightness of colour, with overlain contours of its uncertainty (variance); (**b**) uncertainty represented through successive levels of stippling, with denser stippling representing high uncertainty; (**c**) uncertainty represented as black shading; (**d**) use of a bivariate colour map, with temperature encoded as brightness and uncertainty encoded as colour saturation.

Note that early drafts of WMS-Q attempted to achieve the same goal using a displayable Layer at the variable level, which was allowed to contain children representing its components. However, the WMS standard states that all Styles are inherited by child Layers. It would be inappropriate for Styles that are suitable for whole variables (probably defining compositions of components) to be inherited by Layers representing individual components (for example, a style that plots contours of uncertainty on top of a colour-mapped field would not be suitable for portraying a single component). Because of this, the Layer representing the variable cannot be displayable, therefore we make the Layer representing the whole Variable a displayable child, allowing it to contain a different set of Styles from its individual components.

Example 1: In the CCI-SST dataset described in Section 3 above, there is a "sea surface temperature" variable. The uncertainty of each sample is expressed using two components representing respectively the mean and variance of the variable. This is expressed in WMS-Q using a non-displayable Layer representing the variable, which is tagged with the "qualityCollection" and "statisticsCollection" Keywords. This Layer has three children: the first child represents the variable as a whole, and provides styles that enable the client to visualize the mean and variance together in a single image (e.g., using contours of variance overlain on a colour-mapped image). The remaining two children represent the mean and variance components respectively, and are tagged with Keywords containing the terms "mean" and "variance" from the UncertML vocabulary. This is represented in graphical form in Figure 2. The complete WMS-Q service metadata document can be retrieved using this link: http://ncwms.geoviqua.org/wms?SERVICE=WMS&REQUEST=GetCapabilities&VERSION=1.3.0&DATASET=cci-sst. A demo client showing the data can be found here: http://ncwms.geoviqua.org/godiva2.html.

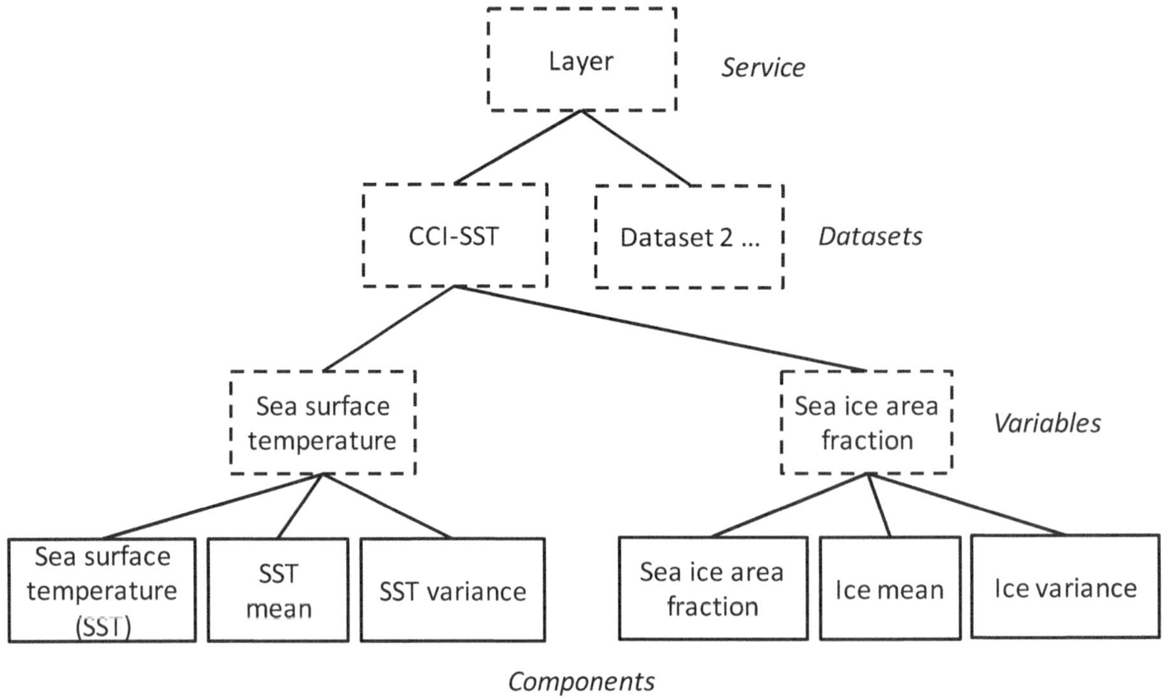

Figure 2. Schematic representation of the structure of Layers in a particular "quality-enabled" profile of Web Map Service (WMS-Q) service instance, illustrating the Service-Dataset-Variable-Component hierarchy. Each box is a Layer in the tree: blue boxes represent non-displayable Layers, whereas orange boxes represent displayable Layers. The derivation of this hierarchy is given in Section 5.5.

Example 2: We consider the "Landcover Classification" variable derived from Landsat in the Spanish National research project DinaCliVe. The classification method applied here combines common remote sensing classification techniques with the cadastre parcels and forces that each complete parcel is considered as a unity allowing statistical treatment of the pixels inside [20]. The variable and its uncertainty are expressed at the sample level using several components. Some of them are categorical components such as the class with the most presence in the parcel and the second and third class present in that parcel. Some other components are continuous such as fidelity, representativity, promiscuity, entropy, *etc.* (see http://qualityml.geoviqua.org for their definition). The variable layer has several children: the first child represents the component that most users will want to see: the class with the most presence in the parcel. This is tagged with the Keyword http://qualityml.geoviqua.org/1.0/values. The remaining classes help the user to understand the variability of the classification in the parcel and the confidence of the classification. All of them are tagged with the Keywords containing the right terms coming from the QualityML vocabulary. The complete WMS-Q service metadata document can be retrieved using this link: http://www.ogc.uab.cat/cgi-bin/GeoViQUA/WMSQ/MiraMon.cgi?SERVICE =WMS&VERSION=1.3.0&REQUEST=GetCapabilities. A demo client showing the data can be found here: http://wms-q-demo.geoviqua.org/geoviqua/wmsq/.

5.7. Behaviour of GetFeatureInfo

As mentioned in Section 5.1 above, the WMS standard does not specify the data returned by the server in response to a GetFeatureInfo request. In WMS-Q, if GetFeatureInfo is requested for a layer that represents a Variable as a whole, we recommend that the server responds with a graphical or structured textual representation of the probability distribution function of the value of the variable at the requested point, if possible. If this is not possible, the server should respond with a document containing the values of each of the sample components at that point. This aspect of WMS-Q has not so far completely defined and will be refined in future versions.

5.8. Extensions to the Symbology Encoding Standard

There are a number of ways of portraying uncertainties in raster data, either as continuous or categorical fields. A full description of all options is beyond the scope of this paper, but common strategies include:

- Portraying the "best estimate" as a colour-mapped image, overlain with contours showing a measure of data uncertainty (top left of Figure 1).
- As above, but uncertainty is represented through varying levels of stippling or texture (top right of Figure 1), as used in the Assessment Reports of the Intergovernmental Panel on Climate Change (e.g., [34]).
- As above, with uncertainty represented using black shading, the opacity of which increases with data uncertainty (bottom left of Figure 1).
- Using a bivariate colour map (e.g., [35]) in which the colour of a pixel is a function of two variables, the "best estimate" and the uncertainty (bottom right of Figure 1).
- Using glyphs (*i.e.*, small icons), the shape, size or colour of which can be mapped to different components of an uncertain variable. A special case of this is the use of "confidence triangles" (e.g., [36]), which visualize the estimated spread of data by dividing the image into squares, each of which is divided into two triangles. The lower triangle is assigned a colour representing the lower bound of the variable, and the upper triangle is coloured according to the upper bound of the variable. The contrast in colours between the two triangles gives a visual estimate of the uncertainty.

The current version of the SE standard allows for only a limited number of means of visualizing raster data using the RasterSymbolizer. Data values can be mapped to pixel colours and the opacity of the image as a whole can be altered. Other functions in SE are oriented around optical remote sensing imagery (with assumed red, green and blue channels) or digital elevation models. Contours, stippling, bivariate colour maps or glyphs are not supported in SE. We therefore propose extensions to SE to support these portrayal types. For reasons of space, a full description of the implementation of these new styles is beyond the scope of this paper, but are described in a draft document at http://www.geoviqua.org/Docs/03_ncWMS_Styling_Specification_1.0.pdf and are implemented in the ncWMS-Q software (see Section 6.1 below and a test client at http://ncwms.geoviqua.org/sldtest.html).

Clients can request that map images be generated using these new styles by crafting a SLD document using these new styles and passing it to the server. Alternatively, a server may offer named styles that implement these new styles, for convenience and simplicity (at the expense of some flexibility).

5.9. Mixing "Quality-Enabled" Data with "Non-Quality-Enabled" Data

In WMS-Q, not all Layers will necessarily have quality information (at the variable or sample level or either) attached. It is permissible to mix "quality-enabled" and "regular" layers in the same WMS service instance. The presence of MetadataURL links to metadata documents with quality indicators and the special keywords "qualityCollection", "qualityComposition" or other keywords in the UncertML or QualityML vocabulary allow the client to detect the quality-enabled layers in a WMS service.

6. Implementations

6.1. Server Implementations

The WMS-Q profile has been implemented in two systems in the context of the GeoViQua project. The ncWMS software [3] has been adapted to be compliant with WMS-Q, providing visualisations of raster data (e.g., numerical simulations and satellite imagery). Output from ncWMS-Q can be seen in Figure 1, showing data from the CCI-SST dataset of sea surface temperature and sea ice [19]. ncWMS is designed to require minimal configuration through automatic detection of metadata from source files in NetCDF format; this automatic configuration has been extended to the recognition of NetCDF-U data files, which contain information about data uncertainty. Therefore system administrators need only to format their data in NetCDF-U and the ncWMS-Q software will automatically provide a WMS-Q compatible endpoint. Manual configuration is possible in cases in which NetCDF-U files cannot be provided.

The Centre for Ecological Research and Forestry Applications (CREAF) has developed a WMS/WMTS server supporting WMS-Q based on MiraMon Map Service technology [37]. This allowed us to test the implementation for both continuous and categorical variables. For continuous variables, the Climatic Atlas of the Iberian Peninsula provides the following quality indicators regarding temperature: mean annual values, extrapolated areas, residual regression, and stability test. For categorical variables, the Landsat classified atlas was complemented by a set of quality indicators regarding land cover: first, second and third most present classes, fidelity, representativity, promiscuity, majority categories, entropy, and uncertainty.

The CREAF implementation of WMS in MiraMon Server is based on a prerendering process that serves both WMS and WMTS endpoints. In practice, this means that symbolization is associated to the layer by the operator publishing the layer. This has the advantage that rich and recommended visualizations can be offered to standard clients not fully aware of the particularities of the WMS extension. Multiple visualizations can be provided but users cannot control the symbolization parameters. The ncWMS-Q software supports the above proposed extensions to the SE standard (see Section 5.1 above) and therefore in this case the client has close control over the result of the image rendering process.

6.2. Client Implementations

ncWMS is packaged with an interactive map-based visualisation client called Godiva2 [38]. Godiva2 has been adapted for WMS-Q compliance, giving users the ability to interactively explore uncertain data on the web.

The MiraMon Map Browser has been enhanced making it aware of WMS-Q. It groups layers to this end following the service metadata document as defined by the layer hierarchy explained within this article. Only modifications on layer grouping and manual reordering by the user were required, proving how easily existing standard WMS clients are to adapt to WMS-Q. In addition, two different tools were added in order to allow users to easily compare parameters associated with a certain layer. The user is able to dynamically modify transparency of layers on runtime to see the layers below; and shift, allowing hiding dynamically one area of the top layer, so the ones below can be seen.

7. Integration of WMS-Q in a Quality Enabled Spatial Data Infrastructure

WMS-Q was developed in the context of the GeoViQua project (http://www.geoviqua.org), which also developed several other components of a "quality-enabled" SDI. Although these components were originally designed and developed as enhancements to the Global Earth Observation System of Systems (GEOSS), they are more widely applicable.

Data cataloguing and discovery was implemented by enhancing the Discover and Access Broker (DAB) to produce a quality-enabled version (DAB-Q, http://essi-lab.eu/do/view/GIcat). This harvests the WMS-Q service metadata ("Capabilities") documents and maps the different Layers there present into ISO 19115 documents, which form the basis of the data catalogue. Those metadata ISO documents can be then used in the GEO Portal (http://www.geoportal.org) to find datasets which are quality enabled. Thus, when a WMS-Q enabled dataset is found in a search, in addition to all the current links presented, links to the different variables and their quality components are given to the user.

WMS-Q services are found through the "WMS-Q" keyword within the "http://www.geoviqua.org/def/doc/conventions/vocabulary" vocabulary (see Section 5.3 above). Variables are detected as having components at pixel-level or not, depending on the presence of the keyword "http://qualityml.geoviqua.org/1.0/qualityCollection", and if they are defined are displayable or not displayable (having the <Name> tag or not, see Section 5.5 above).

A prototype to discover datasets with certain statistical properties was implemented, allowing searches that are aware of the dataset quality (for example, standard deviation less than 1 meter). The final port of this quality enhanced implementation to the GEO Portal search engine is to be done.

8. Discussion and Future Work

This paper introduces WMS-Q, a profile of the WMS standard that provides a mechanism for communicating data quality. The essential features of the profile are:

- Communication of data quality at the level of datasets, variables and individual samples
- Re-use of concepts from related standards and vocabularies, including UncertML and QualityML, using the WMS Keyword tag to communicate the quality and uncertainty of a measured variable.
- Full compatibility with version 1.3.0 of the WMS standard.

- Extensions to the SE specification to give greater control over visualizations of uncertain components.

This profile was developed in the first version as an OGC public engineering report [39]. However, during further test in the GeoViQua project, some problems in the document were detected and solved. The results of these modifications were presented here and also in the WMS Standards Working Group that is now considering it as a possible OGC standard profile.

We have presented examples of use of the profile for two raster datasets: one representing continuous data (sea surface temperature) and one representing categorical data (land use classification). Further extensions to the SE specification could be devised in order to support more ways of visualizing uncertainty, such as the "uncertainty ribbons" and "graduated glyphs" described in [7]. Some extensions to vector data was also tested in the GeoViQua project and will be investigated in future work, building on approaches such as those described in [11].

Parallel work has extended these concepts to the Web Map Tile Service (WMTS) standard and will be described in future publications. WMTS was developed reusing many concepts of WMS but with scalability and performance in mind. Both have the concept of Layers and both deliver maps as a result. An important difference that is important for applying this profile is that layer nesting cannot be done in WMTS. Instead, the concept of *themes* was introduced, decoupling the layer definition from the layer interdependencies. A theme is defined as a tree structure that can contain other themes that can represent variables and layer references that can represent uncertainty components. This way, a quality theme can be defined to describe the sample quality components relation for a variable.

This paper focuses in discovery standards but many concepts could be reused in data access standards and services, such as the Web Coverage Service (WCS) standard. The version 2 [40] of this standard has adopted a modular approach that allows for better extensibility. In WCS, GMLCOV [41] (a GML application schema for coverages) is used to describe coverage offerings and it could be profiled to accurately identify uncertain variables and the components that convey them at the sample level. In addition, WCS operations could be extended to request the necessary quality and uncertainty components.

Recently, we have started to collaborate with weather forecasting agencies including MeteoFrance, the UK MetOffice, KNMI, DWD, ECMWF and AFWA, who are working together on a WMS best practice way to better communicate probability of forecasts resulting model ensembles, through WMS. These organizations are already providing WMS services and they have implemented their own solutions. By considering each ensemble result as a statistical sample and derived products as statistical probability distributions, we recognized the relevance of WMS-Q as a solution for ensemble standardization. We expect that the results of these discussions can converge in the use of an improved version of WMS-Q that fulfils the requirements of the weather and ocean forecasting community.

Acknowledgments

This work was supported in part by the European Commission through the Seventh Framework Programme under grant agreement No. 265178, QUAlity aware VIsualisation for the Global Earth Observation System of Systems (GeoViQua), and in part by the Spanish National project under the reference CGL2012-33927, Spatiotemporal analysis of land cover and vegetation stress in the Iberian P. in light of a half a century (1975–2025) of climate dynamics and its anomalies (DinaCliVe). The first version of the WMS-Q profile was developed in the OGC interoperability experiments series, ninth edition (OWS9). Xavier Pons is recipient of an ICREA Academia Excellence in Research grant (2011–2015).

The authors thank the editors and three anonymous reviewers for constructive comments on earlier versions of this paper.

Author Contributions

Jon D. Blower, Joan Masó, Daniel Díaz and Xiaoyu Yang. designed the WMS-Q profile and wrote the paper. Guy H. Griffiths, Charles J. Roberts and Xiaoyu Yang implemented the profile in the ncWMS software. Daniel Díaz implemented the MiraMon WMS-Q client extensions and Joan Masó and Xavier Pons contributed to the server extensions. Charles J. Roberts, Guy H. Griffiths and Jane P. Lewis designed the extensions to the Symbology Encoding specification.

Conflicts of Interest

The authors declare no conflict of interest.

References

1. De la Beaujardiere, J. *OpenGIS Web Map Service (WMS) Implementation Specification Version 1.3.0*; OGC 06-042; Open Geospatial Consortium: Wayland, MA, USA, 2006.
2. Technical Committee ISO/TC 211, Geographic Information/Geomatics. *ISO 19128:2005 Geographic Information—Web Map Server Interface*; ISO: Geneva, Switzerland, 2005; p. 76.
3. Blower, J.; Gemmell, A.; Griffiths, G.; Haines, K.; Santokhee, A.; Yang, X.A. Web Map Service implementation for the visualization of multidimensional gridded environmental data. *Environ. Model. Softw.* **2013**, *47*, 218–224.
4. Masser, I. *Building European Spatial Data Infrastructures*; Esri Press: Redlands, CA, USA, 2007.
5. Khalsa, S.J.S.; Nativi, S.; Geller, G.N. The GEOSS interoperability process pilot project (IP3). *IEEE Trans. Geosci. Remote Sens.* **2009**, *47*, 80–91.
6. Pebesma, E.; Cornford, D.; Dubois, G.; Heuvelink, G.B.M.; Hristopulos, D.; Pilz, J.; Stohlker, H.; Morin, G.; Skøien, J.O. INTAMAP: The design and implementation of an interoperable automated interpolation web service. *Comput. Geosci.* **2011**, *37*, 343–352.
7. Sanyal, J.; Zhang, S.; Dyer, J.; Mercer, A.; Amburn, P.; Moorhead, R.J. Noodles: A tool for visualization of numerical weather model ensemble uncertainty. *IEEE Trans. Vis. Comput. Graph.* **2010**, *16*, 1421–1430.
8. Potter, K.; Wilson, A.; Bremer, P.T.; Williams, D.; Doutriaux, C.; Pascucci, V.; Johhson, C. Visualization of uncertainty and ensemble data: Exploration of climate modeling and weather forecast data with integrated ViSUS-CDAT systems. *J. Phys. Conf. Ser.* **2009**, *180*, doi:10.1088/1742-6596/180/1/012089.
9. MacEachren, A.M.; Robinson, A.; Hopper, S.; Gardner, S.; Murray, R.; Gahegan, M.; Hetzle, E. Visualizing geospatial information uncertainty: What we know and what we need to know. *Cartogr. Geogr. Inf. Sci.* **2005**, *32*, 139–160.

10. Müller, M. (Ed.) *OpenGIS Symbology Encoding Implementation Specification Version 1.1.0*; OGC 05-077r4; Open Geospatial Consortium: Wayland, MA, USA, 2006.
11. Keijzer, S.C. Visualizing Scale Related Uncertainty in Web Maps for Spatial Planning. Master's Thesis, Faculty of Geosciences Theses, Utrecht University, Utrecht, The Netherlands, 2011.
12. Technical Committee ISO/TC 176/SC 1, Concepts and Terminology. *ISO 8402:1994 Quality Management and Quality Assurance—Vocabulary*; ISO: Geneva, Switzerland, 1994.
13. Devillers, R.; Gervais, M.; Bédard, Y.; Jeansoulin, R. Spatial data quality: From metadata to quality indicators and contextual end-user manual. In Proceedings of the 2002 OEEPE/ISPRS Joint Workshop on Spatial Data Quality Management, Istambul, Turquie, 21–22 March 2002.
14. Technical Committee ISO/TC 211, Geographic Information/Geomatics. *ISO 19157:2013 Geographic Information—Data Quality*; ISO: Geneva, Switzerland, 2013; p. 146.
15. Kresse, W.; Danko, D.M. *Springer Handbook of Geographic Information*; Springer-Verlag: Berlin, Germany, 2012; p. 1120.
16. Veregin, H. Data quality parameters. In *Geographical Information Systems*; Longley, P.A., Goodchild, M.F., Maguire, D.J., Rhind, D.W., Eds.; John Wiley and Sons: New York, NY, USA, 1999; pp. 177–189.
17. Developing Spatial Data Infrastructures: The SDI Cookbook. Available online: http://www.gsdi.org/docs2004/Cookbook/cookbookV2.0.pdf (accessed on 28 September 2015).
18. Lawrence, B.; Lowry, R.; Miller, P.; Snaith, H.; Woolf, A. Information in environmental data grids. *Philos. Trans. R. Soc. A: Math. Phys. Eng. Sci.* **2009**, *367*, 1003–1014.
19. Merchant, C.J.; Embury, O.; Roberts-Jones, J.; Fiedler, E.K.; Bulgin, C.E.; Corlett, G.K.; Good, S.; McLaren, A.; Rayner, N.A.; Donlon, C. *ESA Sea Surface Temperature Climate Change Initiative (ESA SST CCI): Analysis Long Term Product Version 1.0*; NERC Earth Observation Data Centre: Oxford, UK, 2014.
20. Pons, X.; Sevillano, E.; Moré, G.; Serra, P.; Cornford, D.; Ninyerola, M. Distribución espacial de la incertidumbre en mapas de cubiertas obtenidos mediante teledetección. *Rev. Teledetec.* **2014**, doi:10.4995/raet.2014.3059.
21. Devillers, R.; Bédard, Y.; Jeansoulin, R. Multidimensional management of geospatial data quality information for its dynamic use within GIS. *Photogramm. Eng. Remote Sens.* **2005**, *71*, 205–215.
22. Technical Committee ISO/TC 211, Geographic Information/Geomatics. *ISO 19115:2003 Geographic Information—Metadata*; ISO: Geneva, Switzerland, 2003.
23. Technical Committee ISO/TC 211, Geographic Information/Geomatics. *ISO/TS 19139:2007 Geographic Information—Metadata—XML Schema Implementation*; ISO: Geneva, Switzerland, 2007; p. 111.
24. Technical Committee ISO/TC 211, Geographic Information/Geomatics. *ISO 19113:2002 Geographic Information—Quality Principles*; ISO: Geneva, Switzerland, 2002.
25. Di, L.; Yue, P.; Ramapriyan, H.K.; King, R.L. Geoscience data provenance: An overview. *IEEE Trans. Geosci. Remote Sens.* **2013**, *51*, 5065–5072.
26. Aguilar, R.; Pan, J.; Gries, C.; San Gil, I.; Palanisamy, G.A. Fexible online metadata editing and management system. *Ecol. Inform.* **2010**, *5*, 26–31.
27. Aditya, T.; Kraak, M.J. A search interface for an SDI: Implementation and evaluation of metadata visualization strategies. *Trans. GIS* **2007**, *11*, 413–435.

28. Lush, V.; Bastin, L.; Lumsden, J. Developing a GEO label: Providing the GIS community with quality metadata visualisation tools. In Proceedings of the 21st GIS Research, Liverpool, UK, 3–5 April 2013.
29. Ahonen-Rainio, P. Visualization of Geospatial Metadata for Selecting Geographic Datasets. Ph.D. Thesis, Helsinki University of Technology (TKK), Espoo, Finland, 2005.
30. Sevillano, E.; Ninyerola, M.; Zabala, A.; Bastin, L.; Masó, J. *QualityML: A Dictionary for Quality Metadata Encoding*; EGU: Munich, Germany, 2014.
31. Williams, M.; Cornford, D.; Bastin, L. Describing and communicating uncertainty within the semantic web. In Proceedings of the 7th International Semantic Web Conference Uncertainty Reasoning for the Semantic Web Workshop, Karlsruhe, Germany, 26 October 2008.
32. Bigagli, L.; Nativi, S. (Eds.) *NetCDF Uncertainty Conventions (NetCDF-U) OGC Discussion Paper*; OGC 11-163; Open Geospatial Consortium: Wayland, MA, USA, 2013.
33. Lupp M. *Styled Layer Descriptor Profile of the Web Map Service Implementation Specification*; OGC 05-078r4; Open Geospatial Consortium: Wayland, MA, USA, 2007.
34. IPCC. *Climate Change 2013: The Physical Science Basis. Contribution of Working Group I to the Fifth Assessment Report of the Intergovernmental Panel on Climate Change*; Stocker, T.F., Qin, D., Plattner, G.-K., Tignor, M., Allen, S.K., Boschung, J., Nauels, A., Xia, Y., Bex, V., Midgley, P.M., Eds.; Cambridge University Press: Cambridge, UK, 2013.
35. Teuling, A.J.; Stöckli, R.; Seneviratne, S.I. Bivariate colour maps for visualizing climate data. *Int. J. Climatol.* **2011**, *31*, 1408–1412.
36. Pebesma, E.J.; de Kwaadsteniet, J.W. Mapping Groundwater Quality in the Netherlands. *J. Hydrol.* **1997**, *200*, 364–386.
37. Pons, X. *MiraMon.Sistema d'Informació Geogràfica i Software de Teledetecció Centre de Recerca Ecològica i Aplicacions Forestals*; CREAF: Bellaterra, Spain, 2000.
38. Blower, J.D.; Haines, K.; Santokhee, A.; Liu, C.L. GODIVA2: Interactive visualization of environmental data on the Web. *Philos. Trans. R. Soc. A* **2009**, *367*, doi:10.1098/rsta.2008.0180.
39. Blower, J.; Yang, X.; Masó, J.; Thum, S. *OWS 9 Data Quality and Web Mapping Engineering Report*, OGC 12-160r1; Open Geospatial Consortium: Wayland, MA, USA, 2013.
40. Baumann, P. (Ed.) *Web Coverage Service Interface Standard-Core, Version 2.0.1*; OGC 09-110r4; Open Geospatial Consortium: Wayland, MA, USA, 2012.
41. Baumann, P. (Ed.) *GML Application Schema—Coverages, Version 1.0.1*; OGC 09-146r2; Open Geospatial Consortium: Wayland, MA, USA, 2012.

5

Ecological Network Construction Based on Minimum Cumulative Resistance for the City of Nanjing, China

Jihong Dong*, Wenting Dai, Guoquan Shao and Jiren Xu

School of Environment and Spatial Informatics, China University of Mining and Technology, Xuzhou 221116, China; E-Mails: daiwenting@cumt.edu.cn (W.D.); cumtshaogq@126.com (G.S.); jirenxu@126.com (J.X.)

* Author to whom correspondence should be addressed; E-Mail: dongjihong@cumt.edu.cn

Academic Editors: Suzana Dragicevic and Wolfgang Kainz

Abstract: With economic growth and the improvement of the urbanization level, human activities have constantly interfered with landscape patterns, resulting in serious threats to regional ecological security. Therefore, it is of great significance to study the evolution and optimization of the landscape patterns. Based on three TM images from 1990, 2000, and 2010, and selected landscape pattern indexes, the changes in the landscape pattern of Nanjing in the past twenty years were studied based on landscape ecology theory using Remote Sensing (RS) and a Geographical Information System (GIS). The ecological network was built on the basis of extracted ecological nodes and the minimum cumulative resistance. The results show that changes in the landscape pattern of the city of Nanjing were notable. Class-level indexes indicate that the farmland landscape area decreased and the degree of patch fragmentation increased. The construction land area increased, and it tended to show dispersed distribution. The proportion of forest land increased and the shape of patches became more complex. The proportion of water firstly showed a decrease, followed by an increase, and the shape of the water became more regular. Landscape-level indexes indicate that biological diversity and the degree of fragmentation increased. Spatial heterogeneity of the natural landscape increased, and the patch shape of each landscape type developed similarly. The results also call for stepping-stones to enhance the connectivity and optimization of the ecological network, which will help improve ecological services and improve the landscape pattern of the city.

Keywords: landscape pattern; land use change; ecological network; minimum cumulative resistance; Nanjing

1. Introduction

Landscape patterns refer to the spatial structure and characteristics of landscape components, namely the spatial configuration of landscape patches that differ in size and shape [1,2]. They are formed by natural and human factors. As the most densely-distributed area in terms of population, energy, materials, and information, cities with urbanized development are ecosystems that suffer from severe human disturbance [3], and the landscape patterns of cities are also undergoing profound changes. In recent years, many experts and scholars have conducted research on landscape pattern changes in various areas, including cities [4–8]. Most of the research is based on multi-temporal, remote-sensing data sources and uses the landscape pattern index method to measure the change characteristics of the landscape in a given area through a classification-based interpretation of remote sensing images [9–11].

An ecological node is defined as a place that is distributed in a spatial ecosystem with a high service function value; it is a landscape component that connects the adjacent ecological source and plays the key role of the regional ecological flow [12]. An urban ecological corridor is a linear or zonal landscape component that is based on a natural or artificial corridor mainly by virescence. It is the information channel connecting landscape elements, which has the dual function of connector and barrier. Ecological nodes and ecological corridors are connected with each other to form a network system, which is an ecological network. The concept of an ecological network was first proposed during the era of park planning in the late 19th century. Early thought about ecological network construction can be seen in the Boston Park System of Olmsted [13]. In the 1990s, the application range of the ecological network concept became broader, with the research scale expanding from "site" to "local", then to "region". The construction of an ecological network realizes the movement of material, energy, and information in the ecosystem and the migration of species [14,15]. In addition, as a strategic measure, ecological network construction has been valued highly by planners and land managers, and it is widely accepted and recognized all over the world [16,17]. The USA and Canada put the theory of ecological networks into practice and developed region-scale ecological network planning [18,19]. Chinese scholars mostly learned about earlier foreign theories and examples of ecological network construction and used geographic information system (GIS) techniques and the theory of ecological network construction to help build cities. Their studies have focused on urban green-land system planning [20,21], applications of GIS in ecological network construction [22–27], the analysis and evaluation of the structure of urban ecological networks [28], and studies on the ecological network planning model [29]. There are some practical applications of urban ecological network construction in China. For example, Yang used the landscape pattern principle to determine the potential landscape ecological security network in Guangzhou City, and contacted the ecological space planning [30]. In the ecological network planning of Xiamen City, Wang abstracted the green patch and urban ecological corridor as node and link, respectively, by using the network analysis method to analyze the topological relationship between the nodes and links, and finally selected the network model with highest degree of connection as the best ecological network design [31].

This paper presents a study on the landscape pattern changes in the City of Nanjing and the construction of its ecological network using Remote Sensing (RS)/Geographical Information System (GIS) techniques. The aim of this study was to enhance the relationship between the landscape elements and connectivity of the city's complete ecosystem and to promote the ecological service functioning of the urban landscape systems. Both the scientific basis and practical decision support are discussed.

2. Study Area and Methods

2.1. Study Area

Nanjing is a city in China that lies in the central region of the middle and lower Yangtze River and the southwestern part of Jiangsu province. It is a hub city in the western part of the Yangtze River Delta and one of four central cities in the Yangtze River basin. The main rivers in Nanjing City are the Yangtze River and Qinhuai River, as well as the Chunli River and Tianshengqiao River in the south, and the Chu River in the north. Nanjing belongs to the Ningzhenyang region, where low mountains account for 3.5% of the total land area, hilly areas account for 4.3%, hillock areas account for 53%, and low-lying land, river and lack account for 39.2%. Nanjing is surrounded by water and mountains, and the hilly areas are widely distributed. The administrative districts of Nanjing include Gulou District, Xuanwu District, Jianye District, Qinhuai District, Qixia District, Yuhua District, Pukou District, Jiangning District, Liuhe District, Lishui District, and Gaochun District. There are eleven administrative districts, as illustrated in Figure 1.

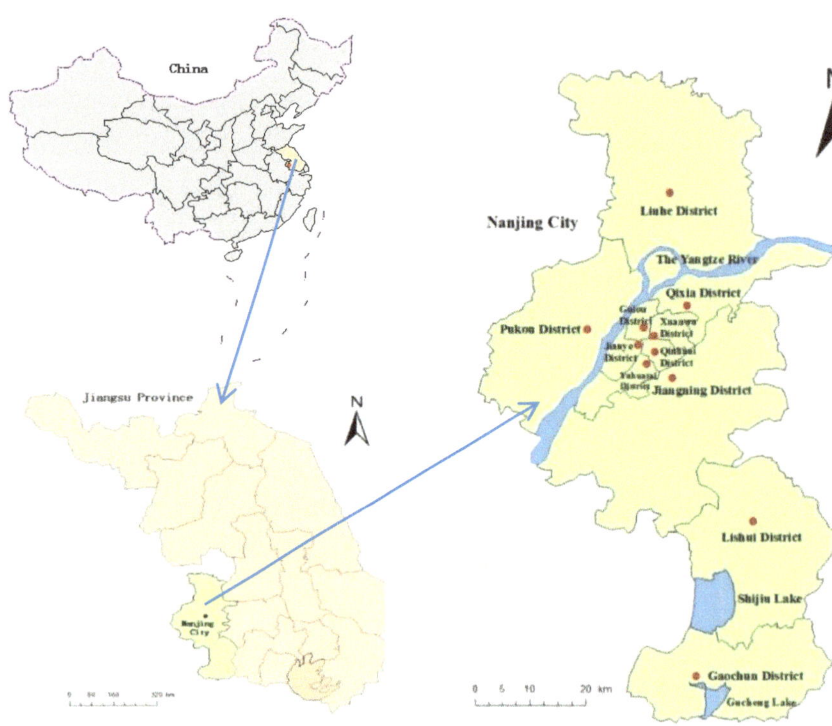

Figure 1. Location and administrative districts of Nanjing in Jiangsu Province, China.

2.2. Data Acquisition and Processing

The three TM images from 1990, 2000, and 2010 were obtained from the Center for Earth Observation and the Digital Earth Chinese Academy of Sciences and were used as the basis for landscape classification. (The center is a subordinate unit of the Chinese Academy of Sciences and is organized based on the Chinese Academy remote sensing satellite ground station and the Chinese Academy aviation remote sensing center and digital earth laboratory. The Center for Earth Observation primarily conducts high-quality operations and data services for the government, industries, and firms in the field of aerospace and aviation). The row-column number of the image track is P122 R38. The three images have approximately 10% cloud coverage, contain six bands (data acquired in 1–5 and 7 bands from the visible to near-infrared) and have a spatial resolution of 30 m. The methods for system radiation correction and geometric correction are based on a ground control point, and terrain correction was used to process the original images. WGS_1984_UTM_Zone_50N was selected as the projected coordinate system. According to the Chinese Land Use Classification System (General Administration of Quality Supervision, Inspection, and Quarantine of the PRC and Standardization Administration of the PRC [32]) and the characteristics of land use status in Nanjing, the four land-cover classes are farmland, construction land, forestland, and water. The images were classified using both supervised and unsupervised classification methods with the help of ENVI software. Figure 2 shows a land cover map from 1990 to 2010 in Nanjing. As shown in Figure 2, from 1990 to 2000, the area of farmland was the largest, while the areas of water, forestland, and construction land were relatively small and dispersed; the area of Nanjing city proper increased, and the construction land areas in the rural and suburban parts tended to decrease. These changes were related to the rural integration and new rural construction of Nanjing. From 2000 to 2010, the areas of water and construction land clearly increased, especially the area of construction land. The fragmentation degree of the Nanjing landscape in 2010 was significantly greater than in 1990 and 2000. According to the Nanjing master urban plan (1991–2010), the government planned to build a historical and cultural city protection zone and scenic tourist area with “mountains, water, city and forest” as the main areas in the city proper and area along the Yangtze River.

2.3. Selection and Calculation of the Landscape Pattern Indexes

The classified remote images were converted into grid data of 30 m × 30 m, and the landscape pattern indexes were calculated using Fragstats 3.3 software to analyze the changes to the landscape patterns in the City of Nanjing during its urbanization process. Fragstats 3.3 is a landscape index calculation software developed by the School of Forest Science at Oregon State University in the USA, in which complicated landscape pattern index calculations can be produced using a simple computer program [33]. Landscape pattern indexes that are less correlated were selected at both the class and landscape levels. The class-level indexes are PLAND, AREA_MN, PD, LSI, and AI; the specific ecological meanings and calculation equations of the class-level indexes are shown in Table 1. The landscape-level indexes are SHDI, SHAPE_AM, FRAC_AM, SI, and SHEI; the specific ecological meanings and calculation equations of the landscape-level indexes are shown in Table 2 [34–36]. A GIS spatial attribute database was constructed using ArcGIS.

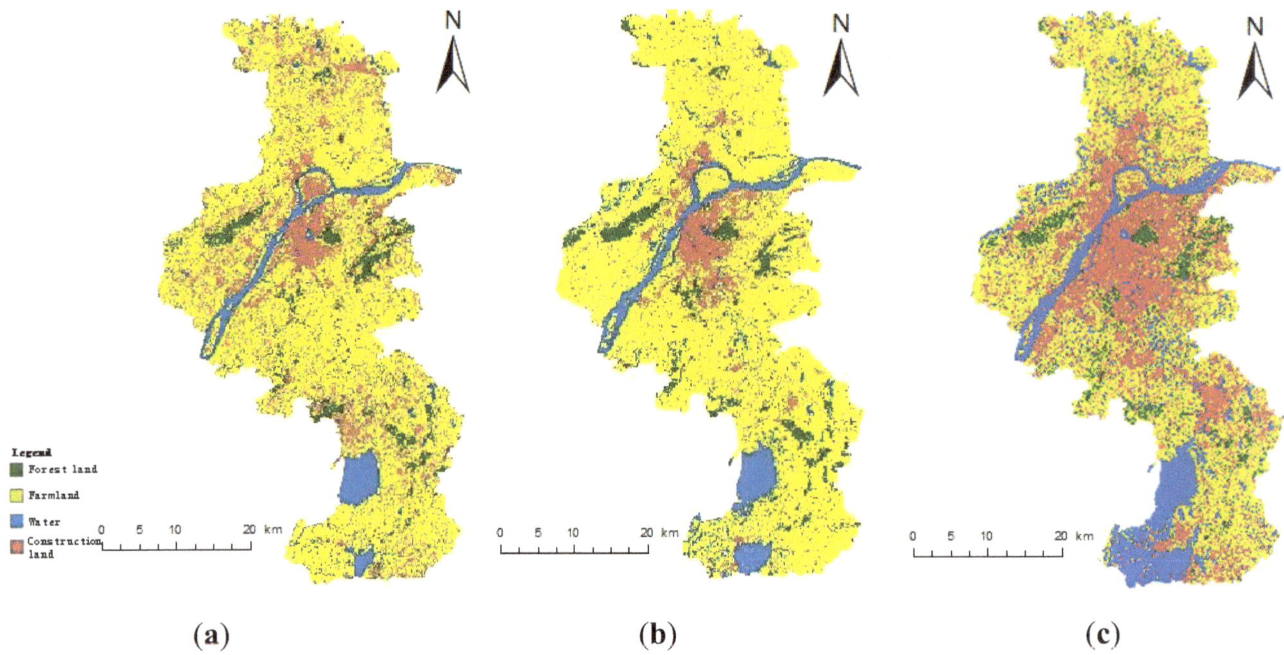

Figure 2. Land cover map from 1990 to 2010 in Nanjing: **(a)** 1990; **(b)** 2000; **(c)** 2010.

Table 1. Specific ecological meanings and calculation of class-level indexes.

Class-Level Indexes	Ecological Meaning	Calculation Equation
PLAND	Reflecting the situation of landscape grouping	(PLAND $= P_i = \frac{\sum a_{ij}}{A}$ (a_{ij} represents the area of patch ij; A represents total area of the landscape)
AREA_MN	Reflecting the aggregation or fragmentation degree of the landscape patches	AREA_MN $= \frac{A}{N}$ (A represents total area of the landscape; N represents the number of the whole patches)
PD	Reflecting the spatial heterogeneity of the landscape	PD $= \frac{n}{A}$ (n represents the number of patches; A represents total area of the landscape)
LSI	Reflecting the variability of the landscape patches	LSI $= \frac{E}{4\sqrt{A}}$ (E represents the total length of all patch boundaries in the landscape; A represents total area of the landscape)
AI	Describing the agglomeration degree of different components in the landscape	AI $= \left[\frac{g_{ii}}{max \to g_{ii}}\right](100)$ (g_{ii} represents the number of adjacent patches in the corresponding landscape types)

2.4. Construction of the Ecological Network

Firstly, ecological nodes in Nanjing were selected as ecological points, including forest parks, scenic areas, and wetlands. Secondly, the resistance value of a landscape type, which refers to the effect of one landscape type on the ecological flow rate, can be determined based on experience and expert investigation. Landscape resistance refers to the impact of the landscape on the ecological flow rate and the creation of the landscape resistance surface using ArcGIS according to the landscape resistance values. Thirdly, the least cost gradient data based on the minimum cumulative resistance (MCR, which is the cumulative distance from a certain landscape unit with the resistance value linked to the ecological source) and the cost distance function in ArcGIS is calculated. The cost distance function is an algorithm of the grid data, moving from the center to the periphery of the eight units. As shown in Equation (1), the cumulative cost that any N_i moves to four vertically-adjacent units $N_{i+1}$is half of the total cost coefficient moving from N_i to N_{i+1}; and as shown in Equation (2), the cumulative cost that any N_i moves to four diagonally-adjacent units N_{i+1} is the product of half of the total cost coefficient moving from N_i to $N_{i+1}$and the square root of 2.

$$N_{i+1} = N_i + (r_i + r_{i+1})/2 \tag{1}$$

$$N_{i+1} = N_i + \sqrt{2} \times (r_i + r_{i+1})/2 \tag{2}$$

In the equations, N_i and r_i represent the cumulative cost and the cost coefficient of unit i, respectively; N_{i+1} and r_{i+1} represent the cumulative cost and the cost coefficient of moving to unit $i+1$.

Table 2. Specific ecological meanings and calculation of the landscape-level indexes.

Landscape-level Indexes	Ecological Meaning	Calculation Equation
SHDI	Investigating the diversity of the landscape pattern	$SHDI = -\sum(P_i lnP_i)$ (P_i represents the ratio that a certain type of patch accounts for in the whole landscape)
SHAPE_AM	Reflecting the complexity of the shape of the patches	$SHAPE\_AM = \frac{0.25P_{ij}}{\sqrt{a_{ij}}}$ (P_{ij} represents the perimeter of the patch ij; a_{ij} represents the area of the patch ij)
FRAC-AM	Reflecting the complexity degree of the shape of average patches in the landscape	$FRAC\_AM = \frac{2\ln(P_{ij}/4)}{ln a_{ij}}$ (P_{ij} represents the perimeter of the patch ij; a_{ij} represents the area of the patch ij)
SI	Reflecting the complexity of the landscape space structure	$SI = \frac{P_{ij}}{4\sqrt{A_{ij}}}$ (P_{ij} represents the perimeter of the patch ij; A_{ij} represents the area of the patch ij)
SHEI	Reflecting the uneven distribution degree of area for each patch in the landscape	$SHEI = \frac{\sum P_k log_2 P_k}{log_2 n}$ (P_k represents the ratio that the area of patch k accounts for in the whole landscape; n represents the total number of patch types in the landscape)

Finally, the direction and route of least cost is determined and the least-cost ecological network is extracted by setting several threshold values of the cost network.

3. Results

3.1. Analysis on the Characteristics of the Landscape Pattern Change

3.1.1. Class-Level

The change characteristics of the landscape types are shown in Table 3.

Table 3. Class-level indexes from 1990 to 2010.

Land Cover Types	Year	PLAND	AREA_MN	PD	LSI	AI
Farmland	1990	67.19	263.01	0.25	82.62	92.51
	2000	65.95	220.42	0.30	73.12	93.42
	2010	54.58	214.73	0.81	87.46	91.32
Water	1990	11.63	19.56	0.91	66.75	79.57
	2000	8.49	21.03	0.63	50.80	80.70
	2010	11.09	22.42	0.34	45.25	86.93
Forest	1990	7.15	14.91	0.37	46.05	87.48
	2000	8.49	15.32	0.40	50.94	87.27
	2010	8.69	16.55	0.90	66.19	83.56
Construction land	1990	14.04	8.60	1.63	104.16	86.78
	2000	15.93	10.29	1.55	104.76	88.09
	2010	25.65	20.38	0.90	90.27	90.13

The variation in the characteristics of PLAND: PLAND reflects the components of the landscape. As shown in Table 1, the proportion of farmland type—mostly composed of natural landscape and managerial landscape—decreased from 67.19% to 54.58%. The proportion of construction land, mainly artificial landscape, increased from 14.04% to 25.65%. The proportion of water area decreased by 3.14% from 1990 to 2000 because of the large-area pond seine culture in the 1990s and the expansion of the city. In 2010, the proportion of water area reached 11.09%; this change was correlated with the construction of urban wetland parks and large-area aquaculture activities in Gaochun District. The proportion of forest and construction land both increased. The increased proportion of forestland reflected a move by the Nanjing government toward the construction of natural reserves, scenic areas, and forest parks. The expansion of construction land was the inevitable outcome of urbanization.

The variation in the characteristics of the AREA_MN: The AREA_MN index represents the degree of aggregation or fragmentation of the landscape patches. We can see from the contemporaneous data that the AREA_MN of farmland showed the highest index value, which means that farmland had a dispersed distribution. The indexes for the other three types of landscape were all at low levels, which indicate that the patches were smaller in size and saw a dispersed distribution. From 1990 to 2010, the AREA_MN value of the farmland in Nanjing decreased by 48.28 *ha*. This shows that the farmland landscape was separated by artificial landscapes and that the degree of fragmentation became deeper. As noted in the master urban plan of Nanjing (1991–2010), the government planned to increase the ecological land area and, thus, the AREA_MN of water and forest increased. The AREA_MN value of construction land increased from 8.6 *ha* to 20.38 *ha* over 20 years. This change of 11.78 *ha* also indicates a widely dispersed distribution of construction land patches.

The variation in the characteristics of PD: PD is defined as the number of patches per unit area; the more that PD increases, the greater the heterogeneity and fragmentation become. The PD of farmland and forestland increased by 0.56 and 0.53 over the past 20 years, respectively, while the PD of water and construction land decreased by 0.57 and 0.73. The results show the increased fragmentation of farmland and forestland and a trend of these lands being converted to small-scattered patches. The anti-risk ability of the eco-function land declined. Water and forestland became small aggregate patches. This can be seen in the classified remote sensing images in Figure 2.

The variation in the characteristics of LSI: LSI reflects the heterogeneity of the landscape patches. Compared with the two other landscape types, the LSI values of construction land and farmland are the highest and the second highest, respectively. The patch shape of the two types was more complex than the shape of the other land types. The LSI of farmland saw an initial decreasing trend followed by an increase. The index decreased by 9.5, which was caused by agricultural land consolidation and the planting of green belts during the first 10 years (1990–2000). In the subsequent 10 years (2000–2010), the index increased by 14.34 due to the separation of farmland with artificial landscapes, which resulted in many patches bearing more complex shapes. The water LSI reduced gradually and the shape of water became more regular. The forestland LSI saw a constant increase, with a more complex shape. The LSI of construction land showed a slight increase from 1990 to 2000, possibly because the Urban Planning Division in Nanjing claimed that development of the city proper should be coordinated with the surrounding towns. Under the guidance of the Urban Planning group, patches of construction land saw a dispersed expansion. Since 2000, the LSI of construction land has decreased sharply from 104.76 to 90.27. The possible reason for this decrease was that the government made local rectifications with the Urban Planning group of Nanjing in 2000: the urban structure of the city proper and the 12 surrounding towns in the Nanjing Metropolitan Area were adapted to the structure of city proper-new city-new urban area. Patches of construction land were combined, and the shape tended to be regular.

The variation in the characteristics of AI: AI is the degree of aggregation of the landscape patches of the different types. From 1990 to 2010, the AI of farmland was 92.51, 93.42, and 91.32, showing a rising-and-falling trend. This trend reflects agricultural development in the 1990s and the later construction encroachment into farmlands up to 2010. The forestland AI fell from 87.48 in 1990 to 83.56 in 2010; this change was due to deforestation by humans. The constantly increasing AI of water and construction land indicated that patches of water and construction land aggregated gradually and tended to distribute like crumbs. This situation was caused by creating planned water conservancy land and extending construction land.

3.1.2. Landscape-Level

The changes at the landscape level are shown in Table 4.

SHDI reflects the complexity and heterogeneity of land use in the paper. The change of SHDI was not notable from 1990 to 2000, during which the index increased only by 0.135 from 2000 to 2010. The index increase of 0.135 from 2000 to 2010 meant that the diversity of the whole landscape increased, the number of dispersed patches became bigger, and spatial heterogeneity was increased because the scattered urban land has broken up the landscape.

SHAPE_AM and FRAC_AM exhibit a decreasing trend, which shows that the shape of the overall city landscape became more regular. The shape of these four landscape patch types became similar. Aggravated disturbance by humans (planning adjustment for land use) was consistent with the result of the analysis.

SPLIT increased from 5.13 to 10.08, which explains the natural landscape fragmentation caused by human landscape disturbance. Landscape fragmentation led to single patch area decreases, and the fragmentation of the territory and loss of connectivity has become a worldwide problem [37–40].

SHEI had not changed significantly between 1990 and 2000. By 2010, however, the SHEI value showed an improvement, indicating that major landscape types no longer played a dominant role, the average patch area became similar and the patches tended to have a uniform distribution.

Table 4. Landscape-level indexes from 1990 to 2010.

Year	SHDI	SHAPE_AM	FRAC_AM	SPLIT	SHEI
1990	1.01	34.23	1.29	5.13	0.73
2000	1	27.80	1.28	5.99	0.72
2010	1.135	22.48	1.27	10.08	0.82

3.2. Creation of the Ecological Network

3.2.1. Extraction of Ecological Nodes

Ecological nodes are the convergence points of materials, energy, and services [41]. In urban and rural areas, ecological nodes include green landscapes, which consist of city green areas, farmland, garden land, forestland, other agricultural land, and blue landscape that includes rivers, lakes, reservoirs, and wetlands.

On the basis of the Ecological Red Line Area Protection Plan of Jiangsu Province, which was developed by the Jiangsu government in 2013 [42], this study selected scenic areas, forest parks, and large water source regions as ecological nodes from the list of ecological Red Line areas in 2010 in Nanjing (see Table 5). Furthermore, these selected ecological nodes were abstracted as ecological points (see Figure 3).

Table 5. Ecological nodes in 2010 of Nanjing.

Type	Name	Location	Function
Forest park	Purple Mountain Forest Park	Jiangning District	Biodiversity, natural and human landscape protection
	Laoshan Forest Park	Pukou District	Biodiversity, natural and human landscape protection
	Fangshan Forest Park	Liuhe District	Water source conservation Natural and human landscape protection
	Pingshan Forest Park	Liuhe District	Natural and human landscape protection
	Wuxiang Temple Forest Park	Lishui District	Biodiversity, natural and human landscape protection
	Dajingshan Forest Park	Gaochun District	Water source and soil conservation
Scenic area	Dragon Mountain Scenic Area	Pukou District	Natural and human landscape protection
	Mufu Mountain Scenic Area	Xiaguan District	Natural and human landscape protection
	Lingyan Hill Scenic Area	Liuhe District	Geo-conservation Biodiversity protection
	Qixia Mountain Scenic Area	Qixia District	Biodiversity, natural and human landscape protection
	Niushou Mountain Forest Park	Yuhuatai District	Natural and human landscape protection

Table 5. *Cont.*

Type	Name	Location	Function
Scenic area	Tang Hill National Geo-park	Jiangning District	Geological relic protection
	Nanshan Scenic Area	Jiangning District	Biodiversity, natural and human landscape protection
	Mao Mountain	Lishui District	Biodiversity protection Water source
	Residual Branch	Gaochun District	conservation
	Youzi Mountain Scenic Area	Gaochun District	Biodiversity, natural and human landscape protection
Wetland	Qiqiaoweng Wetland Park	Qinhuai District	Wetland ecosystem maintenance
	Xuanwu Lake Wetland Park	Xuanwu District	Wetland ecosystem maintenance
	Gucheng Lake Natural Reserve	Gaochun District	Water and aquatic resources
	Shijiu Lake	Lishui District Gaochun District	Natural and human landscape protection
	Wetland by Yangtze river	Xinsheng island, Xinji island, Zaisheng island	Wetland ecosystem maintenance
	Shanhu Reservoir	Liuhe District	Irrigated agriculture
	Daquan Reservoir	Liuhe District	Irrigated agriculture
	Zheshantou Reservoir	Jingqiao Town	Irrigated agriculture
	Yaojia Reservoir	Jingqiao Town	Irrigated agriculture
	Wolong Mountain Reservoir	Lishui District	Irrigated agriculture
	Zhongshan Reservoir	Lishui District	Irrigated agriculture Water source and water quality protection
	Fangbian Reservoir	Lishui District	Irrigated agriculture Water source and water quality protection
	Hewangba Reservoir	Liuhe District	Irrigated agriculture
	Zhaocun Reservoir	Jiangning District	Irrigated agriculture
	Yanziji Water Source	Northcity water plant	Water source and water quality
	Baguazhou Water Source	Yuangu water plant	Water source and water quality
	Jiangpu & Pukou Water Source	Jiangpu water plant	Water source and water quality
	Baguazhou Spare Water Source	Jiangning District	Water source and water quality
	Jiajiang Water Source	Jiangning District	Water source and water quality

3.2.2. Ecological Network Construction

Construction of resistance and resistance trend surface: According to the influence of the landscape units on landscape migration and expert evaluation [43], the landscape units are classified by resistance and allocated the corresponding resistance parameters, which results in the landscape resistance surface with the help of GIS. The specific steps of the resistance surface construction are as follows: firstly, setting the resistance values of water and construction land, where the resistance values of all water and all construction are 1 and 100, respectively; secondly, setting the resistance values of forest land and farmland based on their own proportional values of ecological service; finally, forming the landscape resistance surface of Nanjing with the help of the surface diffusion technology in GIS (see Figure 4a) [44,45]. Using the selected points as targets and the landscape resistance surface for resistance costs, the least ecological function gradient cost data were calculated based on MCR with GIS techniques. The gradient cost data were then classified to generate a cost trend surface according to five levels of degree (see Figure 4b). The different levels represent the different degrees of landscape resistance: Level 1 represents the lowest degree of landscape resistance, which allows for the easiest ecological function flow, while Level 5 represents the highest degree of landscape resistance, which means that the ecological function flow is the most difficult.

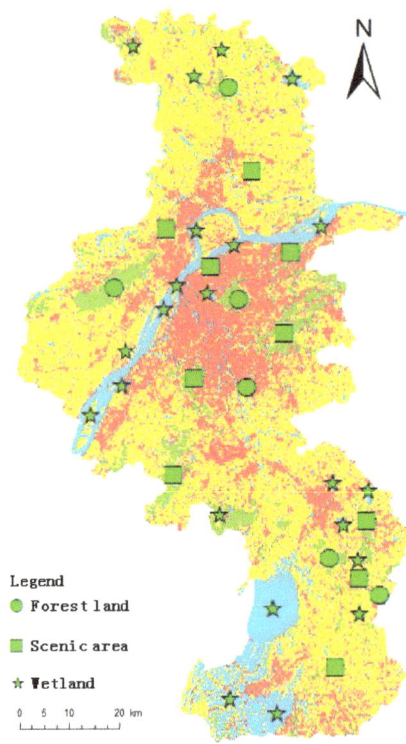

Figure 3. Distribution of ecological nodes of 2010 in Nanjing.

Establishment of four-level linear ecological networks: Based on branch networks and ecological function points, four-level linear ecological networks were constructed in the resistance trend surface, as shown in Figure 5.

It can be concluded from Figure 5 that linear ecological networks at the first and second level have isolated source points. The ecological function flows between source points cannot circulate integrally. Linear ecological networks at the third level are capable of connecting all ecological nodes, but they cannot meet the demand for the development integration of ecological networks and some branch networks that existed in remote rural areas. The linear ecological networks at the fourth level provide the highest coverage rate and are capable of connecting all ecological nodes that showed strong connectivity and form a circular ecological network covering the entire area. Moreover, a wider migration space of ecological function flow was provided.

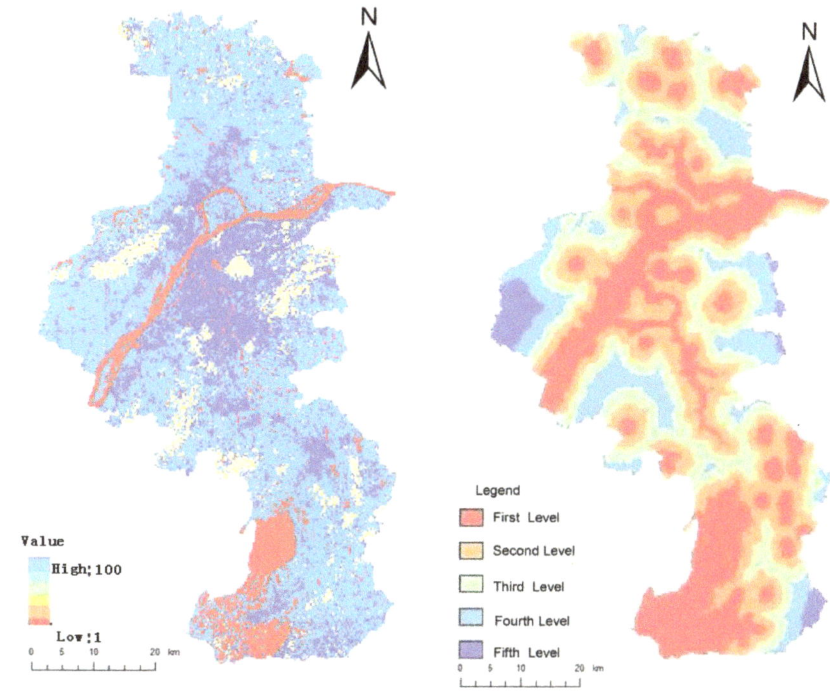

(a) (b)

Figure 4. (**a**) Landscape resistance surface of Nanjing; and (**b**) landscape resistance trend surface of Nanjing.

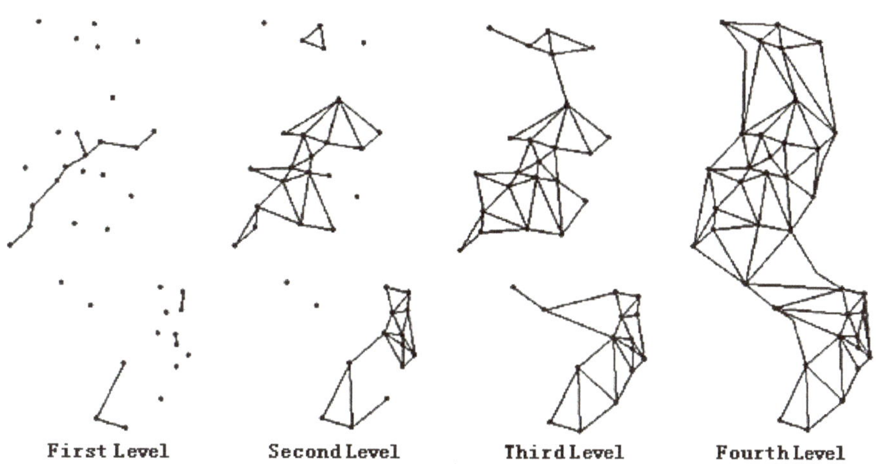

Figure 5. Representation of four-level linear ecological networks.

Construction of the least-cost network: Considering the advantages of the linear ecological network at the fourth level (as stated in the section above), the linear ecological network was analyzed to determine the optimal path. Using MCR from ArcGIS and selecting the direction and path of least cost, the least-cost ecological function network was extracted (see Figure 6).

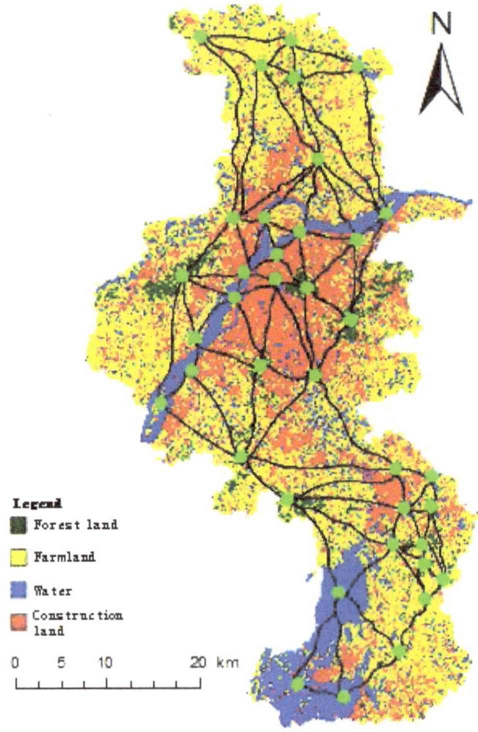

Figure 6. Least-cost ecological network of Nanjing.

4. Conclusions

The effects of the urbanization process on the landscape of the city of Nanjing from 1990 to 2010 were studied using RS, GIS, and the landscape index method. Moreover, an ecological network was generated according to landscape characteristics and MCR. The conclusions were as follows.

(1) The landscape suffered from human disturbance, and natural patches became separated by artificial patches. Landscape fragmentation led to the decrease of farmland patches and the expansion of construction land. SHDI, SPLIT, and SHEI tended to increase, while SHAPE_AM and FRAC_AM decreased, and the shape of the entire landscape became regular.

(2) Laying stepping stones (ecological nodes) between source points and establishing ecological corridors can strengthen landscape connectivity, optimize ecological network structures, and improve ecological services.

In future landscape optimization processes, ecological networks should be built. Moreover, destructive landscapes require rebuilding, wetland parks should be built and protected, and protective green areas need to be planted on both sides of roads and rivers in Nanjing. Confirming inhibitory development scope, such as understanding natural reserves, and scenic areas to avoid damaging the ecological function of the landscape, is necessary. Biological planning should be emphasized to protect internal species and maintain biomass; it is necessary to protect important water sources and reconstruct landscape areas that demonstrate reciprocal inhibition. Finally, MCR theories and methods are expected to be applied in other areas.

Acknowledgments

This work has been funded by the National Science Foundation of China (No.51374208) and the University Advantage Discipline Construction Project in Jiangsu province (No.SZBF2011-6-B53). We want to thank the Jiangsu Key Laboratory of Resources and Environmental Information Engineering at China University of Mining and Technology for its support. Also, we want to thank Songnian Li at Ryerson University, for his guidance and modification of the article.

Author Contributions

Author contributions: Jihong Dong constructed the structure of the article, and wrote the first draft of the manuscript. Wenting Dai was mainly responsible for data and relevant information collection, and finished the initial version of the manuscript. Guoquan Shao completed the images and data processing. Jiren Xu carried out the model validation. All co-authors participated in the field survey, contributed by reviewing the manuscript, and improving the discussion and interpretation of the results.

Conflicts of Interest

The authors declare no conflict of interest.

References

1. Fu, B.J. *Theory and Application of Landscape Ecology*; Science Press: Beijing, China, 2001.
2. Forman, R.T.T.; Godron, M. *Landscape Ecology*; John Wiley and Sons: New York, NY, USA, 1986.
3. Chen, L.D.; Sun, R.H.; Liu, H.L. Eco-environmental effects of urban landscape pattern changes: Progresses, problems and perspectives. *Acta Ecol. Sin.* **2013**, *33*, 1042–1050.
4. Guo, L.; Du, S.H.; Xue, D.Y. Spatio-temporal variation of landscape patterns during rapid urbanization in Guangzhou city. *Acta Sci. Nat. Univ. Pekin.* **2009**, *45*, 129–135.
5. Xu, J.X.; Li, G.; Qu, J.F. Changes of land use and landscape pattern in Hongze Lake basin. *Resour. Environ. Yangtze Basin* **2011**, *20*, 1214–1215.
6. Chen, X.L.; Chen, S.J.; Du, P.J. A study of the landscape pattern change of Longyan City based on multi-temporal remote sensing images. *Remote Sens. L. Res.* **2012**, *2*, 135–136.
7. Sun, F.Y.; Liu, M.; Hu, Y.M. A comprehensive analysis of landscape pattern dynamic change based on multi-methods: A case study of Tieling city, Liaoning province of northeast China. *Chin. J. Ecol.* **2013**, *32*, 2167–2168.
8. Yu, Q.Z.; Zhang, Z.L.; Gao, B. Study on the changes of landscape pattern in Nansihu wetland based on RS and frags tats. *Forest Res. Manag.* **2013**, *1*, 111–112.
9. Chen, A.L.; Sun, R.H.; Chen, L.D. Applicability of traditional landscape metrics in evaluating urban heat island effect. *Chin. J. Appl. Ecol.* **2012**, *23*, 2077–2086.
10. Liu, X.P.; Li, X.; Chen, Y.M.; Qin, Y.; Li, S.Y.; Chen, M.H. Landscape expansion index and its applications to quantitative analysis of urban expansion. *Acta Geogr. Sin.* **2009**, *64*, 1430–1438.
11. Huang, S.Q.; Chen, Y.; Zhang, R.Z.; Wu, W.; Wei, C. Spatial correlation analysis of land fragmentation and agriculture development based on landscape indexes. *Agr. Res. Arid Area.* **2015**, *33*, 238–244.

12. Wu, N.T.; Yue, D.P.; Zhang, Y.C.; Cao, R.; Shang, S. Identification and analysis of ecological nodes of urban and suburban areas based on 3S technology. *Environ. Sci. Tech.* **2014**, *37*, 184–189.
13. Liu, J.; Pu, W.R. *Planning and Design of City Greenland System*; Chinese Construction Industry Press: Beijing, China, 2004.
14. Fu, B.J.; Chen, L.D.; Ma, K.P.; Wang, Y.L. *Principle and Application of Landscape Ecology*; Science Press: Beijing, China, 2001.
15. Jongman, R.H.G. Nature conservation planning in Europe: Developing ecological networks. *Landsc. Urban Plan.* **1995**, *32*, 169–183.
16. Jongman, R.H.G.; Pungetti, G. *Ecological Networks and Greenways: Concept, Design, Implementation*; Cambridge University Press: Cambridge, UK, 2004.
17. Liu, B.Y.; Wang, P. Green ecological network planning development and China research frontier. *Chin. Gard.* **2010**, *26*, 1–5.
18. Liu, B.Y.; Yu, C. Greenway networks planning of USA and its apocalypse to us. *Chin. Gard.* **2001**, *6*, 77–81.
19. Government of Toronto, Canada. Urban Planning and Development Services of Toronto, Canada. Garrison Creek Linkage Plan, 1998. (Unpublished work)
20. Huang, Y.; Chen, H.; Huang, Z.J. Construction of urban green space ecosystem by using corridor networks: A case study in west urban area of Dongying city, Shandong province. *Chin. J. Appl. Ecol.* **2006**, *17*, 1683–1687.
21. Kong, F.H.; Yin, H.W. Developing green space ecological networks in Jinan city. *Acta Ecol. Sin.* **2008**, *28*, 1712–1718.
22. Adriaensen, F.; Chardon, J.P.; Blust, G.D. The application of "least-cost" modeling as a functional landscape model. *Landsc. Urban Plan.* **2003**, *64*, 233–247.
23. Zhang, X.F.; Wang, Y.L.; Li, G.C. Landscape functions network construction and application in watershed scale: A case study on Taiwan Wuxi watershed. *Acta Geogr. Sin.* **2005**, *60*, 974–980.
24. Liu, J.; Yang, W. A GIS-based landscape pattern optimization approach for Lake Dianchi watershed. *J. Nat. Resour.* **2012**, *27*, 802–805.
25. Li, J.H.; Liu, X.H. Research of the nature reserve zonation based on the least-cost distance model. *J. Nat. Resour.* **2006**, *21*, 217–224.
26. Guo, H.B.; Huang, Y.X.; Ye, G.F. A study on the evaluation and optimization of urban ecological function network in Xiamen. *J. Nat. Resour.* **2010**, *25*, 75–76.
27. Liu, C.L. Landscape Ecology Spatial Pattern Change and Optimization Measures in Nanjing. Master's Thesis, Nanjing Agricultural University, Nanjing, China, 2006.
28. Liu, J.J.; Li, J.D.; Liu, H. GIS-based ecological corridor of Shenyang landscape. *J. Shenyang Agr. Univ.* **2012**, *43*, 279–283.
29. Wang, Y.C.; Liu, Y.L. Researches on spatial model application of urban landscape eco-network planning. *Resour. Environ. Yangtze Basin* **2009**, *18*, 819–824.
30. Yang, Z.F.; He, M.C. *Urban Ecological Sustainable Development Planning*; Science Press: Beijing, China, 2004.
31. Wang, H.Z.; Zhang, L.Q. A GIS, landscape pattern and network analysis based planning of ecological networks for Xiamen Island. *Acta Phytoecol. Sin.* **2005**, *29*, 144–152.

32. General Administration of Quality Supervision, Inspection and Quarantine of the PRC, Standardization Administration of the PRC. *Classification of Land Use Present Situation*; Chinese Standardization Press: Beijing, China, 2007.
33. Yang, Z.H. Landscape pattern analysis of Hainan natural meadow in China based on Fragstats and thematic mapper. *J. Landsc. Res.* **2012**, *4*, 13–14.
34. Zheng, X.Q.; Fu, M.C. *Spatial Analysis Techniques and Application of Landscape Pattern*; Science Press: Beijing, China, 2010.
35. McGarigal, K.; Marks, B. *FRAGSTATS: Spatial Pattern Analysis Program for Quantifying Landscape Structure*; Oregon State University: Corvallis, OR, USA, 1994.
36. Chen, W.B.; Xiao, D.N.; Li, X.Z. Classification, application and creation of landscape indices. *Chin. J. Appl. Ecol.* **2002**, *13*, 121–125.
37. Parcerisas, L.; Marull, J.; Pino, J.; Tello, E.; Coll, F.; Basnou, C. Land use changes, landscape ecology and their socioeconomic driving forces in the Spanish Mediterranean coast (El Maresme County, 1850–2005). *J. Environ. Sci. Policy* **2012**, *23*, 120–132.
38. Bhatt, G.D.; Kushwaha, S.P.S.; Nandy, S.; Bargali, K.; Nagar, P.S.; Tadvi, D.M. Analysis of fragmentation and disturbance regimes in South Gujarat forests, India. *Trop. Ecol.* **2015**, *56*, 275–288.
39. Girvetz, E.H.; Thorne, J.H.; Berry, A.M.; Jaeger, J.A.G. Integration of landscape fragmentation analysis into regional planning: A statement multi-scale case study from California, USA. *Landsc. Urban. Plan.* **2008**, *86*, 205–218.
40. Fahrig, L. Effects of habitat fragmentation on biodiversity. *Annu. Rev. Ecol. Evol. Syst.* **2003**, *34*, 487–515.
41. Chen, L.D.; Fu, B.J.; Zhao, W.W. Source-sink landscape theory and its ecological significance. *Acta Ecol. Sin.* **2006**, *26*, 1444–1449.
42. People's Government of Jiangsu Province. Protection Planning of Ecological Red-Line Areas of Jiangsu Province. Available online: http://www.jiangsu.gov.cn/jsgov/tj/bgt/201309/t20130923_400467.html (accessed on 30 August 2013).
43. Jiao, S.; Li, Z.M.; Gao, Q.; Zhou, K.; Wei, C.Y.; He, S.Y. The application of landscape connectivity theory in urban ecology suitability assessment and optimization. *Geogr. Res.* **2013**, *32*, 720–730.
44. Ying, F.N.; Wang, X.L. Ecological landscape planning of the Four-Lake Basin based on accumulative minimum resistance model. *J. Huazhong Agr. Univ.* **2010**, *29*, 231–235.
45. Liu, X.F.; Shu, J.M.; Zhang, L.B. Research on applying minimal cumulative resistance model in urban land ecological suitability assessment: As an example of Xianmen City. *Acta Ecol. Sin.* **2010**, *30*, 421–428.

6

A Novel Spatial-Temporal Voronoi Diagram-Based Heuristic Approach for Large-Scale Vehicle Routing Optimization with Time Constraints

Wei Tu [1,2,3,*], Qingquan Li [2,3,4,*], Zhixiang Fang [4] and Baoding Zhou [2,3]

[1] College of Information Engineering, Shenzhen University, Shenzhen 518060, China

[2] Shenzhen Key Laboratory of Spatial Information Smart Sensing and Services, College of Civil Engineering, Shenzhen University, Shenzhen 518060, China; E-Mail: bdzhou@szu.edu.cn

[3] Key Laboratory for Geo-Environmental Monitoring of Coastal Zone of the National Administration of Surveying, Mapping and GeoInformation, Shenzhen University, Shenzhen 518060, China

[4] State Key Laboratory of Information Engineering in Surveying, Mapping, and Remote Sensing, Wuhan University, Wuhan 430079, China; E-Mail: zxfang@whu.edu.cn

* Authors to whom correspondence should be addressed; E-Mails: tuwei@szu.edu.cn (W.T.); liqq@szu.edu.cn (Q.L.)

Academic Editor: Wolfgang Kainz

Abstract: Vehicle routing optimization (VRO) designs the best routes to reduce travel cost, energy consumption, and carbon emission. Due to non-deterministic polynomial-time hard (NP-hard) complexity, many VROs involved in real-world applications require too much computing effort. Shortening computing time for VRO is a great challenge for state-of-the-art spatial optimization algorithms. From a spatial-temporal perspective, this paper presents a spatial-temporal Voronoi diagram-based heuristic approach for large-scale vehicle routing problems with time windows (VRPTW). Considering time constraints, a spatial-temporal Voronoi distance is derived from the spatial-temporal Voronoi diagram to find near neighbors in the space-time searching context. A Voronoi distance decay strategy that integrates a time warp operation is proposed to accelerate local search procedures. A spatial-temporal feature-guided search is developed to improve unpromising micro route structures. Experiments on VRPTW benchmarks and real-world instances are conducted to verify performance. The results demonstrate that the proposed approach is competitive with state-of-the-art heuristics and achieves high-quality solutions for large-scale instances of

VRPTWs in a short time. This novel approach will contribute to spatial decision support community by developing an effective vehicle routing optimization method for large transportation applications in both public and private sectors.

Keywords: spatial decision support system; vehicle routing problem; Voronoi diagram; heuristics; local search; distance decay

1. Introduction

Flexible transportation service not only reduces travel cost, but also alleviates related energy and environmental concerns, such as traffic congestion, energy consumption, and carbon emission [1,2]. With the advancement of geographic information science (GIS), space-related problem decision-making has been embedded in spatial decision support systems (SDSSs) to provide flexible transportation service in both public and private sectors [3], such as the design of school bus routes [4], collection of solid waste [5,6], distribution of goods in chain business [7], design of police patrol routes [8], and planning express deliveries [9]. Spatial function tools are used for intelligent transportation operation in many services including management of geo-referenced transport network data, geocoding massive human demands, positioning moving vehicles, and navigating routes for drivers [5–9].

Vehicle routing optimization (VRO) aims to design least cost routes to satisfy geographically-distributed human demand through spatial intelligence of SDSSs in the field of transportation [10], including the shortest path problem (SPP) [11–13], traveling salesman problem (TSP) [14], and the vehicle routing problem (VRP) [5–7]. One type of VRO is modeled as the vehicle routing problem with time window (VRPTW), which imposes time constraints on customers, vehicles, and depots to accept or distribute high quality service [4,6,14]. It specifies that a customer only accepts certain transportation-related service in a given time interval [15–17]. The VRPTW is computationally intractable in many real-life applications [18]. The heuristic algorithm is the only justified approach as it can find good solutions in a reasonable time [16,17]. There are two types of heuristic algorithms for the VRPTW, which include local search and evolutionary optimization. Local search iteratively modifies local route structures to improve solution quality and follows a single trajectory to explore the solution space [3,6,7]. Evolutionary optimization simultaneously searches multiple parts of the whole solution space to converge on the best solution [5,19,20]. Recently, state-of-the-art heuristic algorithms have been reported to yield a high quality solution of the VRPTW with hundreds of customers [5,9,18]. More efficient heuristics are needed for even larger VRPTW applications.

Spatial principle has exhibited good performance in space-related problem decision-making [21–23] in problems such as bus stop location [3], hazmat route planning [20], and path covering [24]. To efficiently solve VRO, from a spatial perspective, a few speed-up strategies have been developed to handle spatial issues and extend solvable VRO instances up to thousands of customers, which include the granular neighborhood [25], the *k*th nearest neighbors [26], and the *k*-ring-shaped Voronoi neighbors [27]. The common denominator of these effective strategies is to search only a limited number of spatial neighbors to save computing efforts [28]. However, such useful strategies become ineffective for the VRPTW since time constraint imposes an exceptional influence on the proximity of consecutively served customers in a route. A long wait will occur if two spatially-close customers with quite different time windows (for example, one is 8:00–9:00 a.m. and the other is 3:00–4:00 p.m.) are visited consecutively in a route, which challenges the motivation of spatial proximity-based heuristics approaches. Hence, spatial proximity measures must be extended to address the additional temporal constraints [18]. Moreover, how to accelerate the VRO with the help of spatial-temporal thinking should be investigated.

Voronoi diagrams describe both spatial proximity and topological proximity between spatial objects [29–32]. By extending it to the space-time context, a spatial-temporal Voronoi diagram (STVD) is well suited for measuring the proximity of customers with time constraints. Therefore, this article proposes a spatial-temporal Voronoi diagram-based heuristic approach to solve very large-scale VRPTWs. A derived spatial-temporal Voronoi neighbors from the Voronoi diagram is used to select promising candidates for a local search. Furthermore, motivated by the truth that nearer neighbors have a higher possibility to be consecutively served in a route [28], a Voronoi distance decay strategy is proposed to assign reasonable search efforts on neighbors with different Voronoi distances. A spatial-temporal feature guided search is used to reconstruct unpromising micro route structures. Intensive experiments on benchmark datasets (25–1000 customers) and real-world large scale VRPTWs and its multi-depot variants (2000–10,000 customers) have been implemented to assess the performance of the proposed approach. This efficient and effective approach enables current local search heuristics to solve super large-scale VRPTWs and will contribute to improving spatial intelligence for transportation-oriented SDSSs.

The remainder of this paper is organized as follows. Section 2 reviews related literature. Section 3 introduces the VRPTW, the STVD, the derived spatial-temporal Voronoi distance and neighbors. The proposed spatial-temporal Voronoi diagram-based heuristic for the VRPTW and its multi-depot variants are described in Section 4, and details of the experiments, results, and comparisons are reported in Section 5. Section 6 discusses the impact of the STVD and spatial-temporal proximity behind the best found solutions. The last section draws the conclusions.

2. Literature Review

This section reviews heuristic algorithms for the VRPTW and the integration of vehicle routing optimization and GIS.

2.1. Heuristic Algorithms for the VRPTW

Heuristic algorithms for VRPTWs are divided into two categories: local search and evolutionary optimization. Starting from an initial solution, local search shifts from a current solution to a better neighborhood solution with a simple change of local routes [33]. The change is made iteratively by moving nodes or exchanging arcs within a route or between routes such as 1–0 exchange, 1–1 exchange, 2-Opt, and λ-opt. Usually, the change process becomes trapped in a local optimum. To overcome unpromising results, many intelligent strategies that accept some worse solutions have been developed to further improve the solution quality, such as simulated annealing [34], iterated local search [35], large neighborhood search [36], variable neighborhood search [37], and tabu search [38]. Numerous tests on small- and medium-size VRPTWs with 25–100 customers have demonstrated good performance of local search heuristics [34,36,37].

Evolutionary optimization concurrently improves many solutions and results in high quality solutions. Four major steps are involved: representation, selection, combination, and mutation. Some of the most successful evolutionary optimization algorithms for the VRPTW are provided by Repoussis *et al.* [39] and Vidal *et al.* [40]. Mester and Bräysy [41] used the standard evolutionary optimization framework to guide exploration in the VRPTW solution space with solution initialization and evolution. Gehring and Homberger [42] parallelized the genetic algorithm and first solved VRPTW instances up to 1000 customers. Usually, local search is used as a component in the evolutionary optimization approach. Fast and simplified node or arc exchange-based local search have been used for offspring education in the genetic algorithm [41]. Therefore, more efficient local search heuristics also help to improve evolutionary optimization.

For larger VRPTW instances, recent advances have proposed several accelerating techniques to save computing efforts while still achieving high quality solutions. Following the local search principle, taking spatial distance into account, Cordeau [38] limited the local search candidates with *kth nearest neighbors* to shorten computing time. Toth and Vigo [25] selected the local search evaluated nodes with a fixed distance threshold. Considering time constraints, Ibaraki [43] proposed a time-oriented neighbor list-based strategy to reduce the local search evaluation. Some unpromising search candidates had been excluded with the unmatched time windows. But the spatial proximity has been ignored. Nagata [44] developed a time warp operation for customers with a late service to avoid complex time window analysis. An additional penalty function was added to the original objectives to justify the local search heuristics. However, the spatial dimension and the temporal dimension are still separated in these local search heuristics.

Another effective accelerating approach is to decompose large-scale VRPTW instances into many smaller subproblems. Using spatial clustering, Dondo and Cerdá [45] divided the large scale VRPTW into a set of small-sized problems. Routes were generated for each small problem to provide a high quality initial solution. Qi *et al.* [18] developed an effective spatial-temporal distance measure to deal with both space and time issues. A large-scale VRPTW was partitioned with a *k*-medoid clustering method to generate a good initial solution. A standard genetic algorithm was used to improve the initial solution. Good performance of these effective approaches has been verified by VRPTW instances with no more than 1000 customers.

With regard to the Voronoi diagram, the only work has been provided by Milthers [46] who used two-dimensional Voronoi diagrams to split the VRPTW into many subproblems and then solve them with large neighborhood search heuristics. It showed the effectiveness of Voronoi diagrams in guiding the search process. However, this study dealt only with decomposition of the VRPTW in the solution construction stage. Use of the spatial-temporal Voronoi diagram in speeding up the local search for the VRPTW should still be further considered.

2.2. Integration of Vehicle Routing Optimization and GIS

Effective approaches to find high quality solutions for VRPTWs have been embedded in a GIS-based SDSS to deal with many real world applications. Generally, the integration between GIS and vehicle routing optimization is tightly coupled. Spatial data management, processing, and visualization tools are used to collect customer orders, georeference related data, activate the solving process, and display routes for VROs, as in GIS software such as ArcGIS and TransCAD. In line with local search, Weigel and Cao [7] first reported their efforts in the implementation of a tabu heuristics approach in a GIS environment to deal with VRO for a major American retailer. Following the evolutionary optimization line, Mendoza *et al.* [5] integrated a customized routing module, which improved solution quality with commercial solutions such as SAP/R3 and ArcGIS to cope with VROs in public utilities. Experiments on a real-world case in Bogotá, Colombia with 323–601 customers verified the effectiveness of evolutionary optimization and GIS software.

Recently, progress of GIS related to cloud computing transfers the integration of GIS and VRO to a loose coupling way. Using Google maps, Santos *et al.* [47] developed user-friendly web-based SDSSs that embed VRO to deal with urban trash collection in Coimbra, Portugal. TU *et al.* [48] presented a cloud GIS-based spatial decision support framework with variable neighborhood search heuristics for dynamic vehicle routing using historical traffic information. These practices have proven the dominance of VROs in real-life transportation applications. However, they also indicate that current spatial intelligence should be improved to enhance the ability of SDSSs in transportation departments to cope with increasing numbers of customers.

In summary, state-of-the-art heuristics, including local search and evolutionary optimization, solve small- and medium-size VRPTWs in a reasonable time. They have been verified to be effective in a GIS environment for real-life transportation cases with no more than 1000 customers. For larger-size real-life applications, more efficient heuristic algorithms are needed to reduce computing time. Using the spatial-temporal Voronoi diagram, this paper proposes an efficient local search-based approach to address VRPTWs up to 10,000 customers in a reasonable time. Differing from the neighborhood in the spatial context [25–27,48,49], the spatial-temporal Voronoi neighborhood considering both spatial distance and time windows is proposed to limit the search space. A Voronoi distance decay strategy is developed to assign more searching efforts to nearer neighbors. The spatial-temporal feature guided search is used to escape from local minima. Details of the STVD, Voronoi distance, Voronoi neighbors, and the proposed approach for the VRPTW are described in the next sections.

3. The VRPTW and the Spatial-Temporal Voronoi Diagram

This section introduces the VRPTW, the STVD, and the spatial-temporal Voronoi distance.

3.1. The VRPTW

The VRPTW has five components: customers, depot, vehicle, routes, and the solution.

Customers: a set of customers $V_c = \{v_1, v_2, \dots, v_n\}$ expects to be served. A customer v_i $(I > 0)$ located at (x_i, y_i) has a unique demand, q_i, a service duration, s_i, and a service time window $<e_i, l_i>$. The service is required to arrive after start time, e_i, and before end time, l_i. Such a time window is termed "soft" when it could be violated with a penalty cost, and termed "hard" when no violation is permitted.

Depot: The depot v_0 located at (x_0, y_0) provides service in a time window $<e_0, l_0>$.

Vehicles: A fleet of K identical vehicles with the capacity Q is located at the depot. Each vehicle departs from the depot, serves multiple customers, and returns to the depot. Travel distance and time between nodes (includes customers and the depot) i, j are denoted as d_{ij} and t_{ij}, respectively, where $i \neq j$.

Routes: A route $r = < v_0, v_1, v_2, \ldots, v_R, v_{R+1} >$ is formed by a sequence of visited customers, which represents the service process of a vehicle, where R denotes the number of visited customers. $v_{R+1} = v_0$, so that the vehicle must return to the depot. The earliest departure time at the depot v_0 is a_r, the earliest service time at a customer v_i is a_{v_i}, and the earliest return time at the depot is $a_{v_{R+1}}$, which are defined recursively as follows:

$$\begin{cases} a_{v_0} = a_r \\ a_{v_i} = \max\{a_{v_{i-1}} + s_{v_{i-1}} + t_{i-1,i}, e_{v_i}\}, (i = 1, \ldots, R+1) \end{cases} \quad (1)$$

A route is specified following two restrictions: (1) The route duration time $a_{v_{R+1}} - a_r$ should be no more than the maximum duration T_{max}; (2) The total served demand $\sum_{i=1}^{R} q_i$ should be no more than the vehicle capacity Q. The total travel distance of the route is defined as $\sum_{i=0}^{R} d_{v_i v_{i+1}}$. A route is defined as feasible if the customer's service time window, the route duration time, and the vehicle capacity are satisfied.

The VRPTW solution: A solution for a VRPTW instance is a set of feasible routes $r_1, r_2, r_3, \ldots, r_K$ such that a single vehicle visits each customer once. A hierarchy of objectives of the VRPTW is followed in this paper. The number of used vehicles is first minimized and then the total travel distance is reduced.

3.2. The Spatial-Temporal Voronoi Diagram for VRPTWs

Components of the VRPTW, such as customers, depot, vehicles, and routes, have both spatial and temporal attributes. Spatial attributes refer to customers' locations, the depot location, and the vehicle trajectories. Temporal attributes refer to the time window, service time, travel time, and route duration time. To represent them simultaneously, a spatial-temporal coordinate system is built by taking the temporal dimension to be perpendicular to the planar XY plane. Therefore, time windows of the customers and the depot are modeled as vertical lines rooted in their locations.

Figure 1 illustrates a small-size VRPTW with five customers (v_1 to v_5), one depot (v_0), and two routes (*R1*, *R2*). Customers (v_1–v_5) and the depot (v_0) are modeled as vertical lines from its start time point to end time point, as line segments L_1–L_5 and L_0. Projections on the XY plane denote their spatial locations, whereas projections on the time dimension denote their time windows and actual service time. The transportation between customers denotes the service process in the routes (*R1* or *R2*).

Distance in this coordinate system is defined as Equation (2), where (x_i, y_i) and (x_j, y_j) denote the spatial location of i and j, and t_i and t_j denote the time at i and j, $\bar{v}$ is the mean travel speed:

$$d(v_i, v_j) = \sqrt{(x_i - x_j)^2 + (y_i - y_j)^2 + (\bar{v}(t_i - t_j))^2} \quad (2)$$

Voronoi diagrams partition space into adjacent Voronoi cells with given points [30]. Following this principle, a spatial-temporal Voronoi diagram divides the spatial-temporal space into many cells. To generate a Voronoi diagram for the VRPTW, the spatial-temporal central point of a customer's spatial temporal line, $(x_i, y_i, (e_i + l_i)/2.0)$, which is the red point in Figure 1, is used as the seed point. The STVD for VRPTWs is built by the fast quickhull algorithm [50].

In the STVD, a pair of points whose cells share a boundary (Voronoi face or Voronoi edge) are Voronoi neighbors [30]. The spatial-temporal Voronoi distance from a given point i to the point j, $vd(i, j)$, is defined as the minimum number of crossings of the spatial-temporal Voronoi boundary in any route from *i* to *j* [27,31]. Figure 2 provides an example that illustrates the Voronoi distance. The Voronoi distance from *P1* to *P3* is 2 for the two crossings of Voronoi boundaries (C1–C2). The Voronoi distance from *P1* to *P13* is also 2 for the crossings C3–C4. The Voronoi distance from *P1* to *P12* is 3 for the crossing C5–C7. Thus, Voronoi neighbors with different spatial-temporal proximity can be measured. As both spatial distance and time windows are considered, it is well suited for accelerating local search for the solving of VRPTW. Using these useful definitions, this paper proposes an efficient local search heuristic to address large-scale VRPTWs.

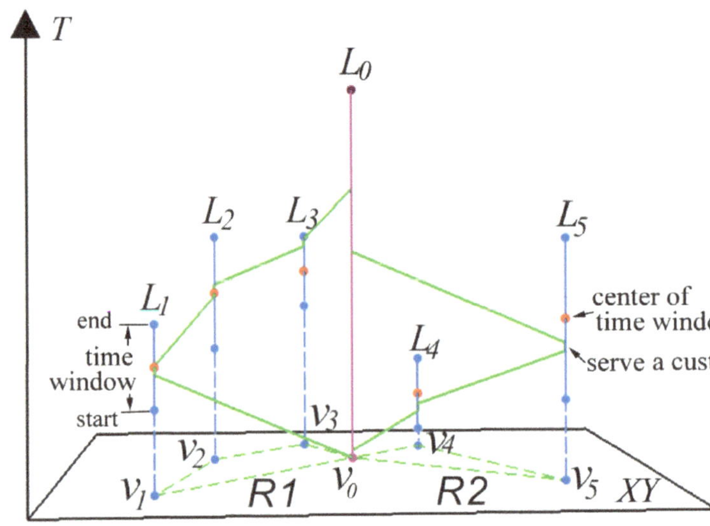

Figure 1. Spatial-temporal representation of VRPTW.

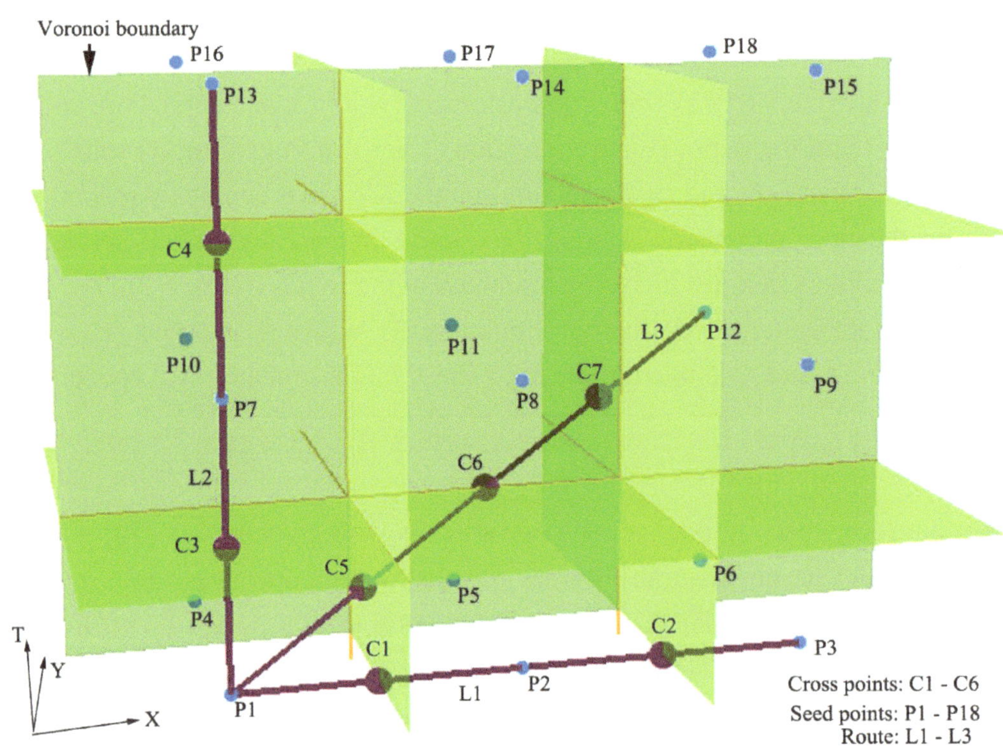

Figure 2. The spatial-temporal Voronoi diagram and the spatial-temporal Voronoi distance.

4. The Spatial-Temporal Voronoi Diagram-Based Heuristic

This section introduces the spatial-temporal Voronoi diagram-based heuristic (STVDH) for large-scale VRPTWs. The principle of the STVDH is to speed up the local search procedure by using useful spatial thinking, including the spatial-temporal Voronoi neighbors, distance decay search strategy, and local route features. The overall framework of the proposed heuristic is summarized in Figure 3. After the construction of an initial solution (step 1-1), a local search heuristic based on spatial-temporal Voronoi neighborhood is followed to improve solution quality (step 1-1 to step 1-4). A spatial-temporal Voronoi distance decay strategy is developed to assign reasonable searching efforts on neighbors with different distances (step 1-2). Spatial-temporal route features are identified and guides the searching process escape from local minima (step 1-3). The stopping condition is a limitation on the number of iterations (step 1-4). Finally, the best found solution is reported (step 1-5).

[The spatial-temporal Voronoi diagram based heuristic (STVDH)]

Input: VRPTW instance,
parameters setting:
N_{max}: maximum number of iterations
Z_{max} : maximum number of non-improved iterations to start spatial-temporal guided local search
$k_{\max}$:maximum local search Voronoi distance
α : parameter controlling the distance decay speed.
Output: the found best solution S_{best}

Step 1-0: Read VRPTW instance and build the spatial-temporal Voronoi diagram with customer nodes.
Step 1-1: Generate an initial solution S_0 with the construction algorithm in section 4.1. Set $S_{best} = S_0$.
Step 1-2: Explore the solution neighborhood by the local search in section 4.2. Update S_{best} if a better solution is found. If the solution quality isn't improved in Z_{max} iterations, record the local minima S_{local}, go to step 1-3; else, repeat this step.
Step 1-3: Implement the spatial-temporal feature guided local search in section 4.3 to disturb S_{local}. Spatial-temporal features in S_{local} are identified and a modification process is followed.
Step 1-4: Check the stopping criteria with N_{max}. If the stopping criterion is accepted, go to step 1-5. Otherwise, go to step 1-2.
Step 1-5: report S_{best} and exit.

Figure 3. The framework of the spatial-temporal Voronoi diagram-based heuristic algorithm.

4.1. The Construction Algorithm

The construction algorithm iteratively inserts an unrouted node into a proper position until all customers are routed. During the insertion, instead of evaluating all unrouted nodes, neighbor customers within a given Voronoi distance are selected as candidates of the next inserted nodes. The details of the construction algorithm are described in Figure 4. First, customers are sorted by the start time window and stored in a queue L (step 2-0). Then, L's front unrouted node is popped out as the operated node μ (step 2-1). An empty route r is initialized with u (step 2-2). Another unrouted customer v is randomly selected from μ' Voronoi neighbors. Node v will be inserted before or after a routed node i in the route

r (step 2-3). If the insertion is performed, u is updated with v. The best insertion place is defined as the one that minimizes the added total travel length and the total value of the end time window violation as Equation (3), where j denotes the next visited node of i. λ is a random value in [0.5, 2]. It should be noted that the arrival time at v, j should not be before the start time, as Equations (4) and (5):

$$\Delta F = \left(d_{iv} + d_{vj} - \lambda d_{ij}\right) + \bar{v}(a_i + s_i + t_{iv} - l_v) \tag{3}$$

$$a_i + s_i + t_{iv} > e_v \tag{4}$$

$$a_i + s_i + t_{iv} + s_v + t_{vj} > e_j \tag{5}$$

[Construction algorithm]

Step 2-0: Label all customers as unrouted. Sort customers with start time window and store them in a queue L. Initialize an empty route r.

Step 2-1: **If** L is empty, go to step 2-4;

Else,

pop L's front node u.

If u is unrouted, go to step 2-2;

Else, repeat step 2-1.

Step 2-2: Insert node u into an empty route r and label u as routed.

Step 2-3: Select an unrouted node v randomly from Voronoi neighbors of the last inserted node u. Find the best insertion place i, insert v before or after i, set u=v, and label v as routed. Repeat this step until there is no place for insertion, go to step 2-1.

Step 2-4: If all customers are routed, report solution and exit.

Figure 4. The construction algorithm.

During the insertion, several constraints including the start time window, route duration time, vehicle capacity, and the maximum route length should be satisfied. If no place is found for insertion, we pop L's front node as u, insert it into a new route, and label it as routed (step 2-2 and step 2-3). The insertion will end if all customers have been labeled as routed (step 2-5). Finally, obtaining all routes is the initial solution.

4.2. Local Search

Local search iteratively explores the solution space to improve solution quality. Usually, the exploration removes or adds edges from the current solution by using local search operators. Three typical local search operators, including the 1-0 exchange move, 1-1 exchange move, and 2-Opt move [51], are used to explore the neighborhood solution in this article.

- *1-0 exchange move* injects a node from the original place and inserts it after another. As Figure 5a shows, node b is removed from its current position and inserted before node i.
- *1-1 exchange move* swaps positions of a pair of nodes. As Figure 5b shows, the places of nodes b and i are swapped.

- *2-Opt move* swaps the ends of two routes after the positions of a pair of nodes. As Figure 5c illustrates, the partial routes after b and i are swapped.

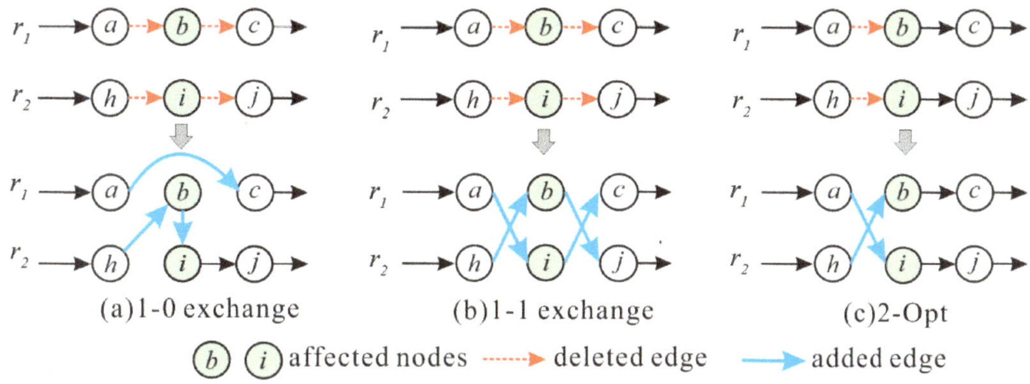

Figure 5. Local search operators (**a**) 1-0 exchange; (**b**) 1-1 exchange; and (**c**) 2-Opt.

The computing complexity of these three operators depends on two aspects: operated nodes and local search evaluation. Two operated nodes, named b and i, are involved as shown in Figure 5. Therefore, the number of operated candidates is $n(n-1)$, where n is the number of customers. The time complexity is $O(n^2)$. For each candidate pair, local search evaluation checks the time window for each customer involved, total duration time, and vehicle capacity for each route. Due to the cascade effect of the arrival time, it will, on average, take $O(n/2R)$ time, where R is the number of routes. To utilize limited computing effort more efficiently, we propose a spatial-temporal Voronoi distance-decay strategy to assign searching efforts on selected candidates. A time warp operation is also used to shorten computing time for each candidate.

4.2.1. The Spatial-Temporal Distance Decay Strategy

Differing from the equal searching intensity of the k nearest neighbors [26] and the k-ring Voronoi neighbors [27,48], the spatial-temporal Voronoi distance decay strategy searches more on nearer neighbors than further neighbors to balance the solution quality and computing time. For spatial-temporal Voronoi neighbors with Voronoi distance k, the searching probability P_k is defined as Equation (6). As the distance k increases, searching probability P_k will decrease. Therefore, most computing effort will focus on near neighbors, but some necessary efforts are left for further neighbors.

$$P_k = \frac{\alpha^{-k}}{\sum_{k=1}^{k_{max}} \alpha^{-k}} \tag{6}$$

where $k = 1, \ldots, k_{max}$, and k_{max} is the maximum searchable Voronoi distance, $\alpha \in [1, \infty)$ is a value that controls the spatial-temporal distance decay speed. As the value of α increases, the decay speed becomes faster.

The probability range $[TP_{k-1}, TP_k]$ ($k = 1, 2, \ldots, k_{max}$) for neighbors no farther than Voronoi distance k is calculated as Equation (7), where $TP_0 = 0$.

$$TP_k = \sum_{i=1}^{k} P_i \tag{7}$$

To implement the Voronoi distance decay strategy, local search randomly generates a value of δ between (0, 1), finds the probability range in which δ is located, and then selects Voronoi neighbors with the corresponding distance *k* as candidates to evaluate. As it avoids evaluating all neighbors, this useful strategy puts more searching efforts on closer nodes and therefore significantly accelerates the local search. As the number of Voronoi neighbors is fixed, the computing complexity reduces from $O(n^2)$ to $O(n)$.

4.2.2. Time Warp Operation

The local search evaluates distance and time constraints for customers and routes, which cost the most computing effort [43]. Nagata [45] proposed a new penalty function in which change can be computed in O(1) time for the VRPTW. When a later service happens for a customer, a time warp operation is performed such that the arrival time is adjusted to the end of the time window, and a penalty is added to the original objective. Figure 6 shows an example of the time warp. The additional penalized cost is defined as the time warp used in Equations (8) and (9), where tw_i is the cost value at customer *i*, and $(a_i - e_i)^+$ is a non-negative value of the service late time at *i*. Therefore, the objective is changed to minimize the total route number, total route length, and total penalized cost *TW(s)*. For O(1) time consumed, the time warp operation is very fast.

$$TW(s) = \sum_{i=1}^{n} tw_i \tag{8}$$

$$tw_i = (a_i - e_i)^+ \tag{9}$$

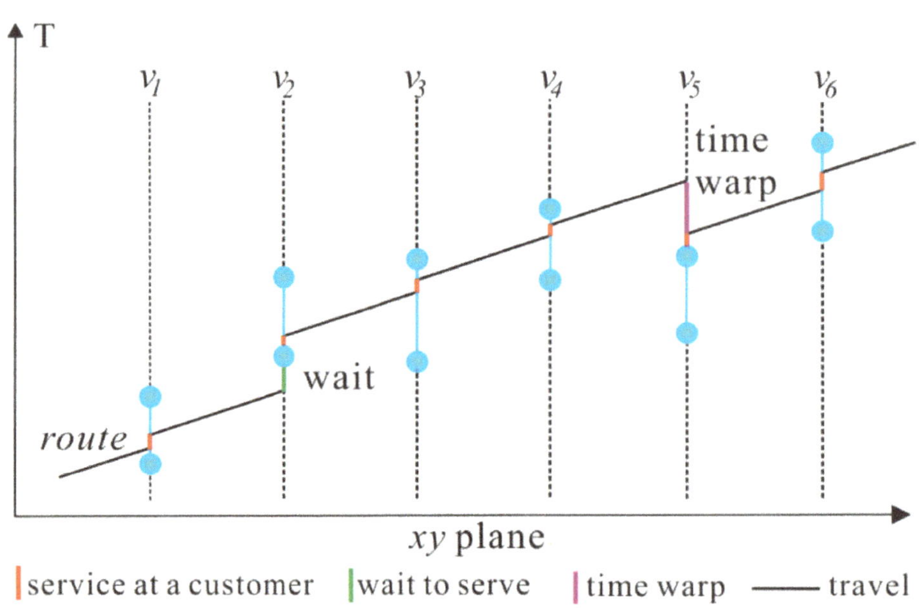

Figure 6. Wait time and time warp at a customer.

4.2.3. Acceptance Criterion

Local search will naturally drop into local minima. A threshold acceptance criterion is used to escape. If the found solution is not 0.5% worse than the current solution, it will be accepted as the new solution.

Otherwise, it will be rejected. If the best solution does not improve in Z_{max} iterations, the spatial-temporal features-guided search will be started.

4.3. Spatial-Temporal Features-Guided Search

Due to the used spatial-temporal Voronoi neighbors, local search explores a relatively small proportion of the neighborhood solutions. Some potential neighborhood solutions will be rejected due to spatial and temporal constraints, such as route duration, time window, and vehicle capacity. Different from traditional random perturbations [37], spatial-temporal features guided search destroys unpromising structures defined by spatial-temporal rules and rebuilds a part of the current solution to escape local minima. It also increases the searchable probability for farther neighborhood solutions. Three components are the spatial-temporal features, the removal algorithm, and the reinsertion algorithm.

4.3.1. Spatial-Temporal Features

Three types of spatial-temporal features are defined in this paper including time window violation nodesφ_t, longest distance nodes φ_l, and smallest route's nodes φ_s. Their definitions are described below.

- Time-window violation nodes

The time warp operation will introduce time window lateness in which the arrival time at a customer may be later than the end time of its time window. However, this route feature should not belong to the best solution. We remove such unpromising nodes from the current solution and reinsert them later. Formally, time window violation nodes φ_t are defined as Equation (10):

$$\varphi_t = \{v_i | t_i > l_i, v_i \in V_c\} \tag{10}$$

- Longest-distance nodes

To cooperate with the time window, some long segments may be kept in routes and difficult to replace in local search, which leads to the increase of total route length [27]. Longest-distance nodes φ_l are end points of the long segment, defined as Equation (11), where S denotes the current solution, e_{ij} or e_{ji} denotes a segment in S. $\bar{e}$ denotes the mean spatial distance of the Voronoi neighbors with Voronoi distance 1. The length of the long segment is more than three times $\bar{e}$.

$$\varphi_l = \{v_i | d_{ij} > 3\bar{e}, e_{ij} \in S || e_{ji} \in S, v_i \in V_c\} \tag{11}$$

- The smallest route's nodes

In some solutions, small routes serve only one, two, or three customers near the depot, which requires more vehicles [48]. Usually, this feature is kept for the suboptimal property. The small route's nodes are defined as φ_s, as Equation (12), where $|r|$ denotes the number of visited customers in the route r:

$$\varphi_l = \{v_i | v_i \in r, |r| \leq 3, v_i \in V_c\} \tag{12}$$

4.3.2. Removal Algorithm

To increase diversification of the guided search, more nodes are removed from the current solution in this study. Additional nodes φ_a are Voronoi neighbors of defined features, just as:

$$\varphi_a = \{v_i | vd(v_i, v_j) = 1, v_j \in (\varphi_t \cup \varphi_l \cup \varphi_s), v_i \notin (\varphi_t \cup \varphi_l \cup \varphi_s), v_i \in V_c\} \quad (13)$$

Therefore, removed nodes φ are spatial-temporal features nodes and their Voronoi neighbors φ_a, as in Equation (14). The removal algorithm sequentially ejects all nodes belonging to φ and leaves a partial solution for the insertion.

$$\varphi_a = \{v_i | v_i \in (\varphi_t \cup \varphi_l \cup \varphi_s \cup \varphi_a), v_i \in V_c\} \quad (14)$$

4.3.3. Insertion Algorithm

This algorithm inserts all removed nodes into new places in the partial solution. First, all removed nodes are randomly pushed into a queue L. Then, the front node of L is popped out and inserted at the best place before or after its Voronoi neighbors to keep spatial-temporal proximity, which minimizes insertion cost as Equation (3), under the limitation of total duration time, time window, and vehicle capacity. It is of note that the end time windows should not be violated. If there is no proper place, a new route is created, and the front node of L is popped out and initialized as the first-served customer. The insertion operation is repeated until all removed nodes are located in proper places. Due to the best choice at each insertion, the final solution is up to the node sequence in the queue. In this article, the sequence is randomly arranged 50 times to obtain different rebuild solutions. Finally, the obtained result of the permutation that minimizes the total objective is accepted as the final solution.

4.4. An Extension of the Solving Algorithm for MDVRPTW

MDVRPTW is a variant of VRPTW, which has more than one depot to serve geographically scattered customers. Compared with the VRPTW, MDVRPTW has an additional problem of how to allocate customers to the right depot. Voronoi partition that divides a study area into many regions with given objects is another typical function of the Voronoi diagram. For the effectiveness, it has been widely used in service area analysis [32]. By using Voronoi partition characteristics of the depots' spatial-temporal Voronoi diagram, we assign customers to be served by the depot of its located Voronoi regions. By that, the presented STVDH is easy to be extended to deal with a large-scale MDVRPTW by decomposing it to many smaller VRPTWs.

Figure 7 illustrates the workflow of the STVDH's adaption for the MDVRPTW. The depots' spatial-temporal Voronoi diagram is built to allocate customers to the located depot so that the MDVRPTW is divided into several VRPTWs, each of which has a depot and many assigned customers. Routes for each VRPTW are first initialized using the construction algorithm in Section 4.1 and then separately improved by the spatial-temporal Voronoi diagram-based local search heuristic in Sections 4.2 and 4.3. To overcome the border effect of Voronoi diagram, an additional improvement between depots is used to move or exchange nodes between routes served by different depots. This is done by using 1-0 and 1-1 exchanges to move nodes between routes served by different depots after the spatial-temporal features guided search. Details of this process are referred to TU *et al.* [49]. The stopping condition is the same as the STDVH. Finally, the best found solution is reported.

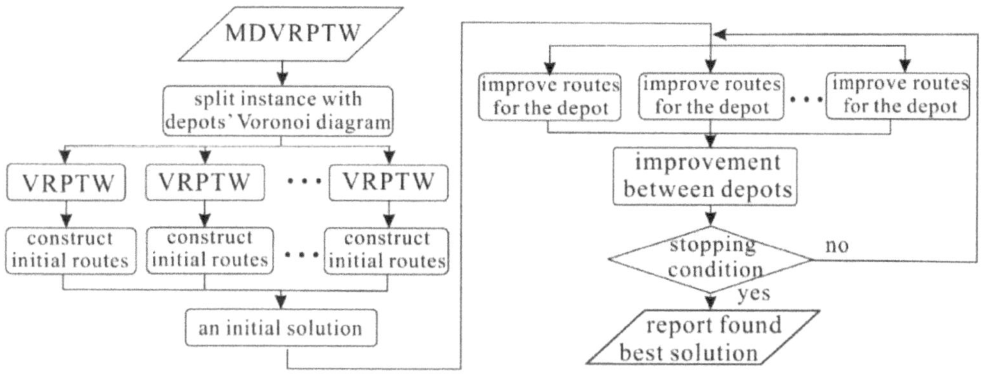

Figure 7. The adaption of the STVDH for the MDVRPTW.

5. Experiment and Comparison

To assess the performance of the STVDH, experiments with benchmark problems and large-scale real-world VRPTW and MDVRPTW problems in Shanghai, China were conducted. The STDVH algorithm was implemented with C++, running on a Windows 7 64-bit system with an Intel Core i7-3.4G processor and 16 GB memory. This section reports test datasets, parameter tuning, computing results, and comparisons with other heuristic algorithms. Computing times of compared algorithms are transformed to the computing platform of STVDH with Dongarra factor [52].

5.1. Test VRPTW and MDVRPTW Problem Dataset

Two types of VRPTW and MDVRPTW problem datasets are tested. The first dataset includes the VRPTW benchmarks of Solomon [15] and Gehring and Homberger [53]. The benchmarks of Solomon have 56 problems with 100 customers. They are divided into six groups with eight to twelve problems, named R1, R2, C1, C2, RC1, and RC2. In the Euclidean plane, customers are randomly distributed (R1 and R2), cluster distributed (C1 and C2) or semi-cluster distributed (RC1 and RC2). Every instance has a single depot within the spatial domain of the customers. The travel time between nodes is equal to the corresponding Euclidean distance. R1, C1, and RC1 have a short time scheduling horizon, whereas R2, C2 and RC2 have a long time scheduling horizon. Similar to Solomon's VRPTW problems, the VRPTW benchmarks of Gehring and Homberger [53] have 300 VRPTW instances with 200, 400, 600, 800 or 1000 customers [54]. Customers have the same distributions.

The second dataset is a large-scale VRPTW and MDVRPTW problem dataset in Shanghai, China [55]. Five VRPTW instances (Sh1a–Sh5a) and 5 MDVRPTW instances (Sh1b–Sh5b) with 2000 to 10,000 customers and one to six depots are generated to simulate the daily parcel delivery service of a logistics company in Shanghai, China. Customers are randomly selected from the commercial points of interest (POI) using a professional digital navigation map to obtain real-world spatial locations. The time window of each customer is a random value in min bounded by 30 and 60, and the service duration is a random value from 1 to 4 min. Demand is a random value in the interval [1, 100]. Depots are located in professional logistic districts in the city. All vehicles have the same capacity, 2000, and maximum route duration, 480 min. The mean travel speed is set to be 45 km/h.

5.2. Parameter Tuning

There are four parameters to be tuned in the presented STVDH algorithm as shown in Figure 3. Following Coy's calibration approach [56], a preliminary experiment is conducted on an instance in the problem dataset to find the best parameter settings. The tuning process initially identifies the first two parameters related to the stopping condition and then the last two parameters related to the Voronoi distance decay strategy. The first two parameters, the maximum number of iterations, N_{max}, and the maximum number of iterations to start the spatial-temporal features guided local search, Z_{max}, control the exit of the STVDH. After intensive experiments with different stopping conditions, N_{max} should be set to 5000N to converge on a stable high-quality solution, where N is the number of customers. Z_{max} is set to 100N to start the spatial-temporal feature guided search.

The last two parameters control the Voronoi diagram-based speedup strategy. The third parameter, k_{max}, indicates the maximum searchable Voronoi neighbors. As the value of k_{max} increases, local search will evaluate more neighbors, which requires more computing efforts. According to the distribution of the Voronoi neighbors, k_{max} should be no more than 5 [49]. The last parameter, α, determines the distance decaying speed. A high value indicates a tendency for fast decay. The proper value of α is within the bound [1,4]. The value space of parameters are summarized in Table 1.

Table 1. Parameter settings of the STDVH algorithm.

Notation	Parameter	Setting
N_{max}	the maximum number of iterations	5000N
Z_{max}	the maximum number of iterations to start the spatial-temporal features guided search	100N
k_{max}	the maximum searching Voronoi neighbors	1, 2, 3, 4, 5
α	Voronoi distance decaying speed	(0, 4)

N is the number of customers.

Experiments on some selected VRPTW instances are conducted to set parameter values for one type of VRPTW dataset. Taking the second VRPTW dataset as an example, we solved Sh2a with all combinations of the last two parameters. Finally, we set k_{max} = 3 and = 2 for the best performance in the experiment. Without loss of generality, identical parameter values were set for each instance. The impact of the Voronoi diagram will be discussed in Section 6. A summary of parameter settings is listed in Table 1. Taking Sh2a as an example, Figure 8 displays the convergence of the number of routes and the total length of routes with the parameter settings used in solving Sh2a.

5.3. The Results of the Solomon [15] and Gehring and Homberger [53] VRPTW Benchmarks

The parameter setting for these VRPTW benchmarks is as: N_{max} = 5000N, Z_{max} = 100N, k_{max} = 2 and α = 1.5. For each problem, the STDVH was run 10 times. The total objective, total routes length, and computing time were recorded. Results of three state-of-the-art methods, including the arc-guided evolutionary algorithm (AGEA) [39], the hybrid genetic algorithm with adaptive diversity management (HGASDC) [40], and the local search based heuristics VRPEJ [57], which limits the searching space with the *kth nearest neighbor* strategy, are compared with the STVDH.

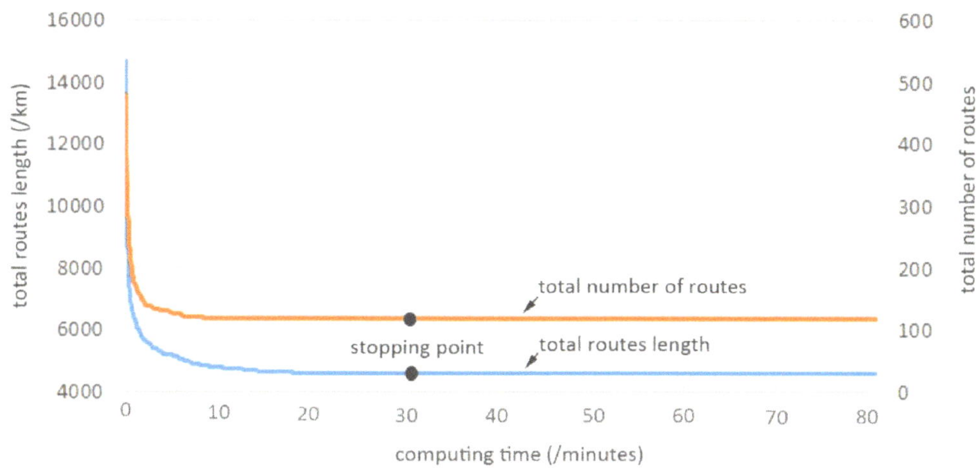

Figure 8. The convergence of the total number of routes and the total routes length (Sh2a).

Tables 2 and 3 present the comparison of the results on the Solomom and Gehring and Homberger VRPTW benchmarks respectively. Each entry reports the average of computing results of a group (average computing time | the average total routes length). They indicate that the presented STVDH is able to achieve high quality solutions of the VRPTWs with no more than 1000 customers in a relatively short computing time. In terms of the number of routes, the STVDH achieves the fewest routes in both the Solomon (total of 405 routes) and Gehring and Homberger (total of 10296 routes) VRPTW benchmarks. In terms of the total route length, the STVDH performs better than the VRPEJ for all size VRPTW instances, only inferiorly to AGEA on Solomon benchmarks, but superiorly to AGEA on the Gehring and Homberger benchmarks. However, the STVDH's results are still worse than that of the HGSADC. For the computing efficiency, the STVDH consumes the least computing time in all six size VRPTW instances (100~1000). Therefore, the proposed STDVH exhibits a good performance on the VRPTW benchmarks of Solomon and Gehring and Homberger.

Table 2. Comparison of results on the Solomon VRPTW benchmarks [15].

Instance	**N**	**AGEA 1 run**	**HGSDAC Best 5 run**	**VRPEJ Best 10 run**	**STVDH AVG 10 run**	**STVDH Best 10 run**
R1	100	11.92|1211.43	11.92|1210.69	11.92|1214.67	11.92|1213.85	11.92|1212.65
R2	100	2.73|954.05	2.73|951.51	2.73|954.10	2.73|954.01	2.73|953.20
C1	100	10.0|828.38	10.0|828.38	10.0|828.38	10.0|828.38	10.0|828.38
C2	100	3.0|589.86	3.0|589.36	3.0|589.86	3.0|589.86	3.0|589.86
RC1	100	11.50|1384.83	11.5|1384.17	11.5|1387.81	11.5|1386.93	11.5|1386.01
RC2	100	3.25|1121.26	3.25|1119.24	3.25|1127.38	3.25|1126.49	3.25|1125.72
CNV		**405**	**405**	**405**	**405**	**405**
CTD		**57,254.73**	**57,195**	**57,366.96**	**57,341.97**	**57,305.10**
T(/min)		**180**	**2.68**	**3.62**	**0.91**	**0.9**
T2(/min)		**52.3**	**1.56**	**3.62**	**0.91**	**0.9**

Instance: group name. **N**: number of customers. **CNV**: cumulative number of vehicles. **CND**: cumulative travel distance. **T**: time reported by metaheuristics. **T2**: time on the computing platform of the STVDH.

Table 3. Comparison of results on the Gehring and Homberger VRPTW benchmarks [53].

Instance	N	AGEA 1 run	HGSDAC Best 5 run	VRPEJ Best 10 run	STVDH AVG 10 run	STVDH Best 10 run
R1	200	18.2|3640.11	18.2|3613.16	18.2|3664.28	18.2|3653.24	18.2|3617.54
R2	200	4.0|2941.99	4.0|2929.41	4.0|2938.53	4.0|2967.87	4.0|2938.53
C1	200	18.9|2721.90	18.9|2718.41	18.9|2749.18	18.9|2758.67	18.9|2721.47
C2	200	6.0|1833.36	6.0|1831.59	6.0|1880.47	6.0|1858.34	6.0|1835.52
RC1	200	18.0|3224.63	18.0|3180.48	18.0|3205.81	18.0|3218.83	18.0|3179.89
RC2	200	4.3|2554.33	4.3|2536.20	4.3|2574.92	4.3|2584.37	4.3|2544.44
CNV		**694**	**694**	**694**	**694**	**694**
CTD		**169,163**	**168,092**	**176,440.8**	**170,415.5**	**168,373.7**
T(/min)		**90**	**8.40**	**5.4**	**1.22**	**1.2**
T2(/min)		**26.7**	**4.9**	**5.4**	**1.22**	**1.2**
R1	400	36.4|8514.11	36.4|8402.57	36.4|8615.29	36.4|8598.39	36.4|8446.36
R2	400	8.0|6258.82	8.0|6152.92	8.0|6274.20	8.0|6279.13	8.0|6160.84
C1	400	37.6|7273.90	37.6|7170.47	37.6|7339.88	37.6|7514.01	37.6|7186.10
C2	400	11.7|3941.70	11.6|3950.95	11.7|4024.82	11.7|3995.21	11.7|3951.71
RC1	400	36.0|8088.46	36.0|7907.14	36.0|8107.86	36.0|8098.32	36.0|7952.00
RC2	400	8.40|5516.59	8.5|5215.21	8.4|5394.54	8.4|5393.49	8.4|5292.92
CNV		**1381**	**1381**	**1381**	**1381**	**1381**
CTD		**395,936**	**388,013**	**397,565.9**	**398,785.5**	**389,385.5**
T(/min)		**180**	**34.1**	**9.8**	**2.30**	**2.3**
T2(/min)		**53.3**	**19.8**	**9.8**	**2.30**	**2.3**
R1	600	54.5|18,781.79	54.5|18,023.18	54.5|18,620.73	54.5|18,587.90	54.5|18,237.74
R2	600	11.0|12,804.60	11.0|12,352.38	11.0|12,615.07	11.0|12,612.60	11.0|12,343.51
C1	600	57.3|14,236.86	57.4|14,058.46	57.3|14,605.53	57.3|14,585.55	57.3|14,271.58
C2	600	17.4|7729.80	17.4|7594.41	17.4|7748.47	17.4|7728.74	17.4|7589.10
RC1	600	55.0|16,767.72	55.0|16,097.05	55.0|16,529.63	55.0|16,524.77	55.0|16,203.93
RC2	600	11.4|11,311.81	11.5|10,511.86	11.4|10,879.26	11.4|10,824.00	11.4|10,626.35
CNV		**2066**	**2068**	**2068**	**2066**	**2066**
CTD		**816,326**	**786,793**	**809,986.9**	**808,635.65**	**792,722.1**
T(/min)		**270**	**99.4**	**16.2**	**3.91**	**3.9**
T2(/min)		**80.0**	**57.8**	**16.2**	**3.91**	**3.9**
R1	800	72.8|32,734.57	72.8|31,311.38	72.8|32,281.48	72.8|32,108.01	72.8|31,540.28
R2	800	15.0|20,618.21	15.0|19,933.39	15.0|20,448.88	15.0|20,339.05	15.0|19,969.61
C1	800	75.2|25,911.44	75.4|24,876.93	75.2|26,097.53	75.1|25,972.63	75.1|25,490.85
C2	800	23.4|11,835.72	23.3|11,475.05	23.4|11,897.31	23.3|11,826.41	23.3|11,621.87
RC1	800	72.0|33,975.61	72.0|29,404.32	72.0|31,071.16	72.0|30,904.01	72.0|30,390.41
RC2	800	15.5|17,536.54	15.4|16,495.82	15.5|16,878.69	15.4|16,733.78	15.4|16,467.01
CNV		**2739**	**2739**	**2739**	**2736**	**2736**
CTD		**1,424,321**	**1,334,963**	**1,386,750.4**	**1,378,838.8**	**1,354,800**
T(/min)		**360**	**215**	**24.8**	**5.82**	**5.8**
T2(/min)		**106.6**	**125.1**	**24.8**	**5.82**	**5.8**

Table 3. *Cont.*

Instance	N	AGEA 1 run	HGSDAC Best 5 run	VRPEJ Best 10 run	STVDH AVG 10 run	STVDH Best 10 run
R1		91.9|51,414.26	91.9|47,759.66	91.9|49,741.43	91.9|49,396.91	91.9|48,523.49
R2		19.0|30,804.79	19.0|29,076.45	19.0|29,871.68	19.0|29,595.30	19.0|29,092.01
C1		94.2|43,111.60	94.1|41,572.86	94.1|43,089.45	94.1|42,682.27	94.1|41,977.06
C2		29.3|16,810.22	28.8|16,796.45	29.0|117,340.13	28.9|17,174.73	28.8|16,877.68
RC1		90.0|46,753.61	90.0|44,333.40	90.0|46,152.97	90.0|45,824.65	90.0|44,974.63
RC2		18.4|25,588.52	18.2|24,131.12	18.4|24,951.31	18.3|24,655.48	18.2|24,248.11
CNV		**3428**	**3420**	**3421**	**3419**	**3419**
CTD		**2,144,830**	**2,036,700**	**2,111,469.7**	**2,093,293.5**	**2,056,930**
T(/min)		**450**	**349**	**34.5**	**7.75**	**7.7**
T2(/min)		**133.3**	**203.1**	**34.5**	**7.75**	**7.7**

Instance: group name. **N**: number of customers. **CNV**: cumulative number of vehicles. **CND**: cumulative travel distance. **T**: time reported by metaheuristics. **T2**: time on the computing platform of the STVDH.

5.4. The Results of Large Scale VRPTW and MDVRPTW Problem in Shanghai, China

The parameter setting for this large scale VRPTW and MDVRPTW problem is as follow: $N_{max} = 5000N$, $Z_{max} = 100N$, $k_{max} = 3$ and $\alpha = 2$. For each problem, the STVDH algorithm was also run 10 times. Total travel distance, number of used vehicles, and computing time were recorded. The obtained results are summarized in Table 4 (https://github.com/spatialsmart/VRPTW/tree/master/Shanghai/Solution).

As Table 4 indicates, the STVDH algorithm reports the best found solution for large-scale VRPTWs up to 10,000 customers in 139.5 min (about 2.32 hours). As the number of customers increases from 2000 to 10,000, the computing time of the STVDH increases from 15.7 min to 120.8 min, whereas for the MDVRPTW, the time increases from 16.1 min to 139.5 min. Compared with the VRPTW, more computing efforts are required for the MDVRPTW due to additional customer allocation between depots. In terms of solution quality, the STVDH utilizes an average of 847.1 total routes that travelled 30,184.725 km to serve all customers in the five VRPTW instances. For the five MDVRPTW instances, 906.4 routes that travelled 26,353.572 km are required to satisfy customer needs. It should be noted that the performance of the STVDH is robust, as indicated by the gap between S_{avg} and S_{best} (−1.60% for the VRPTW and −1.73% for the MDVRPTW). Therefore, the proposed STVDH provides promising results for large-scale VRPTW and VRPTW within reasonable computing times.

Taking problem Sh1a as an example, Figure 9 displays the best found solution and the spatial-temporal vehicle route with ArcScene. It demonstrates the best route compromise between spatial proximity and time constraints. The same vehicle may not yet serve spatially-near customers as Figure 9b shows.

To assess the solution quality of the obtained results, we compare them with solutions reported by two heuristic algorithms. One solution is obtained by solving the simulated instances with the network analysis module in ArcGIS™ 10.2, which uses a tabu heuristics algorithm to solve complex real-world VRPTW problems. Another compared solution is calculated using VRPEJ [55]. Other metaheuristics such as AGEA and HGSDAC are not compared due to unavailability. Each algorithm was run 10 times to find the best result.

Table 4. Results from the large-scale VRPTW and MDVRPTW instances in Shanghai, China.

Instance	Type	N	M	D	Q	STVDH			
						S_{avg} 10 (/km)	T_{avg} 10 (/min)	S_{Best} 10(/km)	T_{Best} 10 (/min)
Sh1a	VRPTW	2000	1	200	2000	59.3|3024.870	15.7	58|2996.104	15.9
Sh2a	VRPTW	4000	1	300	2000	119.8|4674.122	28.3	118|4598.592	30.6
Sh3a	VRPTW	6000	1	400	2000	171.0|6145.245	60.0	169|6052.084	65.9
Sh4a	VRPTW	8000	1	500	2000	220.3|7511.361	93.6	218|7372.345	98.9
Sh5a	VRPTW	10,000	1	600	2000	276.8|8829.127	120.8	274|8681.006	128.8
CNV						**847.1**		**837**	
CTD(/km)						**30,184.725**		**29,701.132**	
Gap to S_{avg}								**−1.60%**	
Total Time (/min)							**318.4**		**340.1**
Sh1b	MDVRPTW	2000	2	200	2000	70.2|3045.601	16.1	69|2999.616	16.9
Sh2b	MDVRPTW	4000	3	300	2000	127.9|4265.365	29.8	126|4174.144	31.5
Sh3b	MDVRPTW	6000	4	400	2000	172.1|6247.152	62.2	169|5260.810	69.6
Sh4b	MDVRPTW	8000	5	500	2000	237.4|6849.848	98.4	235|6238.438	105.4
Sh5b	MDVRPTW	10,000	6	600	2000	298.8|6845.606	138.9	293|6784.751	139.5
CNV						**906.4**		**892**	
CTD (/km)						**26,353.572**		**26,289.418**	
Gap to S_{avg}								**−1.73%**	
Total Time (/min)							**345.4**		**362.9**

Instance: problem name. **Type**: VRPTW or MDVRPTW; **N**: number of customers. **M**: number of depots. **D**: number of vehicles at a depot. **Q**: vehicle capacity. ***S_{avg}* 10:** average result of 10 runs (number of routes| total routes length). ***T_{avg}* 10:** average computing time. ***S_{Best}* 10:** best of 10 runs (number of routes| total routes length). ***T_{Best}* 10:** the computing time of the best solution's run. **CNV**: cumulative number of vehicles. **CND**: cumulative travel distance.

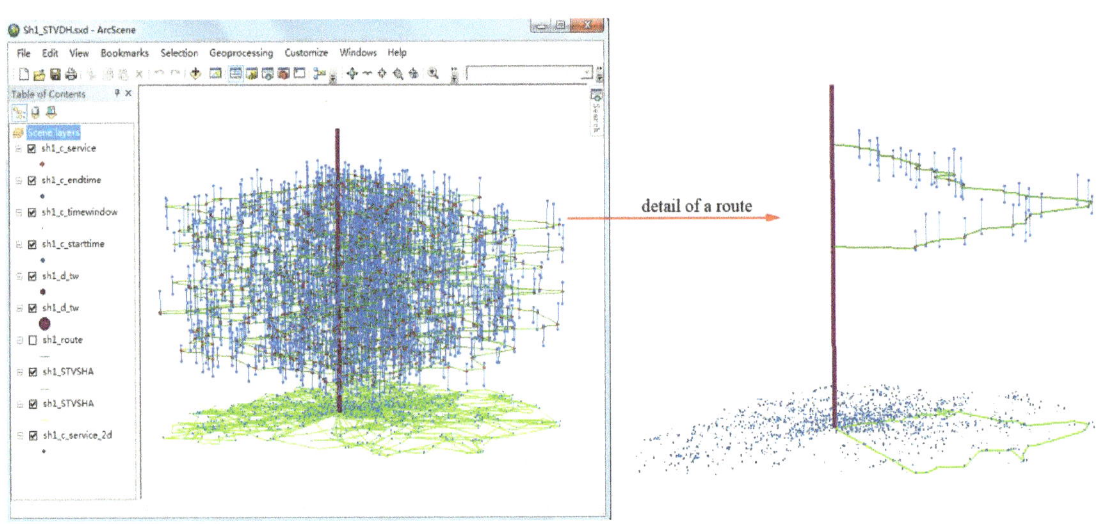

(a) (b)

Figure 9. The result for large scale VRPTW problem Sh1a. (**a**) the final best solution; and (**b**) a spatial-temporal route.

Table 5 compares the obtained results. It indicates that the STVDH presented in this study outperforms both ArcGIS and VRPEJ. In terms of efficiency, comparison with ArcGIS (total of 2141.1 min) and VRPEJ (total of 2355.3 min) indicates that the STVDH costs the least computing effort (total of 703 min). In terms of solution quality, the STVDH requires 1729 routes travelling 55,990.55 km to serve all customers in 10 instances. ArcGIS requires more vehicles (1764 routes) that travel a greater distance (62,047.809 km) to provide the same service. The VRPEJ, which utilizes a spatially *kth nearest neighbors* strategy, requires the most routes (1893 routes) that travelled 63,659.365 km. Hence, it confirms that the spatial-temporal Voronoi neighbor strategy in the presented STVDH is much better than the spatial neighbor strategy for local search to solve large scale VRPTWs.

Table 5. Comparison of the results of the STVDH with other heuristic algorithms.

Instance	Type	N	M	STVDH		ArcGIS		VRPEJ	
				S_{Best} 10(/km)	T_{Best} 10 (/min)	S_{Best} 10(/km)	T_{Best} 10 (/min)	S_{Best} 10(/km)	T_{Best} 10 (/min)
Sh1a	VRPTW	1	2000	58|2996.104	15.9	62|3296.104	60.5	59|3285.404	50.8
Sh2a	VRPTW	1	4000	118|4598.592	30.6	120|4022.248	125.4	126|5234.168	122.2
Sh3a	VRPTW	1	6000	169|6052.084	65.9	173|6738.745	197.2	190|6948.372	242.7
Sh4a	VRPTW	1	8000	218|7372.345	98.9	223|8045.327	254.8	242|8438.982	290.8
Sh5a	VRPTW	1	10,000	274|8681.006	128.8	278|9874.135	407.2	312|9814.247	428.8
Sh1b	MDVRPTW	2	2000	69|2999.616	16.9	72|3308.245	60.5	74|3374.126	58.1
Sh2b	MDVRPTW	3	4000	126|4174.144	31.5	124|4610.283	125.4	132|4878.785	142.3
Sh3b	MDVRPTW	4	6000	169|5260.810	69.6	174|6645.397	197.2	180|7013.522	248.5
Sh4b	MDVRPTW	5	8000	235|6238.438	105.4	239|6982.136	254.8	260|6989.971	280.5
Sh5b	MDVRPTW	6	10,000	293|6784.751	139.5	299|7525.189	568.1	339|7881.788	490.6
CNV				**1729**		**1764**		**1893**	
CTD (/km)				**55,990.550**		**62,047.809**		**63,659.365**	
Gap to STVDH						**10.82%**		**13.70%**	
Total Computing Time (/min)					**703.0**		**2141.1**		**2355.3**

Instance: problem name. **Type**: VRPTW or MDVRPTW; **N**: number of customers. **M**: number of depots. **STVDH**: best of 10 runs (number of routes| total routes length). **ArcGIS:** best result of 10 runs (number of routes| total routes length). **VRPEJ:** best of 10 runs (number of routes| total routes length). **CNV**: cumulative number of vehicles. **CND**: cumulative travel distance.

6. Discussion

Taking the large scale VRPTW and MDVRPTW in Shanghai, China as an example, this section discusses the impact of the spatial-temporal Voronoi diagram and spatial-temporal proximity behind the best found results.

6.1. Impact of the Spatial-Temporal Voronoi Diagram

To evaluate the impact of the Voronoi diagram, this study investigated the balance of solution quality and computing time with different values for parameters k_{max} and α. As Figure 10 illustrates, considering the spatial-temporal Voronoi neighbors, an increase in k_{max} improves solution quality but decreases computing efficiency because more spatial-temporal Voronoi neighbors are involved in local search procedures. With regard to the effect of the Voronoi distance decay strategy, as Figure 10a shows, both the slow decaying speed (α = 1) and the fast decaying speed (α = 3) generate worse results. However, as Figure 10b shows, as a fast decaying speed requires less evaluation on far Voronoi neighbors, the STVDH consumes less computing efforts as the parameter α increases.

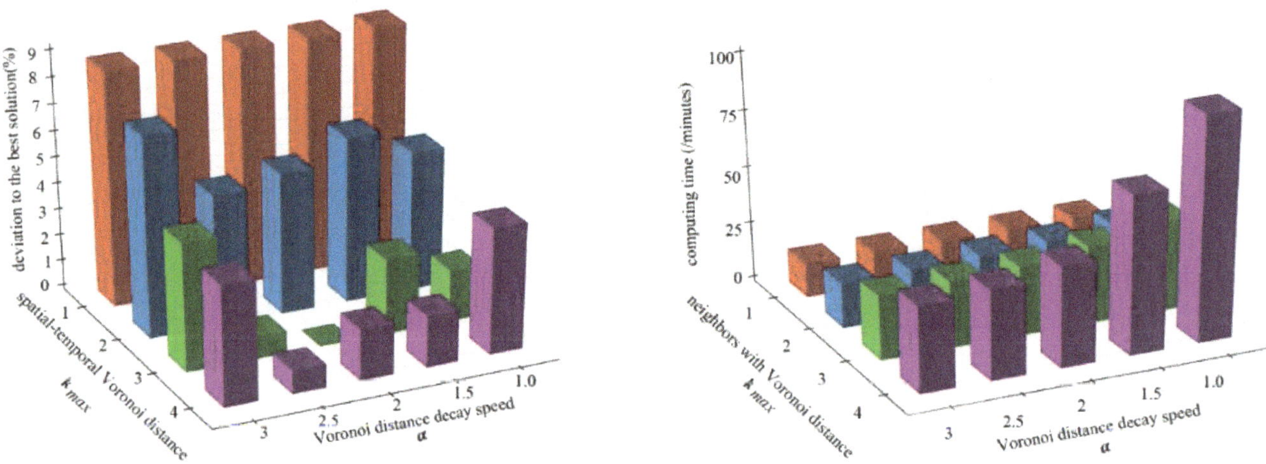

(a) (b)

Figure 10. The impact of the spatial-temporal Voronoi diagram on the STVDH. (**a**) Deviation to the best solution; and (**b**) computing time.

6.2. Spatial-Temporal Proximity Analysis on the Best Found Results

To understand spatial-temporal proximity behind the obtained results, we conducted an analysis on the Voronoi distance between consecutively-visited customers in the route of best found solutions in Section 5.2. Figure 11 displays the average percentage of Voronoi distance in the best found solutions. Two typical features are indicated below.

- The number of larger Voronoi distances is only a small proportion of the best found results. As Figure 11 displays, the percentages of Voronoi distances greater than three in the VRPTW and MDRPTW solutions are only 1.77% and 1.58%, respectively. Such a distribution agrees with the setting of parameter k_{max} ($k_{max} = 3$) in Section 5.2. Therefore, the spatial-temporal Voronoi neighborhood is very typical in the found solution.
- The percentage decreases sharply as the Voronoi distance increases. For the large-scale VRPTW dataset (Sh1a–Sh5a), the percentages for Voronoi distances 1, 2, 3, and >= 4 are 59.44%, 30.13%, 8.66%, and 1.77%, respectively, which is similar to the distribution in the MDVRPTW solution. This verifies that there is a spatial-temporal local compact structure in the routes of best found solutions. Compared with the theoretical searching intensity (as Equation (7) in Section 4.2.1 where $k_{max} = 3, \alpha = 2$) in the Voronoi distance decay strategy, these percentages systematically slightly shift to small Voronoi distances. This result demonstrates the effectiveness of the Voronoi distance-decay strategy, which searches more on near neighbors in the local search but still spends necessary efforts on far neighbors.

In summary, analysis of best found results demonstrates reasons why the spatial-temporal Voronoi diagram is effective in the STVDH algorithm.

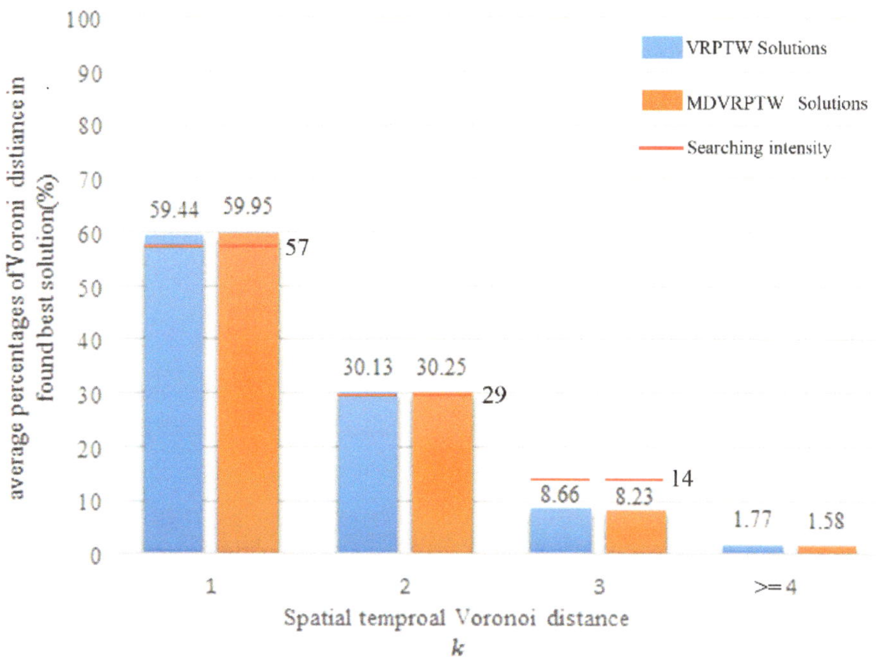

Figure 11. Average percentage of the Voronoi distance in the best found solutions.

7. Conclusions

Vehicle routing designs the least costly routes to satisfy the geographical distribution of human needs. Embedded in the GIS environment, it not only benefits transportation decision-making for both public and private sectors, but also enriches the value of geoinformation with the assistance of location-based service. Inspired by spatial-temporal proximity, this article presents a novel spatial-temporal Voronoi diagram-based heuristic approach to solve large-scale VRPTWs quickly. The derived spatial-temporal Voronoi neighborhood measures proximity considering both spatial and temporal issues. The used Voronoi neighbors limits searching space in the construction algorithm, local search, and spatial-temporal feature-guided search. Moreover, the presented Voronoi distance-decay strategy assigns more searching efforts on spatial-temporal near neighbors.

Experiments on the VRPTW benchmarks and large-scale VRPTW instances in Shanghai, China with 2000–10,000 customers have demonstrated the good performance on VRPTW and MDVRPTW problems of different sizes. The obtained results indicate that the STVDH presented in this study can provide high quality solutions for large-scale VRPTWs in a reasonable time with the help of a spatial-temporal Voronoi diagram. It also verifies that the spatial-temporal proximity is very typical in best found solutions as 98% of segments had a Voronoi distance less than three.

The main contributions of this article are twofold. First, considering both time windows and spatial locations, the spatial-temporal Voronoi diagram is proposed to accelerate the solving process for VROs. In contrast to the spatial proximity based speeding up strategies [25–27], it makes use of spatial-temporal Voronoi neighbor to further reduce computing effort for local search based heuristics. The idea behind the presented approach that makes uses of spatial principles to accelerate complex optimizing processes can be extended to facilitate other space-related decision-making problems such as near real-time emergency response and large-scale facility location. Second, a spatial-temporal Voronoi diagram-based heuristic was developed to solve large-scale VRPTWs. This novel approach exhibits good performance on super large-scale VRPTWs up to 10,000 customers, which can cope with challenges from many complex real-world transportation and logistics applications.

Among future developments that we intend to undertake, we plan to integrate the presented approach with outdoor/indoor ubiquitous positioning and friendly navigating technologies for vehicles with cloud GIS. The presented effective approach will be used to upgrade the network analyst module in SDSSs to provide a more flexible transportation service for daily large-scale logistics and distribution activities. It will further enhance the spatial intelligence of modern transportation and logistics applications.

Acknowledgments

The research is jointly supported by the National Natural Science Foundation of China (No. 41401444; 41371377), China Postdoctoral Science Foundation Funded Project (2014M560671), Shenzhen Dedicated Funding of Strategic Emerging Industry Development Program (No. JCYJ20121019111128 765), Shenzhen Basic Research Program (JCYJ20140828163633980), and the Open Research Fund of State Key Laboratory of Information Engineering in Surveying, Mapping and Remote Sensing (13S02).

Author Contributions

Wei Tu designed this approach, conceived the experiments and wrote the most part of the paper. Qingquan Li contributed result analysis and discussed the idea. Zhixiang Fang provided the framework of the result analysis. Baoding Zhou implemented the experiment.

Conflicts of Interest

The authors declare no conflict of interest.

References

1. Miller, H.J.; Shaw, S.L. *Geographic Information Systems for Transportation: Principles and Applications*; Oxford University Press: Oxford, UK, 2001.
2. Lee, E.; Oduor, P. Using multi-attribute decision factors for a modified all-or-nothing traffic assignment. *ISPRS Int. J. Geo-Inf.* **2015**, *4*, 883–899.
3. Huang, Z.; Liu, X. A hierarchical approach to optimizing bus stop distribution in large and fast developing cities. *ISPRS Int. J. Geo-Inf.* **2014**, *3*, 554–564.
4. Schittekat, P.; Kinable, J.; Sörensen, K.; Sevaux, M.; Spieksma, F.; Springael, J. A metaheuristic for the school bus routing problem with bus stop selection. *Eur. J. Oper. Res.* **2013**, *229*, 518–528.
5. Mendoza, J.E.; Medaglia, A.L.; Velasco, N. An evolutionary-based decision support system for vehicle routing: The case of a public utility. *Decis. Support. Syst.* **2009**, *46*, 730–742.
6. Arribas, C.A.; Blazquez, C.A.; Lamas, A. Urban solid waste collection system using mathematical modelling and tools of geographic information systems. *Waste. Manage. Res.* **2010**, *28*, 355–363.

7. Weigel, D.; Cao, B. Applying GIS and OR techniques to solve sears technician-dispatching and home-delivery problems. *Interfaces* **1999**, *29*, 112–130.
8. Kuo, P.-F.; Lord, D.; Walden, T.D. Using geographical information systems to organize police patrol routes effectively by grouping hotspots of crash and crime data. *J. Transp. Geogr.* **2013**, *30*, 138–148.
9. Janssens, J.; van den Bergh, J.; Sörensen, K.; Cattrysse, D. Multi-objective microzone-based vehicle routing for courier companies: From tactical to operational planning. *Eur. J. Oper. Res.* **2015**, *242*, 222–231.
10. Thill, J.C. Geographic information systems for transportation in perspective. *Transp. Res. C-Emer.* **2000**, *8*, 3–12.
11. Antikainen, H. Using the hierarchical pathfinding A* algorithm in GIS to find paths through rasters with nonuniform traversal cost. *ISPRS. Int. J. Geo.-Inf.* **2013**, *2*, 996–1014.
12. Vanhove, S.; Fack, V. An effective heuristic for computing many shortest path alternatives in road networks. *Int. J. Geogr. Inf. Sci.* **2012**, *23*, 1031–1050.
13. Li, R.; Leung, Y.; Huang, B.; Lin, H. A genetic algorithm for multiobjective dangerous goods route planning. *Int. J. Geogr. Inf. Sci.* **2012**, *27*, 1073–1089.
14. Curtin, K.M.; Voicu, G.; Rice, M.T.; Stefanidis, A. A comparative analysis of traveling salesman solutions from geographic information systems. *Trans. GIS.* **2014**, *18*, 286–301.
15. Solomon, M.M. Algorithms for the vehicle routing and scheduling problems with time window constraints. *Oper. Res.* **1987**, *35*, 254–265.
16. Bräysy, O.; Gendreau, M. Vehicle routing problem with time windows, part I: Route construction and local search algorithms. *Transp. Sci.* **2005**, *39*, 104–118.
17. Braysy, O.; Gendreau, M. Vehicle routing problem with time windows, part II: Metaheuristics. *Transp. Sci.* **2005**, *39*, 119–139.
18. Qi, M.; Lin, W.-H.; Li, N.; Miao, L. A spatiotemporal partitioning approach for large-scale vehicle routing problems with time windows. *Transp. Res. E-Log.* **2012**, *48*, 248–257.
19. Ray, J.J. A web-based spatial decision support system optimizes routes for oversize/overweight vehicles in delaware. *Decis. Support. Syst.* **2007**, *43*, 1171–1185.
20. Huang, B.; Cheu, R.L.; Liew, Y.S. GIS and genetic algorithms for hazmat route planning with security considerations. *Int. J. Geogr. Inf. Sci.* **2004**, *18*, 769–787.
21. Huang, B.; Fery, P.; Xue, L.; Wang, Y. Seeking the pareto front for multiobjective spatial optimization problems. *Int. J. Geogr. Inf. Sci.* **2008**, *22*, 507–526.
22. Wright, D.J.; Wang, S. The emergence of spatial cyberinfrastructure. *P. Natl. Acad. Sci.* **2011**, *108*, 5488–5491.
23. Tong, D.; Murray, A.T. Spatial optimization in geography. *Ann. Amer. Geogr.* **2012**, *102*, 1290–1309.
24. Li, X.; He, J.; Liu, X. Ant intelligence for solving optimal path-covering problems with multi-objectives. *Int. J. Geogr. Inf. Sci.* **2009**, *23*, 839–857.
25. Toth, P.; Vigo, D. The granular tabu search and its application to the vehicle-routing problem. *Inf. J. Comput.* **2003**, *15*, 333–346.
26. Li, F.; Golden, B.; Wasil, E. Very large-scale vehicle routing: New test problems, algorithms, and results. *Comput. Oper. Res.* **2005**, *32*, 1165–1179.

27. Fang, Z.; Tu, W.; Li, Q.; Shaw, S.-L.; Chen, S.; Chen, B.Y. A Voronoi neighborhood-based search heuristic for distance/capacity constrained very large vehicle routing problems. *Int. J. Geogr. Inf. Sci.* **2013**, *27*, 741–764.
28. Tu, W.; Fang, Z.; Li, Q. The empirical Voronoi veighborhood analysis of heuristic solutions for vehicle routing problems. In Proceedings of 2013 International Symposium on Recent Advances in Transport Modelling, Goldcoast, Australia, 21–23 April 2013.
29. Okabe, A.; Boots, B.; Sugihara, K.; Chiu, S.N. *Spatial Tessellations: Concepts and Applications of Voronoi Diagrams*; John Wiley & Sons: Hoboken, NJ, USA, 2000.
30. Chen, J.; Zhao, R.; Li, Z. Voronoi-based k-order neighbour relations for spatial analysis. *ISPRS. J. Photogram. Remote Sens.* **2004**, *59*, 60–72.
31. Zhao, R.; Chen, J.; Li, Z. K-order spatial neighbours based on Voronoi diagram: Description, computation and applications. *Int. Arc. Photogram. Remote Sens. Spat. Inf. Sci.* **2002**, *34*, 10–15.
32. Tu, W.; Li, Q.; Fang, Z. Large scale multi-depot logistics routing optimization based on network Voronoi diagram. *Acta Geod. et Cartogr. Sin.* **2014**, *43*, 1075–1082.
33. Laporte, G. Fifty years of vehicle routing. *Transp. Sci.* **2009**, *43*, 408–416.
34. Baños, R.; Ortega, J.; Gil, C.; Fernández, A.; de Toro, F. A simulated annealing-based parallel multi-objective approach to vehicle routing problems with time windows. *Expert. Syst. Appl.* **2013**, *40*, 1696–1707.
35. Alabas-Uslu, C.; Dengiz, B. A self-adaptive local search algorithm for the classical vehicle routing problem. *Expert. Syst. Appl.* **2011**, *38*, 8990–8998.
36. Ropke, S.; Pisinger, D. An adaptive large neighborhood search heuristic for the pickup and delivery problem with time windows. *Transp. Sci.* **2006**, *40*, 455–472.
37. Polacek, M.; Benkner, S.; Doerner, K.F.; Hartl, R.F. A cooperative and adaptive variable neighborhood search for the multi depot vehicle routing problem with time windows. *BuR-Business Res.* **2008**, *1*, 207–218.
38. Cordeau, J.F.; Laporte, G.; Mercier, A. A unified tabu search heuristic for vehicle routing problems with time windows. *J. Oper. Res. Soc.* **2001**, *52*, 928–936.
39. Repoussis, P.P.; Tarantilis, C.D.; Ioannou, G. Arc-guided evolutionary algorithm for the vehicle routing problem with time windows. *IEEE. Trans. Evol. Comput.* **2009**, *13*, 624–647.
40. Vidal, T.; Crainic, T.G.; Gendreau, M.; Prins, C. A hybrid genetic algorithm with adaptive diversity management for a large class of vehicle routing problems with time-windows. *Comput. Oper. Res.* **2012**, *40*, 475–489.
41. Mester, D.; Bräysy, O. Active guided evolution strategies for large-scale vehicle routing problems with time windows. *Comput. Oper. Res.* **2005**, *32*, 1593–1614.
42. Gehring, H.; Homberger, J. Parallelization of a two-phase metaheuristic for routing problems with time windows. *J. Heuristics.* **2002**, *8*, 251–276.
43. Ibaraki, T.; Imahori, S.; Kubo, M.; Masuda, T.; Uno, T.; Yagiura, M. Effective local search algorithms for routing and scheduling problems with general time-window constraints. *Transp. Sci.* **2005**, *39*, 206–232.
44. Nagata, Y.; Bräysy, O.; Dullaert, W. A penalty-based edge assembly memetic algorithm for the vehicle routing problem with time windows. *Comput. Oper. Res.* **2010**, *37*, 724–737.

45. Dondo, R.; Cerdá, J. A cluster-based optimization approach for the multi-depot heterogeneous fleet vehicle routing problem with time windows. *Eur. J. Oper. Res.* **2007**, *176*, 1478–1507.
46. Milthers, N.P.M. Solving VRP Using Voronoi Diagrams and Adaptive Large Neighborhood Search. Master's thesis, University of Copenhagen, Copenhagen, Denmark, 2009.
47. Santos, L.; Coutinho-Rodrigues, J.; Antunes, C.H. A web spatial decision support system for vehicle routing using google maps. *Deci. Support. Syst.* **2011**, *51*, 1–9.
48. Tu, W.; Li, Q.; Chang, X.; Yue, Y.; Zhu, J. A Spatio-temporal decision support framework for large scale logistics distribution in Metropolitan area. In *Advances in Spatial Data Handling and Anslysis*; Harvey, F., Leung, Y., Eds.; Springer: Berlin, Germany, 2015; pp. 193–206.
49. Tu, W.; Fang, Z.; Li, Q.; Shaw, S.-L.; Chen, B.Y. A bi-level Voronoi diagram based heuristics for large scale mutli-depot vehicle routing problem with time window. *Transp. Res. E-Log* **2014**, *61*, 84–97.
50. Barber, C.B.; Dobkin, D.P.; Huhdanpaa, H.T. The quickhull algorithm for convex hulls. *ACM. T. Math. Softw.* **1996**, *22*, 469–483.
51. Zachariadis, E.E.; Kiranoudis, C.T. A strategy for reducing the computational complexity of local search-based methods for the vehicle routing problem. *Comput. Oper. Res.* **2010**, *37*, 2089–2105.
52. Dongarra, J. Performance of Various Computers Using Standard Linear Equations Software. Available online: http://www.netlib.org/benchmark/performance.ps (accessed on 30 August 2015).
53. Gehring, H.; Homberger, J. A parallel hybrid evolutionary metaheuristic for the vehicle routing problem with time windows. In Proceedings of the EUROGEN99, Jyväskylä, Finland, 30 May–3 June 1999; pp. 57–64.
54. The VRPTW Benchmarks of GEHRING and HOMBERGER. Available Online: http: //www.bernabe.dorronsoro.es/vrp/ (accessed on 10 October 2015).
55. The Large-Scale VRPTW and MDVRPTW Problem Dataset in Shanghai, China. Available Online: https://github.com/spatialsmart/VRPTW/tree/master/Shanghai/Problem (accessed on 10 October 2015).
56. Coy, S.; Golden, B.; Runger, G.; Wasil, E. Using experimental design to find effective parameter settings for heuristics. *J. Heuristics.* **2001**, *7*, 77–97.
57. Groër, C.; Golden, B.; Wasil, E. A library of local search heuristics for the vehicle routing problem. *Math. Program. Comput.* **2011**, *2*, 79–101.

7

Critical Data Source; Tool or Even Infrastructure? Challenges of Geographic Information Systems and Remote Sensing for Disaster Risk Governance

Alexander Fekete [1,*], Katerina Tzavella [1,†], Iuliana Armas [2], Jane Binner [3], Matthias Garschagen [4], Carlo Giupponi [5], Vahid Mojtahed [5], Marcello Pettita [6,7], Stefan Schneiderbauer [7] and Damien Serre [8]

1 Institute of Rescue Engineering and Civil Protection, Cologne University of Applied Sciences, Betzdorfer Str. 2, Köln 50679, Germany; E-Mail: katerina.tzavella@th-koeln.de

2 Faculty of Geography, Department of Geomorphology-Pedology-Geomatics, University of Bucharest, NicolaeBalcescu 1, Sector 1, 010041 Bucharest, Romania; E-Mail: iulia_armas@geo.unibuc.ro

3 Department of Finance, Business School, University House, University of Birmingham, Birmingham B15 2TT, UK; E-Mail: j.m.binner@bham.ac.uk

4 Institute for Environment and Human Security (UNU-EHS), United Nations University, Bonn D-53113, Germany; E-Mail: garschagen@ehs.unu.edu

5 Venice Centre for Climate Studies, Department of Economics, Ca' Foscari University, Venice 30121, Italy; E-Mails: cgiupponi@unive.it (C.G.); v.mojtahed@unive.it (V.M.)

6 Amigo S.R.L., *via* flaminia 48, Roma 00196, Italy; E-Mail: mp@amigoclimate.com

7 European Academy-EURAC, Institute for Applied Remote Sensing, European Academy of Bozen/Bolzano, Viale Druso 1, Bolzano 39100, Italy, E-Mail: stefan.schneiderbauer@eurac.edu

8 UMR ESPACE 7300 CNRS, Université d'Avignon et des Pays du Vaucluse, Case 19, 74 rue Louis Pasteur, 84029 Avignon Cedex, France; E-Mail: damien.serre@univ-avignon.fr

† These authors contributed equally to this work.

* Author to whom correspondence should be addressed; E-Mail: alexander.fekete@th-koeln.de

Academic Editors: Christoph Aubrecht and Wolfgang Kainz

Abstract: Disaster risk information is spatial in nature and Geographic Information Systems (GIS) and Remote Sensing (RS) play an important key role by the services they provide to society. In this context, to risk management and governance, in general, and to civil protection, specifically (termed differently in many countries, and includes, for instance: civil contingencies in the UK, homeland security in the USA, disaster risk reduction at the UN level). The main impetus of this article is to summarize key contributions and challenges in utilizing and accepting GIS and RS methods and data for disaster risk governance, which includes public bodies, but also risk managers in industry and practitioners in search and rescue organizations. The article analyzes certain method developments, such as vulnerability indicators, crowdsourcing, and emerging concepts, such as Volunteered Geographic Information, but also investigates the potential of the topic Critical Infrastructure as it could be applied on spatial assets and GIS and RS itself. Intended to stimulate research on new and emerging fields, this article's main contribution is to move spatial research toward a more reflective stance where opportunities and challenges are equally and transparently addressed in order to gain more scientific quality. As a conclusion, GIS and RS can play a pivotal role not just in delivering data but also in connecting and analyzing data in a more integrative, holistic way.

Keywords: disaster risk management; geographic information systems; remote sensing; volunteered geographic information; crowdsourcing; critical infrastructure; crisis mapping; civil protection

---

1. Introduction

Geographic Information Systems (GIS) and Remote Sensing (RS) have become recognized and utilized as critical tools, methods, and data sources for locating, monitoring, and analyzing human crises and natural hazards in the past decades, particularly boosted by the International Decade for Natural Disaster Reduction (IDNDR). In addition to GIS, data fusion between optical, radar, and thermal imagery is recognized as playing a major role in each of the four phases of disaster management cycle (mitigation, preparedness, response, and recovery), by providing up-to-date geospatial information [1]. The most important application of remote sensing is assessing the extent of damages suffered by an area affected by disasters, and monitoring its recovery. Synthetic Aperture Radar (SAR) data can be used to measure changes in topography and building damage through multi-temporal analysis of pre- and post-disaster imagery [2]. Differential SAR interferometry (DInSAR) is considered one of the best available techniques when ground deformation mapping is needed in earthquake- or landslide-prone areas [3–5]. The accuracy of this technique depends on many parameters, like ground and atmospheric conditions, wave-band, incidence angle, and backscatter variance and intensity [6]. The success of these technologies cannot be gauged only by the proliferation of papers but also by the number of institutions founded for the specific purpose of utilizing these tools for disaster risk management.

In this article, we use GIS and RS both as prominent examples of tools and models for spatial assessment. In many sections, we use them synonymously for data gathering techniques, data sources, and methods of data processing and displaying as well as data analysis (tools) and research on the two. GIS comprises software tools for spatial data processing and RS comprises analysis and a process of data gathering through remotely facilitated sensors. From data types and sources, to processing methods, there are certain overlaps, but also many differing characteristics. It is not the purpose of this article to describe, or even analyze, these differences; it is rather our task to show what they have in common when it comes to certain specific examples of current applications in the field of disaster and risk research, for instance, spatial vulnerability indicators. GIS and RS do not equivalently contribute to processes of spatial assessments and analysis (in terms of processual chains); for example, certain GIS data, such as interview data, is not related to RS sensors but GIS can combine both.

Apart from technical differences, there is also data acquisition and processing that, for both GIS and RS, are highly dependent on financing, and for both, the heterogeneity of availability and accessibility to communities and decision-makers has even widened which would deserve an assessment and article on its own. The (continuous rise in) variety of almost cost-free open access data and acquisition, especially in the fields of humanitarian assistance, is addressed in our article (in a limited way), as are constraints by still highly costly data and software, and educational costs. By broadening the scope from a technical to a more integrative view, we wish to outline how science in the fields of GIS and RS can be connected to the burgeoning fields of Disaster Risk Management (DRM) and Risk Governance, both of which are more integrative in the sense that they go far beyond the data and methods perspective towards involving user needs and decision processes. This article therefore uses GIS and RS often synonymously in order to reveal the possible acceptance perspectives from the field of DRM.

By Disaster Risk Management (DRM) we mean the multidisciplinary process leading to the planning and application of policies, strategies, instruments, and direct intervention measures that favor the prediction, reduction and control of the effects of dangerous physical phenomena on populations, production systems, infrastructure, goods, services, and environment. The sum of actions that favor risk prediction, reduction, and control using prevention, mitigation, preparedness, rehabilitation, reconstruction, and recovery methods [7]. Risk governance is part of the disaster risk management: The totality of actors, rules, conventions and mechanisms relating to the collection, analysis and communication of information on risk and to the processes of management decision making [8]. Many phases of the often seen process cycles of DRM and risk governance overlap at least (compare the process cycles in ISO 31000 [9] and IRGC 2009 [10], for instance, and a further overview on current uses in [11]). We understand Disaster Risk Management as an integrative process(ISO - International Organization for Standardization 2009) consisting of the phases of preparation, analysis, evaluation, decisions, and communication. Risk Governance is a synonym for this process [10], while it points even more to the involvement of political actors. Risk Governance places even more emphasis on communication between all actors, between all phases, and underlines a political aspect. In the field of Disaster Risk Science, or as it is more commonly known, Disaster Risk Reduction, both concepts, DRM and Risk Governance are pivotal to understand the connection between risk processes, resilience and vulnerability, and tools such as GIS and RS.

This article builds upon research and practice of the authors in the field of Disaster Risk Management from the last 15 years. It is thereby a subjective view and comment on a selected number of aspects that prevail in areas that represent only a small fraction of the overall utilization of geo-information and remote sensing. Moreover, within the field of disaster risk science and management, this perspective is further limited to experience with national civil protection authorities in Europe and certain United Nations and international development organizations.

Regarding the broadness of services which GIS and RS can deliver, this article is limited to spatial indicators, early warning mapping/systems and communication, since these are of most direct relevance for DRR applications. Other services for risk reduction of more general type, such as GNSS (global navigation satellite system providing global coverage through a constellation of satellites providing signals from space transmitting positioning and timing data) as well as atmospheric, hydro-meteorological or oceanographic measurements, are excluded since they would exceed the length and purpose of this paper.

The paper will first describe spatial indicators in Chapter 2 as one prominent example of methodology in DRM. In Chapter 3 the paper highlights typical shortcomings of predominant data and methodology-driven approaches prevalent in GIS and RS communities. Chapter 4 discusses Volunteered Geographic Information and Communities in their role and growing use for disaster risk analysis and participation of users. Chapter 5 offers a holistic management approach used in disaster risk management for integrating all the different methods and concepts in GIS and RS. Chapter 6 then investigates whether GIS and RS services do not bear characteristics of other so-called Critical Infrastructure services in terms of having become a key resource for DRM.

2. Spatial Indicators

2.1. Proliferation of Vulnerability and Resilience Indicators

Planning and decision-making for sustainable development, which requires pro-activeness and making informed decisions, is a real challenge for decision makers due to the reactive approach of DRM. Therefore, with the help of GIS, spatial indicators, are widely applied for "measuring" (in a quantitative or semi-quantitative and relative sense) and communicating phenomena such as natural hazards, deforestation, land use change, wildfires, floods, earthquakes, tsunami, climate change. They have all been combined into concise indices with anthropogenic factors and variables, covering social, cultural, institutional, economic, and other sectors. Just to name one example from our experience: the social vulnerability [12,13] and socio-ecological indicators of resilience or vulnerability that have been connected with flood-area maps derived from RS, land use classifications, and precipitation data [14–16]. Drought, earthquake, tsunami, and storms are other areas where such spatial assessments are widely used [17–19].

2.2. Expanding Vulnerability and Resilience Indicators into Fields of Human Assistance, Conflict and Demographic Change

Indicators are not tied to a specific hazard, but some of them are more relevant to one than to another. This special issue for example, uses spatial social indicators to capture the additional impact of social vulnerabilities on the conflict in Yemen (see other article). Many more countries are currently affected by similar humanitarian conflicts and it will be worthwhile to undertake similar assessments and compare them. The Yemen example is not only a very timely application of this methodology, but also helps to provide valuable in-depth information about disaster risks and possible detrimental impacts that go beyond a mapping of hazard hot spots. The recent earthquakes in Nepal (25 April and 12 May 2015) provide another unfortunate case of how RS, GIS but also ground observations (GO) can contribute to disaster management and relief. Information loaded by multi-agency post-disaster efforts on the ground converged with remotely-sensed data on the "Nepal Earthquake 2015: Disaster Relief and Recovery Information Platform" (NDRRIP), a geoportal jointly developed by the Government of Nepal and ICIMOD, to map and assess the consequences of the earthquakes, and in particular landslides, rock falls, and avalanches [20] and orient relief efforts.

Furthermore, crisis mapping, itself, is conducted at an international level within the International Charter Space and Major Disasters initiated in 2000 by CNES (National Centre for Space Studies) and ESA (European Space Agency), joined by CSA (Canadian Space Agency), NOAA (National Oceanic and Atmospheric Administration), ISRO (Indian Space Research Organization), CONAE (Argentina Space Agency), JAXA (Japanese Space Agency), USGS (United States Geological Survey), BNSC/DMCii (Disaster Monitoring Constellation International Imaging Ltd.), CNSA (China National Space Administration), ROSCOSMOS (Russian Federal Space Agency), INPE (National Institute for Space Research of South America), DLR (German Aerospace Center), and KARI (Korean Aerospace Research), and the UN joining in 2003 [20], with the help of UNITAR/UNOSAT (The group of organizations and companies of this Charter coordinate the analysis and mapping of EO based information and provides integrated satellite-based solutions for human security, peace and socio-economic development). Several national space agencies, geoscience centers, institutions, and organizations, such as IPGP (France), INGV (Italy), NUA (Greece), BGS (UK), USGS, GFZ (Germany), DLR-ZKI (Germany), and KOERI (Turkey) alternate in providing rapid response to natural hazards and humanitarian crisis events in the form of satellite maps that designate hazard impact areas, evacuation sites, and other information.

In addition to this charter, similar products are generated by the Emergency Management Service (EMS) of Copernicus. Copernicus is the European Union's Earth Observation program and EMS is one out of six services that integrates satellite-derived data with *in situ* data for the analysis and the monitoring of issues relevant for environment and security. Hand in hand with the development of these Copernicus Services, the European Commission together with the European Space Agency (ESA) launched its own dedicated constellation of satellites—the Sentinels. It is foreseen to give the user—under certain and still-to-be-defined conditions—free access to Sentinel data. If this is going to be realized, one constraint hampering the wider use of satellite-derived information for disaster risk management will be eliminated: the costs for EO data.

Eliminating these costs will enhance the role of emergency services, government, and even international humanitarian organizations. Government and all government agencies have played a major role in saving lives, preserving properties, prevented damage and breakdown of Critical National Infrastructure (CNI). However, the 9/11 incident, Asia Tsunami (2004), complex disaster in Japan (2011) and Christchurch earthquakes (2010, 2011) to mention but a few, have proved that the capacity of CNI, emergency services or governmental agencies are insufficient in facilitating effective response to complex, major or unprecedented incidents. Perhaps this reality or continued CNI failure across the world have forced the general public to self-organize, draw from survival instinct or empathize with one another in the face of death [21], For example, "the Red Cross was already working on mapping Nepal in preparation for an event like the earthquake in April of 2015", says Dale Kunce, senior geospatial engineer for the American Red Cross based in Washington DC. He also mentions that they are using the maps after the earthquake to guide teams on the ground about things like which routes might be prone to landslides, where possible distribution centers could be based, and where banks are.

Remaining challenges are the extension of spatial indicators and assessments on hazard types that are more difficult to map, including rare events, hard-to-spatially-detect information, such as certain supply infrastructure or cyber-attacks, but also many of other forms of sabotage or intentional small-scale and spatially-distributed events. Certainly, spatial information, such as that extracted by RS sensors, offer excellent and unparalleled data and proxy indicators, constraint by definition, play a key role in advancing our understanding about uncertain and complex phenomena, such as vulnerability. However, there remain certain challenges regarding causality and methodological analysis [22]. For example, the area affected by a power failure is often not the area affected by the hazard, such as a flood. There are spatial patterns of relevance for DRM for that direct indicators are difficult to find, namely those of the anthropogenic type. Aspects such as the economic situation, the cultural identity or social networks are only detectibly by remote observations to a limited extent but are the key for understanding vulnerabilities or resilience. In addition, one needs to remember that it remains very hard to directly capture many of the soft components and drivers of vulnerability and resilience through GIS or RS tools. Vulnerability outcomes can often be analyzed with such techniques, e.g., through the mapping of squatter settlements with sub-standard housing quality in areas exposed to floods, storms or landslides. Such methods have proven to be of great support for disaster prevention and relief. However, the institutional and economic driving forces that cause such vulnerability outcomes can, of course, not be captured with GIS and RS tools, limiting their capacity for long-term vulnerability reduction.

Apart from being able to capture the content, another challenge is continuous financing for such GIS or RS services. When it is based upon third-party research projects, then the existence and maintenance of services and platforms often finishes shortly after the project ends. Funding by national bodies of civil protection for example, is also at risk from budget costs and shifts in direction, especially in the field of DRM, which itself is subject to political scrutiny and strategic decisions. Funding can be motivated by certain major disastrous events at one time but at another same time diverted to other events or political processes.

3. The Tool Perspective

3.1. Maslow's Hammer Perspective—Top-down Methodologies of Classifications and Technocentric Perspective of Algorithms

Maslow's Hammer Effect

Research within RS appears often to be driven by a technological perspective, where the sensors, the satellites, the data, the methods to filter them or algorithms to process data often dominate the research carried out.

There is much less focus on the often limited consideration of the impact on potential end users, or evaluations other than statistical sensitivity tests. In certain GIS and RS studies, publication discussion chapters are lacking that are debating the pitfalls in the analysis design, model process or applicability of the results.

While this is a typical characteristic of many other natural science or technical studies, research in RS is typically focused on demonstrating and justifying its power by focusing on technical aspects or applications of algorithms. This observation is based solely on personal observations and it should not necessarily point to critique. However, there appears a need for more studies thinking and discussing

GIS and RS usage in DRM, and, papers analyzing pitfalls within the analyses, as well as in application. This is apparent in the field of spatial vulnerability indicators, where paucity exists on studies critically thinking and discussing challenges and pitfalls of methods or trans-disciplinary operationalization.

GIS and RS are often data-driven research, but also methodology, or algorithm-driven research, and in this respect are subject to Maslow's Hammer effect. Since when we have a GIS at hand not only are spatially-explicit data preselected, but also other types of data are made spatial and conforming to the data representation in a GIS.

On the other hand, this is, at the same time, a necessary prerequisite for data integration, which is a key opportunity offered by GIS technology to the management of RS data and GIS. Different types of scales (temporal, spatial, administrative, *etc.*) and different levels (local, regional, national, international *etc.*) have distinct, specific effects on permitting identification and measurement of, for example, social vulnerability [15,23]. Opportunities and challenges of scale, up-scaling, dis-aggregation, identification of proper research area boundaries, units or auto-correlation effects and tools to analyze and compensate auto-correlation are beyond the scope here and have been intensively discussed elsewhere [17,18,24,25].

3.2. Maslow's Swiss Army Knife—Triangulation of Methods, Embracing Interdisciplinarity

GIS is also not just a hammer, but in fact a Swiss Army Knife version regarding its versatility in combining and analyzing information, gathering and aggregating it. For instance, spatial vulnerability indicators can integrate RS data with demographic, environmental, and structural information, in a GIS environment.

This versatility poses certain problems. One is the risk of overlooking the limitations of the capabilities of GIS and another of disregarding the hazards emerging from the aggregation of data. One is the aggregation problem, which is well known from spatial risk or vulnerability indicators aggregated into concise indices that may blur the representation of reality and diminish the richness of information borne in its variables or individual indicators. Some of these effects are the modifiable areal unit problem, ecological fallacy or spatial autocorrelation, but in this paper we will not go deeper in these issues [26,27]. On the other hand, the advantage of GIS is to map individual indicators, and to allow for multiple layer representation and analyses depending on how the database is structured. Aggregation methods are rarely considered with adequate care by scholars, often overlooking effects of normalization and internal compensation, and neglecting the opportunities offered by the algorithms developed by the literature on Multi-Criteria Analysis (MCA). Another risk stems from data sharing and communication; from an end-user perspective of civil protection bodies that need to take care of possible theft of data that could be misused by saboteurs for planning attacks on the identified vulnerable hotspots, be it people or critical infrastructure.

A greater awareness and transparency on the pitfalls of the usual methodological approach would be useful for the community applying GIS for DRM. Second, a greater awareness of security concerns of end-users or public bodies, or industry would be instrumental in improving the wider acceptance, and effectiveness of GI and RS data and tools. We believe that GIS and RS are indispensable tools and such pitfalls will and should not delimit their potential. However, a certain awareness of pitfalls and maturity in terms of discussion of challenges will be beneficial especially when it comes to interdisciplinary research projects and transdisciplinary collaborations with policy and practice.

In addition to the differences of GIS compared to RS the approach is still technology-driven. Scientists working with these tools, scrutinizing their potential, and developing codes for improved analysis usually do not consider user needs in the first place.

This fact hinders the wider use of RS and GIS products and has the potential to frustrate users by a lack of appropriate "expectation management" [28].

3.3. Bottom-Up Methodologies Based on GIS and RS

The potential role of GIS and RS in DRM must be augmented when they are used with certain models that describe an underlying spatial process. Together, they can be an invaluable tool for assisting policy makers to build adaptive or coping capacity *ex ante* or *ex post* disasters.

GIS and RS can be coupled with other methodologies to simulate and study the socio-economic behavior of individual or group of agents in a complex system, such as cities, in response to climate change or different natural and manmade disasters. An important methodology of coupling Bottom-Up research with the help of GIS maps is the spatial agent-based (ABM) models and as an example we recommend the article of Darvishi and Gholamreza [29]. ABMs are becoming the dominant paradigm in social simulation since they suggest that complex systems emerge from the bottom-up, are highly decentralized, and are composed of a multitude of heterogeneous objects called agents. Sources of heterogeneity can come from the multi-layer geographical characteristics of agents that are provided by the GIS maps. The agents act given their objectives and they interact, usually through time and space, which generates emergent order, often at higher levels than those at which such agents operate. There have been many applications of the spatial ABMs in DRM. For instance, authors in [30,31] provided models of emergency evacuation for urban area. These models provide estimates for the evacuation time given GIS inputs on roads, buildings, and population density. Kwan and Lee [32] used 3D GIS for an intelligent emergency response system based on ABM and network theory aiming at making quick emergency response easier during terrorist's attacks such as 9/11.

A different attempt has been made to make the use of GIS maps collaborative, multimodal, and interactive during the emergency management situations. Rauschert [33] developed a new GIS interface that overcomes problems with unimodality, personal, and unidirectional usage of GIS. The development of these interfaces brings the complex GIS databases directly to the emergency control rooms without the need of the GIS analysts.

4. Volunteered or Commanded Geographic Information?

4.1. Top-down Dissemination of Results and Gaps between Knowledge Production and Decision Making—False Expectations on the Decision Makers and End-Users

Another development is crowdsourcing of Geo-Information (GI). "It had been used in previous emergencies, such as the Wikis created to map Hurricane Katrina and bird flu, but none seemed to have a life beyond the particular incident", said Microsoft's Nigel Snoad, an adviser to the ICT4Peace Foundation, "but in Haiti, Ushahidi and its partners seemed to have a real impact on the way the humanitarian response worked," (NAIROBI, 5 July 2010, IRIN). By far the most known online platform is Ushahidi, which was initially developed to map reports of violence in Kenya after the post-election fallout at the beginning of 2008. Some others such as OpenDRI (https://www.gfdrr.org/opendri), ArcGIS.com (http://www.arcgis.com/features/), Sahana (http://sahanafoundation.org/), Google Crisis Response (https://www.google.org/crisisresponse/), as well as the UN Secretary-General's innovative Global Pulse project (http://www.unglobalpulse.org/) and the Humanitarian ID (http://humanitarian.id/) provide outlets for individuals and organizations to engage with each other to capture real-time issues and risks—ultimately to save lives and strengthen resilience to shocks and disasters.

Improvement of web-based mapping, invention of cell phones and devices that are equipped with Global Positioning System (GPS), PDAs, and digital cameras have made it possible for ordinary people to collect spatial data, which are then shared and disseminated on the Internet using web map services, especially Web 2.0. Producing maps and Geographic Information by non-expert people without academic studies and with local knowledge about their environment and, generally, the world, is preparing a phenomena that is named in general as "citizen science" and, more specifically, by different terms in geographic researches such as Neogeographic, Public Participation GIS, Ubiquitous cartography and, in general, the collection of spatial data and dissemination of them on the Internet by citizens were named by Goodchild [34] as Volunteered Geographic Information (VGI). Web sites such as OpenStreetMap (OSM) and Wikimapia, aiming to produce a free and editable map of the world, are examples of VGI.

For the most part, research on disaster response has assumed that states or other quasi-governmental entities (e.g., the United Nations) would be the primary actors in disaster relief, with NGOs playing a secondary role. Therefore, it comes as no surprise that the role of IT was primarily viewed as a means to enhance the command, control, and dissemination of information [35–38]. First and foremost, there is a need to consider which scientific questions can be answered by citizen science according to the patterns of data collection, the ability to recruit and train volunteers, the suitable participation level, and other aspects of VGI. Second, there is a need to overcome the cultural issues and to develop an understanding and acceptance of citizen science within the scientific community. This will require challenging some of the deeply held views in science, such as viewing uncertainty not as something that can be eliminated through tighter protocols but as an integral part of any data collection and, therefore, developing appropriate methods to deal with it during analysis. Moreover, the view of science as separate from societal and ethical concerns is also a challenge especially at higher levels of engagement between scientists and participants [39].

"It is sobering to be reminded that one of the basic instincts of human nature mutual cooperation for no cost—is thriving on a global scale", mentions Keegan [40].

VGI data are sometimes called "asserted" because there is no standard for checking their quality and there is no reference or citation for them, in divers the official data are called "authoritative" because their quality is checked with standards [34]. Although the quality of VGI data might not be clear, in emergency situations, such as forest fires, where we have no official data, using volunteered data with quality vagueness is better than waiting for better data to arrive [41]. The most significant advantage of VGI is that they can be up-to-date in less time than traditional data, so for the projects that have time limitations, such as updating the streets of a city in few weeks, the local citizens are the best source to collect data and update information about the streets [42]. On the other hand, some monitoring systems are more and more exposed to the risks of budget cuts and volunteered systems are liable to fluctuations of people and their motivations.

Crowdsourced data, in general, and volunteered geographic information, in particular, are becoming the huge source of data. VGI has enormous advantages such as: it's free, has the ability to produce large amounts of data in short period of time, and collect local data that are, in some cases, impossible to obtain by traditional methods of mapping. Despite its benefits, volunteered geographic information cannot be used in many applications because its quality is not determined and there is vagueness about it. Therefore, much research has been carried out so far to determine the quality of VGI [43].

In some specific cases, the problem of the VGI quality could be not crucial and the lack of precision can be used for other purposes. For measuring environmental parameters, VGI data usually do not satisfy the requirements of precision, accuracy, and error. Nevertheless, with appropriate means, VGI can become usable and produce relevant environmental data. For example, in case measures from conventional instrumentation are available and, at the same time, a large amount of VGI can be collected, despite the scarce accuracy of the latter, the high accuracy of the standard measurements tool can be used to correct the bias and to use the VGI as a measure of the spatial variability of the measure. In this case, the accuracy does not play a relevant role because we are interested in the spatial variability of the measured field.

Similarly, VGI data have the problem of "False Observation", which are wrong data collected by users (deliberately or accidentally), but also in those cases, when the concurrent number of VGI is large, False Observations can be easily invalidated with standard techniques of large-deviation filtering.

Participatory GIS (PGIS), Public Participation GIS (PPGIS), and community-integrated GIS are newer and issue-driven approaches of volunteer-based community participation (insiders) for the creation of geographical information to be fed into a GIS. The difference lies in the coordination of this participative process by the researchers (outsiders) through various methods: sketch mapping, transect walk, Internet-based mapping, scale mapping and enhancing spatial accuracy of end-products [44,45]. The concept of participative mapping has been around since the 1980s, as a fast way of extracting indigenous spatial knowledge when adequate cartographic materials were missing. PGIS is currently used on a wider scale in democratic spatial decision-making, because it loads a given space with socio-cultural, economic, or ecological value. PGIS is enhanced VGI because it helps the community identify and better define a common interest problem and generate possible solutions, while supporting self-confidence and determination, community cohesion and identity consolidation [46,47]. PGIS produces maps that represent interactive vehicles for learning spatial analysis [48], helping information exchange and analysis, decision-making, and promotion of community interests through various forms of lobby and advocacy. PGIS is, at the same time, a method of involving local communities in hazard management, a precondition for sustainable disaster risk reduction, bringing with it the deep local knowledge that may differ from scientific of official knowledge [49].

4.2. Crowd-Sourcing as a New Risk?

The proliferation of crowd sourcing, particularly loose networks in disaster communication, may even increase risk. In Section 1.21 of the UK National Security Strategy the potential impact of a new "mass of connections" upon security was highlighted.

It was argued that networks, including social networking technologies and 24 hour news media, could impact security as interest groups become more able to pressurize governments and a wide range of ideas easily proliferate, globally [50]. A recent article on ZDNet covered national "unfriend day", which argued that loose connections on Facebook lead to increased risk of terrorism [51]. Loose networks can lead to the propagation of both intentional and unintentional rumors. In January 2010 a Twitter rumor led to the evacuation of Grand Central Station in Manhattan [52]. Finally, at The Red Cross-hosted Emergency Social Data Summit a key conclusion was that "the major obstacle to the use of social media in crisis situation is the same obstacle to adoption we've seen since the beginning of the technology: a hesitation to shift from broadcasting information to engaging," [53]. Getting emergency managers in particular to embrace and adapt to these new technologies in an age of not only uncertainty but resource scarcity may be a key challenge. There is also the danger that feedback loops between new technologies and the media escalate rumor and speculation as is evident in the work of [54] through the development of the notion of "spectacle" in the media coverage of disaster events (see also [55]).

However, despite reservations around particular technologies, findings in Preston *et al.* [56] show how social networking technologies around transport attacks are potentially transmedia orientated and make use of sentiment. In particular, authors in [56] specifically looked at the characteristics of daily social media conversation and how this may react to specific events through automated content and sentiment analysis of Twitter data. Geo-tagged social media data was used by Schmidt and Binner [57] as a proof of concept for the development of an early warning tool to help first responders and emergency management personnel to quickly assess the scope and location of a current crisis, and to quickly summarize the state of affairs [57]. A related study by Preston *et al.* Authors in [56] carried out an initial manual content analysis on a Tweet dataset to understand how the attack on Domodedovo airport had been discussed in the UK Twitterverse. From a dataset of over 300,000 tweets posted from 24 to 27 January 2011, we extracted tweets that contained the following keywords: "bomb", "explosion", "Moscow", "Domodedovo", "airport". Of the 198 posts retrieved, 61 were directly related to the attack, 17 referred to other attacks and 120 were of "conversational" nature (e.g., "calm as a bomb"). It was noted that all the data directly related to the attack could be grouped into the following four categories: (1) Broadcasting, (2) Fact-finding, (3) Reacting, and (4) Projecting. The categories closely match the three-step cognitive process of SA theory [58] as each step seems to emerge in chronological order and it coincides with a deeper understanding of the event. For example, first the news is broadcasted "Russian media reporting that at least 23 people were killed and 100 injured in Moscow airport bombing". As the news is being broadcasted, people look for more information—a transmedia interaction between old (broadcast) and new (social) media—"Any word on number of bombs? News reports saying possible multiple". Once the gravity of the situation is understood (comprehended), the public react with emotional posts: "Very sad about bombings at Moscow Domodedovo airport (...)". A few final tweets contain references to past experiences and "projections" of future threats "Need to find my blog from years ago about suicide bombs (...)" "Have FIFA said anything on the Moscow bombing? Tragedy all round. Questions over airport security ahead of WC that relies on them so heavily".

Although these were initial observations, it is clear that there is scope for applying automated content and sentiment analysis, within a transmedia context, to assist researchers and stakeholders in gaining a deeper understanding of the content and modalities of daily conversation and crisis communication. Our short-term plans include the incorporation of a variety of semantic algorithms, permitting the communications and graphs to be connected in more meaningful ways. This enhancement would also allow the graphs to be tagged to support, e.g., semantic search operations.

Adding a search tool or a watch-list of interesting terms would enable the utility to be used to display the results of simple searches. One simple extension to this concept is to collect geo-tagged results from commercial search engines, to better display those results. It would be interesting to revise the displays such that these technologies could be used in real time and be widely accessible. It would also be useful to experiment with adding these graphical concepts to small displays, such as cell phones and the latest generation of tablet computers.

5. Holistic Sensing, Knowledge and Management

Chapter 4, as well as Chapters 2 and 3, have only highlighted a few key tool characteristics and a relatively new field of research, crowdsourcing. These examples, spatial indicators, the general prevailing tool-pursuit in conceptual designs of spatial assessment research, and its current combination in the field of what is coined by some as Volunteered Geographic Information, only pinpoint some of the features by which GIS and RS tools and related models are currently used in relation to DRM and Risk Governance. The next question is how are these related to each other and which could be a framework for such tools and models? The following chapter and concurrent chapters will try to outline certain important fields and concepts that (a) might be an option and (b) are already used quite pervasively, for instance in the fields of resilience and vulnerability studies [11].

5.1. Lacking Integration of Product Thinking and Risk Analyses into a Holistic Risk Management Concept

Many risk analyses only cover individual processes of a risk management or governance framework and in most cases they directly start with the analysis itself. Not many papers using GIS or RS in the context of DRM are aware of integrated risk management frameworks, such as the one in ISO 31000 [9] or IRGC 2009 [10]. Building upon knowledge acquired from past events, forward-looking activities and related feedback loops set the context for decision-making and policy development in a strategic planning framework. Carried out by subject experts, forecasting and prediction in that concept have a strong modeling component and are focused on quantitatively assessing and measuring a certain near-future condition. Foresight processes strongly consider inter-related communication of various stakeholders and multidisciplinary experts with the objective of creating common visions and consistent scenarios and thus shape and construct long-term future developments in a favorable manner. Disaster risk mitigation, irrespective of hazard-specific characteristics, can be seen as the common overall goal spanning the entire timeline [59]. In such a risk management framework, the analysis is just one part, embedded in an overall process starting with a pre-analysis phase of identifying the objective possible stakeholders to involve methods to use. After the analysis, the phase of evaluation is separated in order to stress out the need to validate but also in order to highlight that the decision making process is often done by other persons and methods than the ones in the analysis. After that, the phase of identifying and then implementing measures to handle risks follows, and while in some, especially of the older versions of such frameworks, communication and dissemination of results is often the last phase, before the risk management process starts all over again. However, this top-down thinking has been criticized and communication in some frameworks is now put at the core of risk management, taking part in all phases of the process. Applying this risk management process to many GIS and RS studies which are dealing with risks will help to align it with other stakeholders. Such stakeholders can be the decision makers who can be better involved in the process right from the phase of pre-analysis.

5.2. Technical and Social Sensors: Geo-Information as a Vital Part of Risk Management Culture (and not Just as a Tool)?

Another idea to transform the top-down and technocentric culture of GIS and RS, as is often still the case, is offered by bottom-up decentralized crowd-mapping and use of social media (see previous sections of this paper). This is a strong trend and many scientists are eager to pick up the opportunity to integrate feedback on data they disseminated in the classic top-down dissemination chains of products, and others integrate the data itself that is fed back by people texting and mapping disaster information on the spot.

This is certainly becoming a new culture and it might be interesting to undertake studies on how to join remote sensing and local sensing, but also technological sensing and human or social sensing [60]. A new look at technical and social sensors could help to include the perspectives of people and dramatically expand the range of context information to any point of interest (POI) of a crisis area, for instance. In addition, it would help to better integrate social scientists and get their acceptance of this methodology. Overall, it will tremendously increase the accuracy of information on the ground, the context to human factors, and thereby the richness and validity of a risk analysis and thereby risk management process.

5.3. Not another Science Policy Platform—From DSS to Knowledge Management

One unsolved problem and challenge remains in how all the different results, data, and method developments are managed. This calls for a knowledge management concept for knowledge preservation, back-up, but also for a service provision how to make this knowledge available to a large variety of stakeholders and users. While there is an abundance of platforms designed for certain experts, or other vaguely-specified "end-users" for which an evaluation of their sustainability, efficiency, and acceptance would be valuable. Partly driven by the logic of funding institutions, be it national or EU, communication platforms and Decision Support Systems (DSS) proliferated to an extent where many of those end-users might lose oversight and interest. Research is needed to develop an overall knowledge management concept first rather than creating or connecting more and more products and communication platforms, once more.

6. GIS and RS as a "Critical Infrastructure" for DRM

RS offers a great variety of services to DRM, but overviews and analyses of such services and their range are often wanting. While such overviews are sometimes used to justify funding of RS projects or services, so far RS services are, to a lesser degree, analyzed in a scientific manner. One way to address RS could be to analyze the range and importance of services they offer for DRM. As an example, RS information systems could be scrutinized whether they are a Critical Infrastructure (CI), in the same sense as CIs are analyzed for the purposes of civil protection. CI policy and, concurrently, academic research, were stimulated by activities in the United States, such as the Presidential Commission on CI.

Many other countries followed and classified supply infrastructure such as electricity, gas, mineral oil, water, food and also other services such as the financial system, government, school, media or national heritage and symbols as CI. Is RS a CI of national significance? Well, maybe it is wise to be more modest and start to analyze whether RS carries the characteristics to be critical for DRM. For this purpose, methods such as criticality assessment should be used, that identify the significance of a CI service to a certain customer, in this case, DRM.

RS is certainly interesting in this respect, since it is both a provider of critical services, such as GPS for navigation of rescue teams but, on the other hand, is critically dependent on many other CI services, such as electricity, IT and telecommunications, manpower, roads, *etc.* Such an assessment could help to identify the range and importance of RS for DRM in a more structured and in the end more visible manner.

Table 1 ([61]) gives and exemplary overview on just a few selected services that Earth Observation/Remote Sensing and Geographic Information Systems offer to risk governance bodies and people affected by risks and disasters. While certain qualities of these services are specific, overall, their character that could allow them to be classified as "critical" is based on the observation on the ubiquity and pervasiveness of GPS and similar navigation services amongst public and official navigation devices and practices.

Certainly, it remains open and debatable whether RS should bear the name tag "Critical Infrastructure", since this term is tightly coupled with the politically-driven field of Critical Infrastructure Protection (CIP). In fact, even within our consortium of authors, this terminology is debated. However, this is the purpose of this article, to stimulate discussion and trigger assessments, whether RS carries or not all the ingredients and is according to the criteria of a CI. It could be argued that such a scientific assessment bears its own merits while the political connotations and, therefore, also the usability of RS services for DRM and Risk Governance is another task for debate.

The selection of services builds up on our subjective observations of GIS and RS tools and data utilized by local and national disaster risk governance bodies, including ambulance and firefighting. For a thorough analysis of all services upon their criticality, we refer to national and international strategy papers and guidelines [62] and our previous publications where we have outlined a set of methods to analyze the goals of identifying a service as one of being of a critical infrastructure [63,64]. In this process that exceeds the scope and purpose of this sections, the societal values that are of interest would be identified (saving lives, health, the environment, economic interests, legal requirements, *etc.*), then the services that are related to it, then criticality criteria, such as number of people supplied by it, urgency, and other time dimensions and the specific quality that makes a service critical [65]. For example, the value "human lives" would be at risk from an interruption of navigation data when an ambulance is on the way to a major emergency. When a certain number of ambulances are reliant on such navigation data and time is critical to reach the target position, a certain number of victims of an accident are at risk. The specific quality of navigation data is its accuracy and availability, even for non-local helpers or locations that are difficult to find. It is exactly this combination of critical volume (of people using it and are dependent on it) with critical time frames and service quality that makes a service critical. From a disaster risk governance or economic management point of view it might be questionable whether it makes sense to identify more and more "critical infrastructures", but from a scientific point of view, this is more a question of whether the criteria we deploy and the concept behind it identify certain services as critical and sort out others sufficiently. This can be done for GIS and RS but demands for a more complete analysis than we suggest.

Table 1. Usefulness and critical infrastructure characteristics of selected GIS and RS services for civil protection and disaster risk management/governance.

GIS and RS Services (Selected Examples)	Usability for Civil Protection and Disaster Risk Management	Critical Quality *
Navigation: GPS, *etc*	Ambulance and firefighting services, police, location of disaster areas, people trapped under rubble, *etc.*	Life-saving reduction of response times
Communication: Satellite services, internet connectivity, telecommunications	Early warning services, communication between control rooms and remote teams and affected people, social media devices	Number of people reached by warning, better quality of information based on feedback possibility of people affected, more timely coordination of operations and therefore loves saved, effective risk and crisis communication improving public awareness, knowledge and learning
Monitoring: Natural phenomena and recovery progress Weather and climate monitoring	✓ Space Charter Calls for rapid production of disaster event overviews, long-term planning maps on hazard and risk zonation. ✓ Weather forecasts, Climate scenarios disaggregated on regions where civil protection forces are responsible for	Lives and health conditions improved for a large number of populations by better land use zonation and urban planning. Unparalleled overview and accessibility of data saving time to assess a risk or disaster
Surveillance: Humanitarian relief and refugee camps as well as forced displacement	✓ Rapid assessment of impacts and relief needs. ✓ Identification of refugee streams, possible locations suitable for refugee camps	Knowledge on quantity and geographical distribution of impacts is prerequisite for all other steps of response and relief.

* critical infrastructure meaning of such relevance for society that an interruption could cause human losses, serious health implications or devastating effects on the continuity of a community or society (based on and shortened after definition of national CI Criticality criteria in general linked to critical number, time dimensions and quality to be affected [61].

Satellites are not always ideally positioned at the time of the disaster event. However, some natural disasters may be, to some extent, forecasted, and thus satellites can be alerted and prepared to examine areas at risk. In some cases, core data and systems are unavailable. In others, Disaster Managers are unaware of the existence and utility of RS/GIS. If the data cannot be easily understood, processed, integrated, and presented in a timely manner, Disaster Managers will find other alternatives. Moreover, if society prevents Disaster Managers from acquiring RS/GIS data for public policy reasons (e.g., cost, privacy, security), the benefits will also remain illusory.

Table 2. Dependencies and interdependencies of GIS and RS with traditional National Critical Infrastructure.

GIS and RS Assets and Processes (Nodes and Vectors)	National CI Critical ** in the Sense of Causing Serious Impediment in Case of Failure	Mutual Feedback Loops (Physical, Geographic, Cyber and Logical) ***
Navigation: GPS satellite	Electrical power Oil and Gas Banking & Finance Continuity of Government	Physical, cyber, geographic, logical
Communication	Electrical power Oil and Gas Continuity of Government Transportation	Physical, cyber, geographic
Monitoring	Electrical power Oil and Gas Transportation Water Telecommunication Banking & Finance Emergency Services Continuity of Government	Physical, cyber, geographic, logical
Surveillance	Electrical power Oil and Gas Banking & Finance Continuity of Government Emergency Services	Physical, cyber, geographic, logical

** Infrastructures from the Executive Order 13010; *** interdependency criteria based on [61].

Table 2 provides an insight on the dependencies of GIS and RS services on other infrastructure services, often classified as National Critical Infrastructure (NCI). For a better general understanding of the classifying interdependency criteria listed in Table 2 consider the summer vacationers flocking to the highways when gasoline prices are low, resulting in increased traffic congestion. In this case, the logical interdependency between the petroleum and transportation infrastructures is due to human decisions and actions and is not the result of a physical process [61].

This section displays the ambivalent character of criticality—a service becoming so useful and used that if it fails will cause serious delays, impediments or even worse detrimental effects on all users who got accustomed to rely on it.

7. Conclusions

Geographic Information Systems (GIS) and Remote Sensing (RS) tools and services have become recognized and utilized as major tools for monitoring and analyzing human crises and natural hazards in the past decades. The compounded and complex difficulties described above in this use of GIS and RS tools in practice are both a communication and co-ordination problem.

Our conclusion is that digital technologies, such as these, have obvious capabilities in both dispersing information and coordinating responses in the event of a serious impediment or failure in the national critical infrastructure. The implementation of prototype models to serve as leading indicators and early warning systems in disaster management such as that described in detail in [57] are, at present, in their infancy and in need of further research to tackle situations of abuse and misuse, although we have provided examples of good practice. Ultimately, an interdisciplinary, and not simply technology-led, approach is required to investigate this issue in detail. Despite the technological advances in this area, we also need to understand both the mathematical structure of networks and the inter-relatedness of agent behavior, including the limitations of the models and associated barriers and obstacles in practical implementation. As our research to date shows, we also need to consider the role of sentiment and interactions between new and old media (transmedia) in solving these complex problems.

In this article, we could not address a range of issues that would deserve much more detailed explanation and analysis. As an example, the differential characteristics of GIS and RS and the concurrent implications for consideration within the DRM and Risk Governance fields we have indicated in Chapter 1 and have tried to showcase a few examples throughout the following chapters. One other issue related to this and, in fact, one of the major crosscutting themes of the applicability of spatial information is the question of scale. We did not have the space to cover the different implications of spatial, temporal and other conceptual scales such as administrative scales in this paper. In our previous research [66], based on key literature [23] we have shown at the example of spatial vulnerability indicators how differently in some and similarly in other respects data, assessment methodologies, and the relation to the analyzed objects on the ground interact regarding choice of scale, level, unit, and research area. Since this, itself, is nothing new to spatial sciences, even when related to disaster and risk topics, we have not elaborated on this. However, specifically, crowd-mapping and mobility tracking of individual mobile phone user feedback are adding very scale-specific opportunities for analysis, as well as decision making. The paper has, therefore, focused on these aspects but we acknowledge that we ourselves need and want to carry out more research on the scale effects. Specifically, it would be worthwhile to undertake scientific assessments on the reality of how different stakeholders and the so-called "decision-makers", as well as so-called "affected people" truly utilize and perceive the products from GIS and RS assessments. While certain products are provided and considered for a specific user-group, it would be interesting to analyze if this is, in fact, useful and whether it is the appropriate group to target.

Just as one outlook, the task of developing user-centered and scale-specific concepts is in keeping with other similar developments such as the national alert system which is set to begin in New York City to alert the public to emergencies via cell phones. This new Personal Localized Alert Network (PLAN) will enable presidential and local emergency messages as well as Amber Alerts to appear on cell phones equipped with special chips and software. The Federal Communications Commission and the Federal Emergency Management Agency confirms that the system will also warn about terrorist attacks and natural disasters. There is clearly much work to be done. The goal is far more important than the mere display of message data on a graph or map. The ultimate objective is to create a reliable tool that allows first responders and others to leverage digital technologies to protect the public at large. The testimony of the value of such a tool occurs when those who use the prototypes designed for their work areas are able to claim these tools are directly responsible for lives being saved.

In order to achieve this, it is necessary to include users already in the design phase of a service in order to (1) be able to do expectation management in the sense that it is realistically described what EO technology can offer and what it cannot—and (2) to best fulfill the user requirements. These are ingredients towards a more participative DSS and processes, where GIS can play an important role with participative simulations, exercises and more coordinated decision based on collective spatial analysis.

Concluding, GIS and RS will be able to express and enhance their capabilities into the field of DRM and Risk Governance much more pointedly, when taking into account the user side. The user may be an individual affected by a disaster situation or a governmental institution responsible for crisis prevention or a risk manager in a company. In many of the examples we have shown, such as vulnerability indicators, crowdsourced or mapped from space, or spatial assessments of mobile phone users, the participation of the users or, people affected by disaster, is a paradigm changer that has prompted many scientists to embrace the opportunities provided. This is not only about delivering better data quality and resolution (spatial, temporal, and causal context quality) for the GIS and RS analyst. It turns the perspective from a product delivery mentality and therefore top-down data processing tool and model view towards an integrative, feedback-up taking concept view.

GIS and RS can play a pivotal role not just in delivering data but also in connecting and analyzing data in a more integrative, holistic way. There might be a path in development from a tool to a concept to even a perspective of infrastructure vital (or "critical") for improving understanding and actions of handling risks and disasters.

Author Contributions

Alexander Fekete has provided the majority of the research and text body for this paper. Katerina Tzavella wrote Section 4 and provided to Sections 1, 2, 3, 6 and coordinated the reviews. Iuliana Armas contributed mainly to Section 4.1. Jane Binner wrote a large section in Section 4.2 and wrote part of the conclusion. Damien Serre advised on sections in Section 3.2 and the overall document. Matthias Garschagen helped to focus the general title and added to Section 2.2. Carlo Giupponi contributed to Section 2.2 and Vahid Mojtahed contributed to Section 3.3. Marcello Pettita added to Section 4.1 and Stefan Schneiderbauer added to Section 2 and Section 3.2. All co-authors provided guidance and in joint meetings we discussed the research idea and structure to investigate this topic from a broader, multi-author perspective.

Conflicts of Interest

The authors declare no conflict of interest.

References

1. Cartwright, S. *National Civil Defense Emergency Management Plan Order 2005*; New Zealand Government: Wellington, New Zealand, 2005; p. 68.
2. Matsuoka, M.; Yamazaki, F. Building damage mapping of the 2003 Bam Iran Earthquake using Envisat/ASAR intensity imagery. *Earthq. Spectra.* **2005**, *21*, doi:10.1193/1191.2101027.

3. Choudhury, S.; Dasgupta, S.; Saraf, A.K.; Panda, S. Remote sensing observations of pre-earthquake thermal anomalies in Iran. *Int. J. Remote Sens.* **2006**, *27*, 4381–4396.
4. Rott, H.; Nagler, T. The contribution of radar interferometry to the assessment of landslide hazards. *Adv. Space Res.* **2006**, *37*, 710–719.
5. Necsoiu, M.; McGinnis, R.; Hooper, D. New insights on the Salmon Falls Creek Canyon landslide complex based on geomorphological analysis and multitemporal satellite InSAR techniques. *Landslides* **2014**, *11*, 1141–1153.
6. Joyce, K.E.; Belliss, S.; Samsonov, S.; McNeill, S.; Glassey, P.J. A review of the status of satellite remote sensing and image processing techniques for mapping natural hazards and disasters. *Prog. Phys. Geogr.* **2009**, *33*, 183–207.
7. International Risk Governance Council. *Risk Governance: Towards an Integrative Approach*; International Risk Governance Council: Geneva, Switzerland, 2005.
8. IADB/IDEA Inter-American Development Bank/Instituto de Estudios Ambientales. The Notion of Disaster Risk: Conceptual Framework for Integrated Management. Available online: http://idea.unalmzl.edu.co (accessed on 13 July 2015).
9. International Organization for Standardization. *ISO/Iec 31010:2009: Risk Management—Risk Assessment Techniques*; International Organization for Standardization: Geneva, Switzerland, 2009.
10. International Risk Governance Council. *An Introduction to the IRGC Risk Governance Framework*; International Risk Governance Council: Geneva, Switzerland, 2008.
11. President's Commission on Critical Infrastructure Protection. *Executive Order 13010—Critical Infrastructure Protection*. President's Commission on Critical Infrastructure Protection: Washington, DC, USA, 1996.
12. Rajeev, K.M.; Ankita, S. Perception, awareness and behavior based analysis in relation to environmental issues: A case study. *Int. J. Geol. Earth Environ. Sci.* **2013**, *3*, 223–239.
13. Cutter, S.L. Vulnerability to environmental hazards. *Prog. Hum. Geogr.* **1996**, *20*, 529–539.
14. Damm, M. Mapping Social-Ecological Vulnerability to Flooding: A Sub-National Approach for Germany. Ph.D.Thesis, University of Bonn, Bonn, Germany, 2010.
15. Fekete, A. Assessment of Social Vulnerability to River-Floods in Germany. Ph.D.Thesis, University of Bonn, Bonn, Germany, 2010.
16. Giupponi, C.; Mojtahed, V.; Gain, A.G.; Biscaro C.; Balbi, S. Integrated risk assessment of water-related disasters. In *Hydro-Meteorological Hazards, Risks, and Disasters*; Paron, P., di Baldassarre, G., Shroder, J.F., Jr., Eds.; Elsviser: New York, NY, USA, 2015; pp. 163–200.
17. De Sherbinin, A.M. Mapping the Unmeasurable? Spatial Analysis of Vulnerability to Climate Change and Climate Variability. Ph.D. Thesis, University of Twente, Enschede, The Netherlands, December 2014.
18. Hagenlocher, M. Integrated Spatial Indicators for Modeling, Exploring and Visualizing Vulnerability to Vector Borne Diseases. Ph.D. Thesis, University of Salzburg, Salzburg, Austria, December 2014.
19. Taubenböck, H.; Esch, T. Remote Sensing—An Effective Data Source for Urban Monitoring. Avalable online: http://earthzine.org/2011/07/20/remote-sensing-an-effective-data-source-for-urban-monitoring/ (accessed on 10 June 2015).

20. National Disaster Relief and Recovery Information Platform. Available online: http://apps.geoportal.icimod.org/ndrrip/ (accessed on 17 September 2015).
21. Fakuade, D. Mass Population Response to Major Incidents and Critical National Infrastructure Failure. Ph.D. Thesis, University of Canterbury, Christchurch, New Zealand, 2014.
22. Fekete, A. Spatial disaster vulnerability and risk assessments: Challenges in their quality and acceptance. *Nat. Hazard* **2012**, *61*, 1161–1178.
23. Gibson, C.; Clark, C.; Ostrom, E.; Ahn, T.K. The concept of scale and the human dimensions of global change: A survey. *Ecol. Econ.* **2000**, *32*, 217–239.
24. Kienberger, S. Assessing the vulnerability to natural hazards on the provincial/community level in Mozambique: The contribution of GIscience and remote sensing. In Proceedings of 2007 Joint CIG/ISPRS Conference on Geomatics for Disaster and Risk Management, Toronto, ON, Canada, 26 May 2007.
25. Meentemeyer, V. Geographical perspectives of space, time, and scale. *Landsc. Ecol.* **1989**, 3, 163–173.
26. Cao, C. Lam, N.S.N. Understanding the scale and resolution effects in remote sensing and GIS. In *Scale in Remote Sensing and GIS*; Quattrochi, D.A., Goodchild, M.F., Eds.; Lewis Publishers: Boca Raton, FL, USA, 1997; pp. 57–72.
27. Openshaw, S. *The Modifiable Areal Unit Problem*; Geo Books: Norwich, UK, 2012.
28. Schneiderbauer, S. What "on" the world is going on? *Nat. Hazards* **2013**, *68*, 199–200.
29. Mehdi, D.; Gholamreza, A. Validation techniques of agent based modelling or geospatial simulations. *Int. Arch. Photogramm. Remote Sens. Spat. Inform. Sci.* **2014**, *XL-2/W3*, 91–95.
30. Crooks, A.; Christian, C.; Michael, B. Key challenges in agent-based modelling for geo-spatial simulation. *Comput. Environ. Urban Syst.* **2008**, *32*, 417–430.
31. Chen, X.; John, W.M.; Benjamin, Z. Agent-based modeling and analysis of hurricane evacuation procedures for the Florida Keys. *Nat. Hazards* **2006**, *38*, 321–338.
32. Kwan, M.; Jiyeong, L. Emergency response after 9/11: The potential of real-time 3D GIS for quick emergency response in micro-spatial environments. *Comput. Environ. Urban Syst.* **2005**, *29*, 93–113.
33. Rauschert, I.; Agrawal, P.; Pyush, I.R.; Sharma, R. Designing a human-centered, multimodal GIS interface to support emergency management. In Proceedings of the 10th ACM International Symposium on Advances in Geographic Information Systems, McLean, VA, USA, 8–9 November 2002.
34. Goodchild, M.F. Citizens as sensors: The world of volunteered geography. *GeoJournal* **2007**, *69*, 211–221.
35. Alexander, D.E. Advanced information networks: Management and real-time decision during natural and ecological disasters. In Proceedings of 1991 Safety and Technological Innovation, Turin, Italy, 5 September–5 October 1991; pp. 8–26.
36. Comfort, L.K. Integrating information technology into international crisis management and policy. *J. Conting. Cris. Manag.* **1993**, *1*, 17–29.
37. Eve, G.; Weber, M. Internet and emergency management: Prospects for the future. *Int. J. Mass Emerg. Disasters* **1998**, *16*, 55–72.
38. Quarantelli, E.L. Problematical aspects of the information communication revolution for disaster planning and research: Ten nontechnical issues and questions. *Disaster Prev. Manag.* **1997**, *6*, 94–106.

39. Zook, M.; Graham, M.; Shelton, T.; Gorman, S. Volunteered geographic information and crowdsourcing disaster relief: A case study of the Haitian Earthquake. *World Med. Health Policy* **2010**, *2*, doi:10.2202/19484682.1069.
40. Keegan, V. Meet the Wikipedia of the Mapping World. Guardian Unlimited. Available online: http://www.guardian.co.uk/technology/2010/feb/04/mapping-opensource-victor-keegan (accessed on 16 June 2015).
41. Elwood, S.; Goodchild, M.F.; Sui, D.Z. Researching volunteered geographic information: Spatial data, geographic research, and new social practice. *Ann. Assoc. Am. Geogr.* **2012**, *102*, 571–590.
42. Haklay, M.; Weber, P. OpenStreetMap: User-generated street maps. *IEEE Perv. Comput.* **2008**, *7*, 12–18.
43. Esmaili, R.; Naseri, F.; Esmaili, A. Quality assessment of volunteered geographic information. *Am. J. Geogr. Inf. Syst.* **2013**, *2*, 19–26.
44. Abbot, J.; Chambers, R.; Dunn, C.E.; Harris, T.; Merode, E.D.; Porter, G.; Weiner, D. Participatory GIS: Opportunity or oxymoron? *PLA Notes* **1998**, *33*, 27–34.
45. Rambaldi, G. Introduction to PGIS Practice. Available online: http://pgis-tk-en.cta.int/m04/u01.html (accessed on the 17 September 2015).
46. Braun, R.A.; Hocde, H. Farmers participatory research in Latin America: Four cases. In *Working with Farmers: The Key to Adoption of Forage Technologies*; Australian Centre for International Agricultural Research: Bruce, ACT, Australia, 2003; pp. 32–52.
47. Rambaldi, G.; Bugna, S.; Tiangco, A.; de Vera, D. Bringing the vertical dimension to the negotiating table: Preliminary assessment of a conflict resolution case in the Philippines. *ASEAN Biodivers.* **2002**, *2*, 17–26.
48. Rambaldi, G.; Kyem, P.K.; McCall, M.; Weiner, D. Participatory spacial information management and communication in developing countries. *Ejisdc* **2006**, *25*, 1–9.
49. Gaillard, J.C.; Maceda, E.A. Participatory three-dimensional mapping for disaster risk reduction. *Particip. Learn. Action J.* **2009**, *60*, 109–118.
50. UK National Security Strategy, 2010. Available online: http://www.direct.gov.uk/prod_consum_dg/groups/dg_digitalassets/@dg/@en/documents/digitalasset/dg_191639.pdf (accessed on 1 December 2010).
51. Whittaker, Z. How "National Unfriend Day" Can Prevent Terrorism. Available online: http://www.zdnet.com/blog/igeneration/how-national-unfriend-day-can-prevent-terrorism/6696 (accessed on 2 December 2010).
52. Taylor, C. Evacuation at Grand Central? Anatomy of a Twitter Rumour. Available online: http://www.bnet.com/blog/new-media/evacuation-at-grand-central-anatomy-of-a-twitter-rumor/4615 (accessed on 2 December 2010).
53. Emergency Social Data Summit. Available online: http://redcrosschat.org/about-the-emergency-social-data-summit/ (accessed on 2 December 2010).
54. Gotham K.F. Critical theory and katrina: Disaster, spectacle and immanent critique. *City Anal. Urban Trends Cult Theory Policy Action* **2007**, *11*, 81–99.
55. Cottle, S. Global crises in the news: Staging new wars, disasters, and climate change. *Int. J. Commun.* **2009**, *3*, 494–516.

56. Preston, J.; Binner, J.M.; Branicki, L.; Ferrario, M.A.; Kolokitha, M. Multiple Attacks on Transport Infrastructure: An Inter-Disciplinary Exploration of the Impact of Social Networking Technologies upon Real Time Information Sharing, Response and Recovery. Available online: http://www.csap.cam.ac.uk/media/uploads/files/2/preston-et-al-unpublished-social-networking-on-info-sharing.pdf (accessed on 11 May 2015).
57. Schmidt, V.A.; Binner, J.M. A Semi-automated display for geotagged text. In *City Evacuations: An Interdisciplinary Approach*; Springer: New York, NY, USA, 2014; pp. 107–116.
58. Endsley, M.R. Measurement of situation awareness in dynamic systems. *Hum. Factors* **1995**, *37*, 65–84.
59. Christoph, A.; Dilek, Ö.A.; Joachim, K.; Sérgio, F. Freire future-oriented activities as a concept for improved disaster risk management. *Disaster Adv.* **2013**, *6*, 1–10.
60. Preston, J. (University of East London, London, UK). Private communication, 20 April 2015.
61. Rinaldi, S.; Peerenboom, J.; Kelly, T. Identifying, understanding, and analyzing critical infrastructure interdependencies. *IEEE Control Syst.* **2001**, *21*, 11–25.
62. Federal Ministry of the Interior of Germany. *National Strategy for Critical Infrastructure Protection (CIP Strategy)*; Federal Ministry of the Interior of Germany: Berlin, Germany, 2009.
63. Fekete, A. Safety and security target levels: Opportunities and challenges for risk management and risk communication. *Int. J. Disaster Risk Reduct.* **2012**, *2*, 67–76.
64. Fekete, A.; Lauwe, P.; Geier, W. Risk management goals and identification of critical infrastructures. *Int. J. Crit. Infrastruct.* **2012**, *8*, 336–353.
65. Fekete, A. Common criteria for the assessment of critical infrastructures. *Int. J. Disaster Risk Sci.* **2011**, *2*, 15–24.
66. Fekete, A.; Damm, M.; Birkmann, J. Scales as a challenge for vulnerability assessment. *Nat. Hazards* **2010**, *55*, 729–747.

8

Multi-Scale Analysis of Regional Inequality based on Spatial Field Model: A Case Study of China from 2000 to 2012

Shasha Lu [1,*], Xingliang Guan [2], Danlin Yu [3], Yu Deng [4,*] and Linjie Zhou [2]

[1] School of Economics and Management, Beijing Forestry University, No. 35, Tsinghua East Road, Beijing 100083, China

[2] National Academy for Mayors of China, Huixin West Street, Beijing 100029, China; E-Mails: guanxl.09b@igsnrr.ac (X.G.), zhoulinjie2008@126.com (L.Z.)

[3] Department of Earth and Environmental Studies, Montclair State University, Montclair, NJ 07043, USA; E-Mail: yud@mail.montclair.edu

[4] Institute of Geographic Sciences and Natural Resources Research, Chinese Academy of Sciences, Datun Road, Beijing 100101, China

* Authors to whom correspondence should be addressed; E-Mails: lushasha@bjfu.edu.cn (S.L.), dengy@igsnrr.ac.cn (Y.D.)

Academic Editor: Wolfgang Kainz

Abstract: A large body of recent studies—from both inside and outside of China—are devoted to the understanding of China's regional inequality. The current study introduces "the spatial field model" to achieve comprehensive evaluation and multi-scale analysis of regional inequality. The model is based on the growth pole theory, regional interaction theory, and energy space theory. The spatial field is an abstract concept that defines the potential energy difference that is formed in the process of a regional growth pole driving the economic development of peripheral areas through transportation and communication corridors. The model is able to provide potentially more precise regional inequality estimates and generates isarithmic maps that will provide highly intuitive and visualized presentations. The model is applied to evaluate the spatiotemporal pattern of economic inequality in China from 2000 to 2012 amongst internal eastern-central-western regions as well as north-south regions at three geographical scales—*i.e.*, inter-province, inter-city, and inter-county. The results indicate that the spatial field model could comprehensively evaluate regional inequality, provide aesthetically pleasing and highly adaptable presentations based on a pixel-based raster, and realise the multi-scale analyses of the regional inequality. The paper also investigates the limitations and extensions of the spatial field model in future application.

Keywords: regional inequality; spatial field; nodality index; accessibility; China

1. Introduction

China has achieved great progress in economic and social development since the "open-door policy" of 1978 and subsequent economic reforms. During the period following the transformations of the late 1970s, the country adopted industrial and urban priority development strategies which focused specifically on economic growth based on regional comparative advantage, foreign direct investment, international interaction and technological innovation. In so doing, regional inequality is deemed as an inevitable result of such transformation process. Under such strategies, cities and coastal areas in China have gained substantially development during the reform period due largely to their comparative advantages, either as a result of geographic location or historical development. Consequently, regional economic inequality in China has increased drastically during the past three decades [1,2]. Rising regional inequality—such as that seen in China at present—may cause social and economic instability within a country and compromise the prospect of achieving long-term sustainable growth through a variety of social, political, and economic mechanisms [3]. Regional inequality has thus become an important issue, attracting considerable attention from both the government and researchers.

Since the second half of the 1990s, the Chinese government has attached more emphasis to mapping out programmes and national development strategies to alleviate regional inequality. For instance, in 1999 the central government of China formulated the Western Development Strategy (*xibu da kaifa*) to boost the economic development in relatively underdeveloped western China. In 2003 and 2006, Reinvigorating Old Industrial Bases of Northeast China (*zhenxing dongbei*) program, and the Central Region Grow-Up Strategy (*zhongbu jueqi*) were launched in order to boost up economic growth in the north, as well as in central regions [4]. Recently, the central government has compiled the Twelfth Five-Year Plan (2011–2015), through which more attention has been paid to strengthening the coordination of regional development, promoting a more balanced development of basic public services, and gradually narrowing the gap between rural and urban residents' living standards, as well as striving to solve the unbalanced, uncoordinated, and unsustainable problems associated with existing economic and social development patterns.

Regional inequality has also come to be seen as a serious issue amongst researchers, attracting considerable attention from the academic community. The problem has been extensively documented both inside and outside of China, generating a growing literature that can be characterised by five key features. Firstly, the term "regional inequality" tends to clearly denote a regional economic difference that can be divided into "absolute" and "relative" difference, the former reflecting the difference between regions' total economic outputs, the latter representing the difference in their economic development levels [5]. Secondly, the spatiotemporal pattern of the evolution of regional economic difference—that is, the spatial pattern of economic gaps between regions, as well as the features of the evolution of those gaps over a sufficiently long time period—is still a core area of study within research addressing regional inequality [6]. Thirdly, we note that the scale addressed by recent research has been diverse: the analysis scale of regional inequality has in fact shifted between the "macroscopic" scale—e.g., addressing interior–coastal, north-south, rural–urban, inter-provincial or intra-provincial inequality [6–9]; the "mesoscale"—e.g., addressing inter-rural and inter-urban, and inter-regional inequality [7]; and the "microscale"—e.g., inequality at the county level as well as the township level [10]. Fourth, increasingly vigorous technological and methodological contributions have been made in measuring regional inequality. Inequality indices now include the Gini coefficient, the coefficient of variation, the Theil index, the generalized entropy index, and the Atkinson index [6,11,12]. More recently, spatial analysis models using geographic information systems (GIS) and geographically weighted regression, and economic growth models have gradually been introduced to this field [13,14]. Finally, explanations—as well as corresponding policy prescriptions—addressing the factors that are responsible for regional inequality have become both the focus and the main foothold of current research efforts. The effects of physical conditions [15], geographical location [16], labour migration [6], marketization, globalisation [17], foreign direct investment [15], social capital [18], total factor productivity (TFP) and factor inputs, national policies [19] and regional development strategies have been explored as important factors that (jointly) trigger regional inequality. Further, some scholars propose that human capital investment in less-developed areas, factor market reform, infrastructure investment, social protection investment, and governance can all contribute to promoting economic growth in lagging regions and reducing regional gaps [8,15].

Despite these contributions, at least three aspects of study on regional inequality deserve far greater attention. First of all, most of the previous studies use only a single indicator or a simple composite indicator system to evaluate regional inequality—for instance, by considering per capita GDP, per capita GNP, and/or per capita income. The development of the indicator plays an important role in the degree and even the direction of change seen in the inequality being investigated; as such, conclusions may be contingent upon the selection of specific measuring indicators [12,20]. Due to the limitation of a single indicator, comprehensive measurement with respect to regional inequality (e.g., using comprehensive, multiple, or composite indicators, introduced through the spatial field model developed in this study) should garner greater attention in future research [20,21]. Secondly, though more studies have taken a spatial data analytical approach, the majority of the existing work undertaken on the subject has applied traditional statistical models in their analysis of regional inequality. In this regard, we believe that the use of intuitive and visualised presentations of regional inequality, graphically presented through isarithmic maps, should be further enhanced [22]. The third issue evidenced in previous work on regional inequality lies in the different—and even opposite—conclusions yielded by the diversity of analysis scales, measuring indicators and methods, statistical data, as well as urban definitions [6,7]. Such variation is perceived as being serious problem, and can lead to a variety of diverse policy implications, rather than a targeted response [23].

The objective of this study was to develop a "spatial field model" in response to the aforementioned challenges. The model that we put forward has several advantages, including: (i) proposing a comprehensive evaluation of regional inequality in terms of the current features of a region, and its future potential for economic development; (ii) presenting evaluation results as intuitive and highly visualised isarithmic maps; and (iii) achieving a multi-scale analysis of regional inequality. The spatial field model was applied to evaluate intraregional inequality amongst the three often-referred-to economic regions (the eastern, central, and western regions) in China, as well as the broader division of "the north" and "the south" of China, in the period 2000–2012. Each evaluation is done at the provincial-, the prefecture- and county- levels. Results indicated that the spatial field model is a useful instrument in the evaluation of regional inequality. A discussion of the results also highlights the limitations of the spatial field model, and expands upon a series of matters requiring further attention in future applications of the model.

In detail, the eastern region includes 11 provinces: Beijing, Tianjin, Hebei, Liaoning, Shandong, Jiangsu, Shanghai, Zhejiang, Fujian, Guangdong and Hainan; the central region consists of 8 provinces: Heilongjiang, Jilin, Shanxi, Henan, Hunan, Hubei, Jiangxi and Anhui; the western region comprises 12 provinces: Inner Mongolia, Shaanxi, Gansu, Ningxia, Qinghai, Xinjiang, Sichuan, Chongqing, Guangxi, Yunnan, Guizhou and Tibet. Moreover, the country is divided into "the north" and "the south" based on the natural geographical boundaries, especially the climate boundary—*i.e.*, Qinling-Huaihe Line. South China includes 16 provinces: Shanghai, Jiangsu, Zhejiang, Fujian, Anhui, Jiangxi, Hunan, Hubei, Guangdong, Guangxi, Hainan, Sichuan, Chongqing, Yunnan, Guizhou and Tibet; north China consists of the remainder provinces.

The remainder of this paper proceeds as follows. Section 2 describes the spatial field model employed and presents the data acquisition and processing. Section 3 presents the results of regional inequality in China from 2000 to 2012 at three geographical scales. Discussions and concluding remarks are presented in Sections 4 and 5 respectively.

2. The Spatial Field Model

2.1. Defining the Spatial Field Model

The "central place theory", originally proposed by the German geographer Walter Christaller in 1933, constitutes both a doctrine on urban scale and urban hierarchies and a theoretical model for urban planning and urban construction [24]. Subsequently, Perroux in the 1950s proposed the growth pole theory in which growth poles serve as "centres from which centrifugal forces emanate and to which centripetal forces are attracted" [25]. In general, a growth pole might exert two distinctly different types of influence on its neighbours through strong attractive forces (concentration) and radial forces (diffusion): these are termed as "backwash effects" and "spreading effects" [26–28]. A stable core-periphery pattern and an integration of regional development would be expected to achieve under the impact of backwash and spreading effects of a well-devised growth pole strategy [29]. In the early stages of a growth pole, polarisation effects lead the central city to absorb adjacent regions' population, capital, information, technology, and various other production elements. As the pole develops, the "spreading effects" tend to dominate, leading to the spread of various factors and elements from central cities to adjacent regions through various contact "channels" [30–32], which in real world mainly involves the trunk lines of transportation and communication as well as channels for energy and water movement [16].

Since a growth pole is always more developed than its surrounding peripheral regions, from a pure energy perspective (*i.e.*, everything is converted to some types of energy), we can imagine that there exists energy difference from the pole to the peripheral. We can designate such energy decays from the pole to the farthest peripheral with values between 1 and 0. The diffusion distance of the energy will likely be a function of the concentration of regional transportation and communication infrastructure facilities. The "spatial field" is an abstract concept that we coined in an attempt to measure the sphere of influence of any growth poles at any given level [33]. Using energy as a universal measuring unit, the growth pole shall have the largest spatial field (high energy level) than any other locations in its sphere of influence. Such energy declines from the pole to the peripherals following typical distance decaying mechanism.

2.2. Constructing the Spatial Field Model

In general, a growth pole could drive the economic development of peripheral areas through the spreading and backwash effects [26]. If a central city (as regional growth pole) has strong control ability to aggregate and diffuse resources and production factors, and has good regional accessibility, the central city would have a strong spatial field energy, and vice versa. Given that the value of the spatial field energy gradually decreases as distance from the growth pole increases, a peripheral point that is far away from the growth pole would have weak spatial field energy. Thus, the spatial field can be described with the gravity-based potential model which is the most widely used types of interaction model [34,35]. In physics, gravitation is a force that decreases with distance and increases with mass [35]. The gravity-based potential model aims at measuring the influence exerted by a set of masses on a unit of mass located at a given point in space [36–38]. The formula of the potential at point $[i, j]$ is given as follows:

$$E_{ij}^{k} = \frac{Z_k}{\left(D_{ij}^{k}\right)^{a}} \tag{1}$$

where $[i, j]$ indicates the horizontal and vertical coordinates of the point; E_{ij}^{k} indicates the intensity of the potential at point $[i, j]$ deriving from the mass k; Z_k is usually represented by population size of the mass k; D_{ij}^{k} is the distance between the mass k and point $[i, j]$ (e.g., Euclidean distance, economic cost, time); a is the distance friction parameter (normally the value of a is set as 1). In this paper, Z_k is represented by the ability of central city k to aggregate and diffuse resources and the factors of production, and D_{ij}^{k} is represented by the time cost between the mass k and point $[i, j]$.

In general, a spatial point has a composite spatial field for that it is simultaneously influenced by all of the surrounding central cities. The composite spatial field constitutes the superposition of the spatial field (potential energy value) from all of the surrounding central cities. The intensity of the composite spatial field can be described as follows:

$$E_{ij} = \sum_{k=1}^{K} E_{ij}^{k} \tag{2}$$

where E_{ij} is the composite spatial field of the point $[i, j]$ deriving from all of the central cities (k=1, …, K).

It should be noted that when D_{ij}^{k} equals zero, the value of spatial field in the point $[i, j]$ would be not in conformity with the mathematical rules. In this situation, we set E_{ij}^{k} as the largest spatial field value of its surrounding points.

The intensity of a point's composite spatial field is determined by the control ability of its regional central cities (growth poles) to aggregate and spread resources and factors of production (Z_k, k=1, …, K), as well as the regional accessibility of those cities (D_{ij}^{k}, k=1, …, K). In this paper, the ability of regional central cities (growth poles) to aggregate and spread resources and factors of production is reflected by the nodality index, which is a composite indicator system. The distance cost between a central city and its neighbouring area is reflected by regional accessibility based on the Cost-Weighted Distance Method.

2.2.1. The Indicator System of the Nodality Index

Each city has an absolute importance for it provides residents with goods and services. Preston put forward nodality index and used it to describe the absolute importance of a city [39]. In the study, the nodality index is used to evaluate the capability of central cities (growth poles) in aggregating and spreading resources and factors of production from/to their neighbours. The nodality index therefore reflects the aggregation ability and spread ability of central cities in driving the economic development of peripheral areas. Based upon the considerations of manoeuvrability, dynamics, scientific validity, and comprehensiveness, a critical indicator system was established using the expert consultation method, in accordance with the scientific definition of a nodality index and with reference to related literature, feasibility reports, and official reports. The index consists of the target layer, the criterion layer, and the index layer, and includes 29 identified specific indicators to reflect economic development, social development, infrastructure development, investment environment, science and technology, and the condition of informatization (Table 1).

Table 1. The indicator system of nodality index.

Target Layer	Criterion Layer	Specific Indicators
Nodality Index	Economic development condition	Gross domestic product(X_1, 10, 000 RMB), per capita gross domestic product (X_2, RMB), total population (X_3, 10, 000 persons), gross industrial output value (X_4, 10, 000 RMB), local government bugetary revenue(X_5, 10, 000 RMB), total retail sales of social consumer goods (X_6, 10, 000 RMB), savings deposit of urban and rural household (X_7, 10, 000 RMB)
	Social development condition	Urbanisation rate of population (X_8, %), average wage of staff and workers(X_9, RMB), mobile phone users (X_{10}, 10, 000 number), number of theater and cinemas (X_{11}, unit), total collections in public libraries (X_{12}, 1000 copies), number of licensed/assistant doctors (X_{13}, 10, 000 persons)
	Infrastructure development condition	Built-up area (X_{14}, km^2), vegetation cover in respect to built-up area (X_{24}, %), total passenger traffic•(X_{20}, 10, 000 persons), total freight traffic• (X_{21}, 10, 000 tons), area of paved roads at year-end (X_{22}, 10, 000 m^2), number of public vehicles under operation at year-end (X_{23}, unit)
	Investment environment condition	Number of fully-employed staff and workers (X_{15}, 10, 000 persons), total investment in fixed assets (X_{16}, 10, 000 RMB), amount of foreign capital actually utilized (X_{17}, 10, 000 dollar), water supply (X_{18}, 10, 000 tons), electricity consumption (X_{19}, 10, 000 kwh)
	Science & technology and informatization condition	Postal and telecommunications services (X_{25}, 10, 000 RMB), total research & development personnel (X_{26}, persons), total educational funds (X_{27}, 10, 000 RMB), national regular institutions of higher education (X_{28}, unit), number of students enrollment in higher education (X_{29}, persons)

Note: • indicates the indicators in relation to statistical data of municipal district is missing and are replaced with the statistical data from the whole city.

Due to the difficulty in obtaining data, this study made use of data in two years, 2000 and 2012 for prefecture-level or higher level cities. It should be noted that the evaluation result would be not accurate due to the study ignored the county-centers for the difficulty in obtaining data. In fact, the county-centers are always more developed than its surrounding peripheral regions and they might exert influence on its neighbor through strong attractive forces (concentration) and radial forces (diffusion). Thus, it is necessary to comprehensively consider the city-centers as well as the county-centers in the concessional conditions.

The data in these two special years were more complete and representative than others. In terms of 2000, it was not only the first year of the tenth "Five-Year Plan Period", but also the beginning for the central government began to accelerate the urbanisation process, strengthen the leading role and improve the socio-economic function of the regional central cities. In the same year, the country initially constructed a safe and effective regional transportation framework. In 2012, it was the beginning of the twelfth "Five-Year Plan Period". In addition, the socio-economic development and transportation infrastructure construction in the year have not been affected by the country's four trillion RMB economic stimulus plan. In hence, these two years were very significant and typical for us to research the spatio-temporal changes of regional inequality in China.

Since the economic reforms initiated in 1978, China has significantly restructured its urban administrative/spatial system in order to empower central cities to play a leading role in driving national and regional economic development. The country has constructed 260 prefecture- or higher level cities (including 4 municipalities—Beijing, Tianjin, Shanghai and Chongqing, and 256 prefecture level cities) in 2000 and 287 cities (including 4 municipalities and 283 prefecture level cities) in 2012, respectively. This study selected those cities as the growth poles across the country, and derived their nodality index.

In order to ensure the comparability of the nodality index, the raw data of the various periods was standardised using a Deviation Standardisation Method. Due to the difficulty in determining the weight of each indicator as well as the high correlation between the indicators, principal component analysis (PCA) was adopted to calculate the nodality index. Kaiser (1974) suggests adopting the Kaiser–Meyer–Olkin (KMO) measure of sampling adequacy to test the appropriateness of the data reduction techniques such as PCA. A KMO value over 0.9 represents strong suitability for PCA; a value of 0.8 indicates suitability; a value of 0.7 shows moderate suitability. According to our preliminary analysis, the KMO value of the 29 indicators was 0.938 in 2000 and 0.946 in 2012, respectively, indicating that the sample was suitable for PCA. The first four principal components of the PCA were found to have eigen values above 1. These four components explained 85.02% and 86.39% respectively of total variance in 2000 and 2012, which meant that they represent reasonably well the 29 original indicators with limited loss of information.

Figure 1 provides the calculation results of the central cities in 2000 (Panel A) and 2012 (Panel B). It could be noted that the cities evidenced a drop in their numbers and density when comparing eastern coastal China to central China, and then central China to western China. The results were correlated with the overall level of economic development in those three areas; the urban systems in the three agglomeration area (Yangtze River Delta, Pearl River Delta, and Bohai Rim) were most integrated and well-developed in the country [40,41].

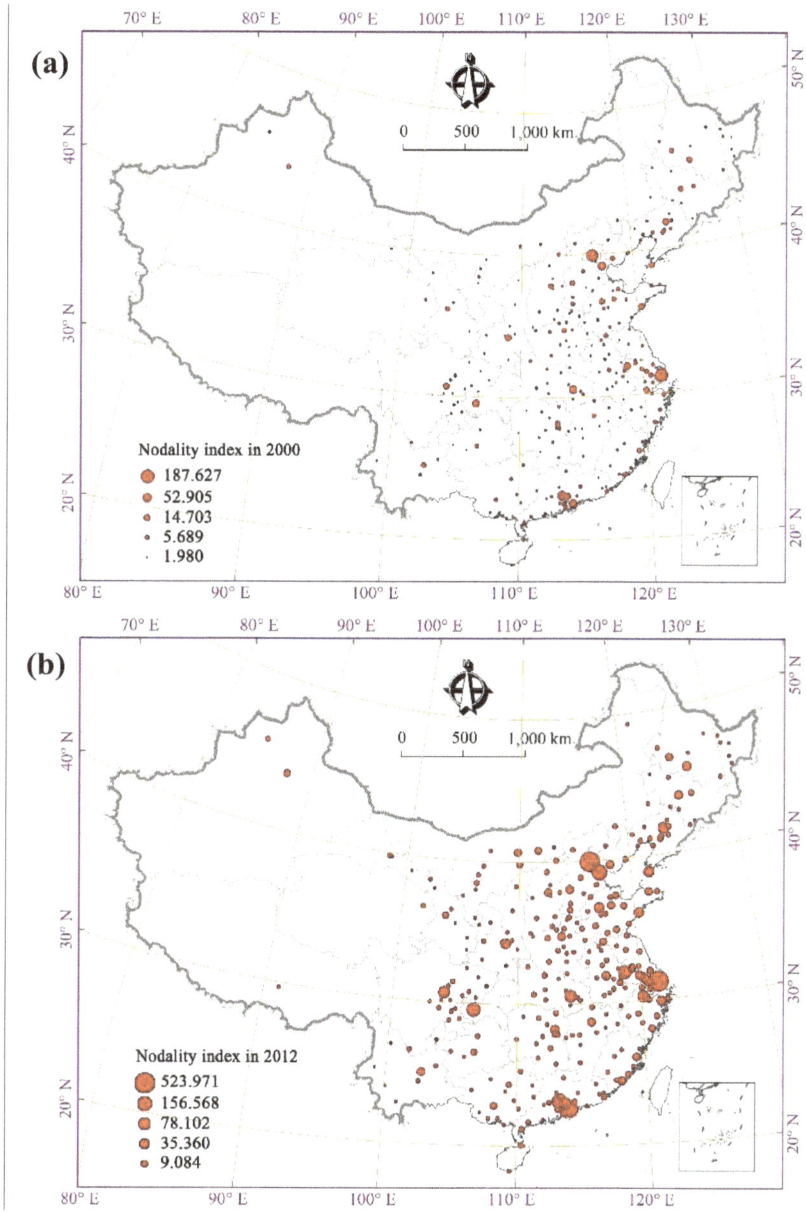

Figure 1. Nodality index in relation to the central cities for different years in China for year (**a**) 2000 and (**b**) 2012.

2.2.2. Regional Accessibility

Regional accessibility is defined areas as the time required for a specific location to travel to its central city [42,43]. This study uses the Cost-Weighted Distance Method (*i.e.*, the shortest path method) to calculate the time cost between a central city and its neighbouring area [44].

The main calculation steps followed in this study in calculating regional accessibility were as follows (following [45]): (a) 1000 m × 1000 m grid cells were generated using the Grid Module of ArcGIS software. The whole of China was divided into 5515 × 4833 grid cells. (b) The transportation speed was set according to different road types in different periods based on the designed road speed provided by Highway Engineering Technical Standard of the People's Republic of China (JTGB01-2003). The average speeds of high-speed railways, normal railways, highways, national highways, provincial highways, and normal roads in 2000 were set at 250 km/h, 80 km/h, 100 km/h, 70 km/h, 50 km/h, and

40 km/h respectively; the corresponding average speeds in 2012 were set at 250 km/h, 100 km/h, 120 km/h, 80 km/h, 60 km/h, and 50 km/h respectively. In addition, the default value with respect to the speed of land was set at 10 km/h. Water areas incur higher time costs than land, and thus the average speed over water was set at 1 km/h. It is noted that the rapid development of the high-speed railway forms an important characteristic in China's transportation profile, as such the study also took into account the 12 high-speed lines that were in operation by the end of 2012. (c) A spatial diffusion map of accessibility was generated using the Spatial Analysis Module of ArcGIS software. Figure 2 provides the time cost of each grid cell in terms of distance to the nearest central cities in 2000 (Panel A) and in 2012 (Panel B).

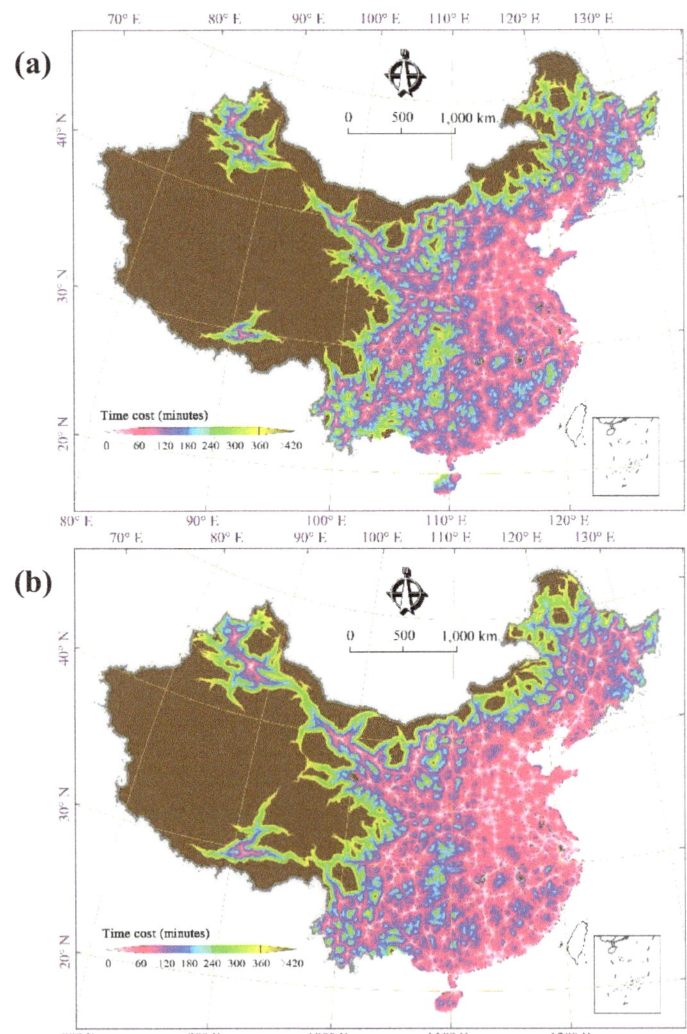

Figure 2. Spatial pattern of regional accessibility in relation to time cost in China for the year (**a**) 2000 and (**b**) 2012.

2.3. Data Sources and Processing

(i) Spatial data: Vector data, in the form of spatial administrative boundaries of 1:1,000,000 and China's 1 km resolution digital elevation dataset, were acquired from the National Geomatics Centre of China (NGCC). Transportation datasets for 2000 and 2012 (which include data on high-speed railways, normal railways, highways, national highways, provincial highways, and normal roads) were acquired from the Resources and Environment Data Centre at the Chinese Academy of Sciences (CAS). The digital map 1: 400, 0000 Highway Transport Version, also used in this study, was published by the Ministry of Transport of the People's Republic of China.

(ii) Attribute data: All socio-economic data was obtained from the annual reports of the National Bureau of Statistics of China, including the Chinese City Statistical Yearbook of 2001 and 2013, the Chinese City Construction Statistical Yearbook of 2001 and 2013, as well as the yearbooks of 32 Chinese provinces (autonomous regions and municipalities) relating to the relevant years. Because we aimed to describe and compare all of these data for the selected cities, we made sure that any data used were from unified statistical sources and that the economic data were based on comparable pricing (*i.e.*, using the constant price in 2000).

It should also be noted that population data as well as any "average" per capita indicator does have a great influence on the evaluation of regional inequality. Due to the unique Chinese *hukou* systems, there were two different types of population statistics (*i.e.*, *de jure* and *de facto* population) [46,47]. The *de jure* population (*huji renkou*) includes people who are registered but actually do not live in the unit and exclude those who live in the unit but lack local *hukou* registration, which might tend to significantly overstate the regional inequality [7]; the *de facto* population (*changzhu renkou*) includes local residents and migrants without local *hukou*, which can depict an actual picture of regional inequality. In this study, the population data as well as per capita indicators was based on *de facto* population statistics. The *de facto* population of the year 2000 came from the fifth national population census of China; the *de facto* population of the year 2012 was substituted with the sixth national population census of China in 2010, which could not be obtained from the relevant statistical yearbook. During the rapid urbanization process in China in recent years, the population and industry scales of different size cities present a polarization trend—*i.e.*, large and mega cities swiftly expanding and medium and small-sized cities relatively shrinking. The substitution would narrow the gap of population size between large cities and small-sized cities, which would reduce the disparity on nodality index between the different size cities. Therefore, the corresponding consequence of the substitution would slightly reduce the regional development disparity in China.

3. Results Analysis

By feeding our data in ArcGIS, we obtained the value of the spatial field across the 5515 × 4833 grid cells. The dasymetric map of the pattern of the spatial field in China in 2000 (Panel A) and in 2012 (Panel B) is provided at Figure 3. Following this first step, we then computed the mean of the spatial field values based on the grid cells located in the specific administrative units—including provincial level, prefecture level, and county level. The coefficient of deviation (CV) was employed in measuring regional inequality across China, as well as intra-regional inequality in the eastern, central, and western regions, and in the north and the south.

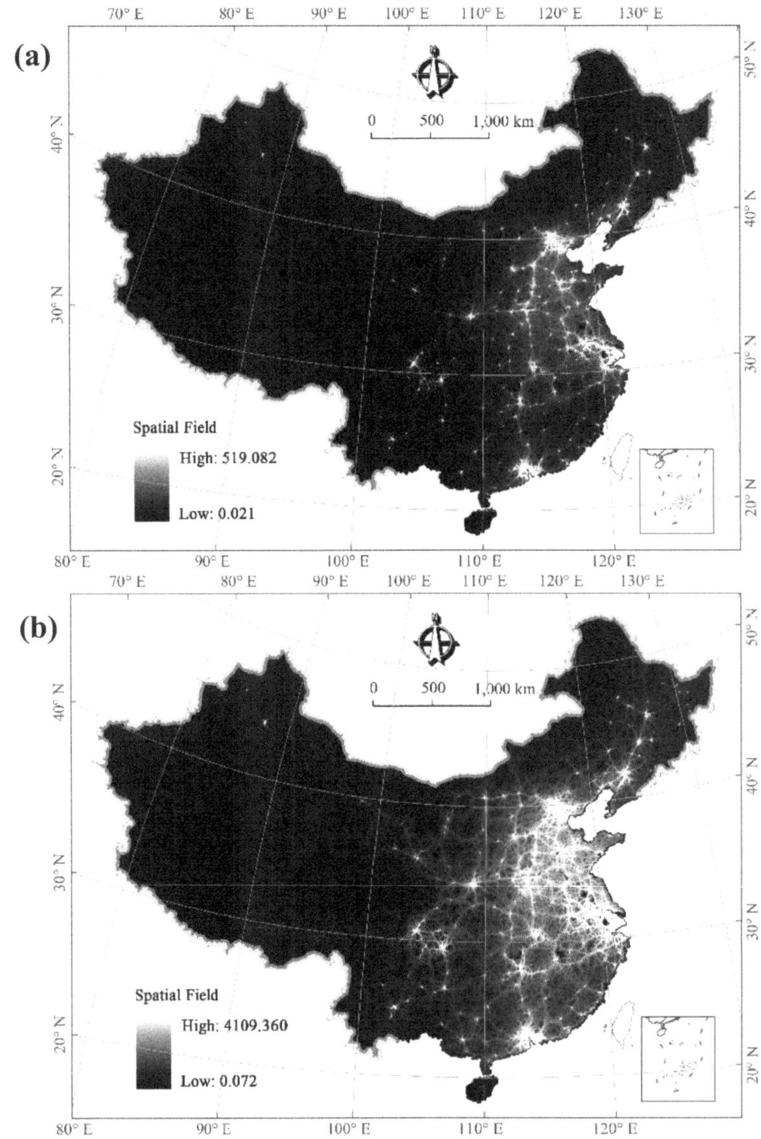

Figure 3. Distribution pattern of spatial field in China for the year of (**a**) 2000 and (**b**) 2012.

3.1. Inequality in China at Three Geographical Scales

This paper details an investigation of China's regional development pattern, and the economic inequality it revealed in the period 2000–2012. The results of the study suggest that China has attained remarkable achievements in the socio-economic development since the turn of the century, seen in rapid increases in the spatial fields (Table 2). Meanwhile, the study also showed that the unevenness of the process of economic development across the country had intensified, in line with the rapid economic growth. In fact, the coefficient of variation (CV) of the spatial field in relation to inter-province, inter-city, and inter-county inequality rose significantly, from 1.336 in 2000 to 1.421 in 2012, from 0.888 to 0.999, and from 1.131 to 1.496, respectively. The spatial field's deviation coefficient showed, at all three geographical scales, that regional inequality in China increased significantly within the period studied (Table 2).

Table 2. Difference in the spatial field at three geographical scales in China.

Year	Indicators	The Whole Country of China		
		Provincial-Level Unit	Prefecture-Level Unit	County-Level Unit
	Mean	0.295	0.247	0.254
2000	Std. Dev	0.394	0.219	0.287
	Coefficient•	1.336	0.888	1.131
	Mean	1.653	1.316	1.355
2012	Std. Dev	2.349	1.314	2.028
	Coefficient•	1.421	0.999	1.496

Note: • Coefficient is defined as the ratio of the standard deviation to the mean, which is a statistical index commonly employed in measuring regional inequality.

It is worthwhile to point out that the degree of regional development inequality seen at the inter-province and inter-county levels were more serious than that at the inter-city level. In addition, inter-county inequality across the country was shown to experience a more conspicuous increase than inter-province and inter-city inequality over the study period. Observed levels of regional inequality and change within those levels can therefore be described as being contingent upon the geographical scales [20]. As such, the effect of the geographical scale of a given inquiry needs to be accounted in the design and assessment of studies addressing regional equality, since the sensitivity of the degree and direction of change in the inequality measured shifted as the choice of geographic scale changes. Such shifts have the potential to generate quite different policy implications.

3.2. Intra-Regional Inequality in the Three Regions Studied

Table 3 reports the averages of the spatial field of the three regions at the three geographical scales. It is noted that the mean values of the spatial field evidenced a drop from eastern coastal China to central China, and then to western China in both 2000 and 2012. These results correlate with the overall level of economic development in those three areas, with respect to which the eastern region is still far ahead of the central and western regions [22].

Table 3. Difference in the spatial field of administrative units at various levels in the three regions of China.

Year	Indicators	Provincial-Level Unit			Prefecture-Level Unit			County-Level Unit		
		Eastern	Central	Western	Eastern	Central	Western	Eastern	Central	Western
	Mean	0.526	0.213	0.091	0.380	0.247	0.106	0.393	0.265	0.112
2000	Std. Dev	0.560	0.098	0.055	0.296	0.121	0.079	0.388	0.220	0.119
	Coefficient•	1.064	0.457	0.604	0.780	0.490	0.741	0.989	0.831	1.062
	Mean	3.064	1.085	0.472	2.100	1.277	0.534	2.204	1.344	0.564
2012	Std. Dev	3.323	0.565	0.301	1.833	0.711	0.390	3.114	1.165	0.553
	Coefficient•	1.084	0.520	0.638	0.873	0.557	0.730	1.413	0.867	0.981

Note: • Coefficient is defined as the ratio of the standard deviation to the mean, which is a statistical index commonly employed in measuring regional inequality.

Table 3 also presents the development disparity in the three regions at the three geographical scales. Overall, the estimated values of inequality evidenced at the provincial, the prefecture and county levels indicates that intraregional inequality of the three regions presents impressive characteristics of "reverse-to-natural gradient", that is the interior parts of the eastern region has experienced more uneven development than the western region, and the western region has experienced a more uneven development process than the central region.

The level of development inequality recorded in the internal eastern coastal region was particularly high, and had increased significantly over the study period, an upward change that may stem from the fact that a number of the wealthiest provinces as well a number of the poorest provinces in China are both located in this region. Hainan and Guangxi, for example, which are located in the coastal region, are significantly more impoverished than Beijing, Tianjin, and Shanghai. In addition, the largest three urban agglomerations in China—the Yangtze River Delta, the Pearl River Delta, and Beijing-Tianjin-Hebei—are also located in the eastern region, which is constructive in stimulating high-speed socioeconomic growth in the region, but which also poses significant threats in terms of increasing disparity. A number of scholars argue that the distribution of municipalities, urban agglomerations, provincial capital cities, and special resource cities have a fundamental influence on the overall regional inequality of China [22,48].

In contrast, the interior parts of the central region experienced the lowest degree of inequality and an unimpressive degree of upward change. In fact, the provinces of the central region were found to enjoy similar resource endowment and development levels in relation to natural conditions, geographical location, and industrial structure, making the inequality in the central region significantly lower than that of the eastern region and the western region. Currently, in order to stimulate the development of the central region and balance the development of interior and coastal parts of the country, the Chinese government has decided to engage in efforts to foster several urban agglomerations–Shijiazhuang-Taiyuan, Wuhan, Zhongyuan, Changsha-Zhuzhou-Xiangtan, and Nanchang-Jiujiang—and develop them as regional growth poles. These urban agglomerations will gradually become aggregation sites, absorbing a wide range of resources and factors of production, which may ultimately result in increases in the region's internal regional inequality in the future.

Due to a series of factors—in particular, less favourable natural conditions, lagging infrastructure, a lack of hometown connections amongst overseas Chinese, a poor transportation network, and poor accessibility to international markets—economic development in Western China at present lags far behind the eastern and central regions. The intraregional inequality in this region is still higher than that of the central region. One possible explanation could be that since 1999, the central government has pursued a Western Development Strategy, through which it has fostered several regional central cities—Chongqing, Chengdu, and Xi'an—in order to boost the economy of the western region. After a decade of efforts, these large metropolitan areas can be seen to have basically reached the development level of the central cities located in the coastal areas of the other regions. Widening regional inequality within the western region thus also represents a direction sequence of national strategies.

3.3. Inequality in Relation to the Interior of North and South China

Table 4 illustrates the mean as well as the coefficient of variation (CV) in relation to the spatial field of the interior, expressed through the division between the north and the south of China. The mean value accorded to the spatial field of the south approximately equalled that of the north in 2000. However, the value of the spatial field of the south was found to be significantly larger than that of the north in 2012. This shift implies that the south experienced more significant growth than its northern counterpart in the period 2000–2012. This remarkable development may stem from a number of factors. Firstly, it is noted that the "open-door policies" benefited the south much earlier and much more strongly than they contributed to the north. The Special Economic Zones were first established in the south (in Shenzhen, Zhuhai, Shantou in Guangdong province, and Xiamen in Fujian province). Subsequently, the Yangtze Delta, the Pearl River Delta, and coastal Fujian, as well as Hainan Island, joined the troops entitled with special privilege policies and other economic incentives. Secondly, for historical and geographical reasons, most overseas Chinese have origins in southern areas, especially in Fujian, Guangdong, and Zhejiang. The strong hometown connections of overseas Chinese have led to the south to become the main recipient of foreign investment. Thirdly, the south has enjoyed the advantages of a geographical location adjacent to Hong Kong, Macao, and Taiwan, and as a result is more competitive than the north in terms of international trade, foreign investment, and technology spillovers.

Table 4 demonstrates the widening nature of intraregional inequalities in both the south and the north. Despite this, it is noted that the extent of the change in regional equality varies at different analysis scales. Table 4 also shows that the inequality evidenced in interior parts of the south region is more significantly severe than in interior parts of the north.

Table 4. Difference in spatial field of administrative units at various levels in "the north" and "the south" of China.

Year	Indicators	Provincial-Level Unit		Prefecture-Level Unit		County-Level Unit	
		South	North	South	North	South	North
	Mean	0.317	0.271	0.267	0.226	0.245	0.224
2000	Std. Dev	0.492	0.269	0.258	0.166	0.298	0.234
	Coefficient•	1.552	0.991	0.966	0.734	1.212	1.046
	Mean	1.804	1.493	1.492	1.127	1.399	1.307
2012	Std. Dev	2.863	1.730	1.567	0.935	2.484	1.346
	Coefficient•	1.587	1.158	1.050	0.829	1.776	1.030

Note: • Coefficient is defined as the ratio of the standard deviation to the mean, which is a statistical index commonly employed in measuring regional inequality.

3.4. Summary of an Evaluation of Inequality in China

Since the turn of the new millennium, China has made great advancements in socioeconomic development, evidenced in rapid increases in the values of the spatial field of its cities (Table 2). Moreover, it is noted that the mean value of the spatial field and its change are significantly higher in the eastern region and in the south than in the central or western regions and in the north (Tables 2 and 3); this demonstrates that the eastern region are still far ahead of the central and western regions, and that the south is more developed than the north.

Meanwhile, in line with China's rapid economic growth, the uneven process of economic development across the country intensified in the period studied (Table 5). In considering the direction of changes in intraregional inequality in China between 2000 and 2012, we noted that the development of the eastern region was characterised by greater unevenness than the development of the central and western regions; at the same time, the interior parts of the south has experienced a greater inequality than the interior parts of the north. It is, however, also worthwhile noting that the extent of changes in inter-regional inequality, as well as in intra-regional inequality, varied at different scales of analysis.

Observed levels of regional inequality and its change have been previously seen to be sensitive to the geographical scales [20], a characteristic which also became apparent in this study. Since the late 1990s, the central government has attempted to readjust its regional development strategies in order to improve the economy of inland areas and promote coordinated economic development across the country. Whilst the "open-door policies", the Western Development Strategy, and the Central Region Grow-Up Strategy have improved the socioeconomic performance of inland areas, they also pose significant threats to inter- and intra-region inequality. Development theory and practice within the development field prove that balanced development constitutes a long-term process and, further, that "absolute balance" is simply not feasible between and within regions which have different natural conditions, geographic positions, and historical and economic development bases [16].

Table 5. The development level and inequality of internal eastern-central-western region and north-south region from 2000 to 2012.

Indicators	Period	Three Economic Regions			North-South Region	
		Eastern	**Central**	**Western**	**South**	**North**
Development level°	2000 and 2012	High	Medium	Low	High	Low
Development speed°	From 2000 to 2012	High	Medium	Low	High	Low
Degree of inequality°	2000 and 2012	High	Low	Medium	High	Low
Increase of inequality•	From 2000 to 2012	Low	Medium	High	Low	High

Note: ° indicates the overall conclusion was drawn from the results at the three geographical scales; • indicates the conclusion was drawn from the results on the inter-provincial inequality. In fact, the pattern in relation to the increase of inequality drawn from the results at the three analysis scales is different even opposite.

4. Discussions

4.1. Advantages of the Spatial Field Model

4.1.1. Comprehensive Evaluation of Regional Inequality

The spatial field model developed through this paper has the advantage of being able to realise "comprehensive" evaluations of regional inequality, offering an alternative to the "single indicator" evaluations that have been employed in previous studies. Regional inequality, we argue, requires a comprehensive index that is able to reflect a number of different aspects, including economic output, industrial structure, residents' income, social services, livelihood standards, and living environment suitability.

The construction of the proposed model was based on a "nodality index"—a measure of the ability of central cities (growth poles) to aggregate and to spread (and thus to drive the economic development of periphery areas), and "regional accessibility"—a measure indicating the convenience of regional transportation and communication infrastructure. The prosperity and development of the economy of a specific region is closely related to the spillovers, dispersion, and spread effects of its regional growth pole; the model reflects this relation. The results of the evaluation undertaken in this study comprehensively demonstrate not only the current features of economic development but also future economic development potential.

With regard to territorial exploitation and regional development, a higher value in terms of a regional central city's spatial field means that that city not only possesses strong abilities in terms of aggregation and spreading(and therefore in driving the economic development of periphery areas), and convenient and efficient contact channels, but also enjoys remarkable development potential. By contrast, a lower value in terms of a city's spatial field signals a city with weaker economic strength and spread effects, inconvenient and inefficient contact channels, and ultimately unpromising development potential.

4.1.2. Provision of Aesthetically Pleasing and Highly Adaptable Presentations based on a Pixel-Based Raster

It is commonly known that socioeconomic datasets (e.g., data on gross domestic product, or demographic datasets) are routinely collected through censuses and surveys, and are compiled for political or administrative units [49]. These datasets are aggregated to areal units, such as counties and census tracts [50]. All previous research on regional inequality has been based on aggregated datasets, and has thus been subject to the typically arbitrary nature of the administrative units. Choropleth maps are usually used to present the results of evaluations of regional inequality, giving the impression that socioeconomic elements are distributed homogeneously within each political or administrative unit [50]. In reality, however, the actual location of boundaries is arbitrary, and in most cases bears no logical relationship to the location of genuine discontinuities in the property being mapped. As such, the spatially discontinuous choropleth map has been considered a poor representation of the underlying continuous but not uniform distribution of various socioeconomic elements, especially where these elements are concentrated in relatively small numbers of villages, towns, and cities [51].

The spatial field energy concept proposed through this study is calculated based on an exhaustive tessellation of square grid cells, in which socioeconomic elements are modelled as a continuous density surface instead of an irregular partitioning into arbitrary administrative units. The results of evaluations of regional economy using the model are graphically represented as dasymetric and isarithmic maps, which retain a high degree of spatial accuracy and simultaneously provide aesthetically pleasing and highly adaptable presentations [51]. The results are actual density measures, which are recorded at a consistent level of generalisation by grid cells that are easily modifiable. Moreover, the model can yield precise spatial inequality estimates, providing the reader with a graphic representation and, in comparison to the conventional choroplethic method, a more accurate impression of various socioeconomic elements [51].

4.1.3. Realisation of Multi-Scale Analyses of Regional Inequality

The spatial field model can realise multi-scale analyses of regional inequality. Regional inequality has been proved to be sensitive to geographical scale; as such, a single-scale investigation might hide important characteristics of regional inequality [20,52]. In fact, there are several different geographic scales in relation to the patterns of regional inequality in China—*i.e.*, inter-region, between the three regions; inter-province, between all provinces; inter-municipal-level city, between all cities; inter-county/district, between all counties/districts; inter-township, between all townships; as well as intra-region/province/city/county/township.

In previous studies, scholars have had to obtain evaluation indicators within various administrative units if they have wanted to realise multi-scale analysis. Undoubtedly, obtaining the socioeconomic datasets is both complicated and time-consuming, especially when the administrative units change frequently. The results of the evaluation of inequality presented in this paper are represented as isarithmic maps that are calculated from a pixel-based raster. Analysis can therefore appraise regional inequality at various geographical scales, by calculating the sum or the mean of pixel values within the corresponding administrative units. As shown in Figure 4, in order to evaluate the inequality present at inter-province, inter-city, and inter-county levels, we calculated the sum of pixel values within the corresponding administrative units. As a pixel-based evaluation mode, the spatial field model provides a very convenient method to realise multi-scale analysis. This study evaluated regional inequality in the eastern-central-western economic regions and the northern-southern region of China at three geographical scales. The results indicate that the observed levels of regional inequality are contingent upon the geographic scales.

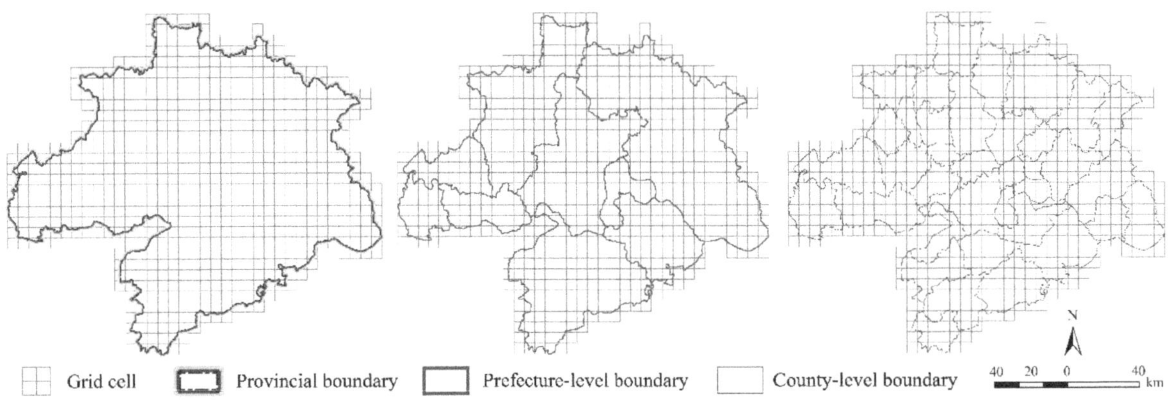

Figure 4. Inequality evaluation based on calculating the sum of pixel values within the administrative units at different geographical scales.

4.2. Limitations and Extensions of the Spatial Field Model

4.2.1. Limitations in the Construction of the Spatial Field

To begin with, greater importance should be attached to the construction of the indicator system related to the spatial field. In this paper, the spatial field was defined based on nodality index and regional accessibility. However, the choice of elements to measure within the nodality index (*i.e.*, indicator system) was largely subjective. As such, it is necessary to construct more scientific and accurate indicator systems.

Another deficiency in the indicator system used in calculating spatial field lay in the uncertainty surrounding the act of setting the distance frictional coefficient (*i.e.*, a), a value which had a significant influence on the results. It is therefore necessary to conduct sensitivity analyses in the future so as to determine a reasonable coefficient.

Thirdly, the regional accessibility intended to capture the transportation infrastructure and reflect the degree of convenience of contact and interaction between growth poles (central cities) and the periphery areas. However, the transportation modes considered were limited to surface road networks without accounting for air and waterway transportation. It was not consistent with the reality through integrating these surface networks by their speed without taking into account the traffic capacity and location of nodes/stations. Consequently, more scientific evaluation on regional accessibility remains to be explored in depth in future.

4.2.2. Limitations and Extensions with Respect to the Solution Algorithm and Computational Requirements

The spatial field model can easily be implemented in an integrated raster-based GIS. It is, however, noted that the model required large amounts of running time due to the complexity of regional accessibility as well as the vast territory of China. In fact, in order to calculate the distance from the 5515 × 4833 grid cells to the 260 central cities in 2000, and to the 287 central cities in 2012, with the aid of an ArcGIS secondary development tool (*i.e.*, ArcEngine) under the c# 2005 development environment, the model consumed approximately 1245 min and 1438 min respectively. It is in reality necessary to calculate the shortest distance between any spatial location and all of the central cities. In this study, the grid cells were set as large as 1000 m × 1000 m (the area of a grid is about 1 square kilometres) in order to shorten the computation time. It would, however, be much better to set the size of the grid cells as small as 100 m × 100 m in order to obtain a higher precision result (a shift requiring strengthened computational facilities). While the spatial field model was used in this study in order to evaluate the development inequality with respect to a small region (e.g., city/county), we suggest setting the grid cells at 50 m × 50 m (*i.e.*, about the width of a road) in order to generate a more accurate and scientific evaluation result.

4.2.3. Extensions Related to the Inequality Evaluation in Other Dimensions of Human Development

Another potential extension lies in using the spatial field model to evaluate development inequality in relation to other dimensions of human development—e.g., production, consumption, wage, education, and healthcare, as well as living environments. In fact, substantial inequality exists between the eastern-central-western regions, the rural-urban, and the provinces in China in these dimensions [12,53]. Further research should attach more attention to the crucial discrepancies evidenced in the diverse regions of the country.

4.2.4. Extensions Related to the Identification of Urban Agglomerations

In the past decade, central and provincial governments have sought to foster urban agglomerations (metropolitan areas) as regional growth poles and as strategic core areas with the potential to drive continuous economic growth. An incipient structural system of urban agglomerations has indeed formed in China [41], however the concept and standard definition of urban agglomerations remains a subject of great dispute in the country. The value of a region's spatial field reflects the level and the future potential of economic development of that region, and can be adapted to determine the spatial location and boundary of an urban agglomeration based on the dasymetric spatial field map. Figure 3 provides evidence of at least four mature urban agglomerations (the Yangtze River Delta, the Pearl River Delta, Beijing-Tianjin-Hebei, and the Shandong Peninsula) located in coastal regions, and several incipient metropolitan areas (Chongqing-Chengdu, Wuhan, Xi'an, Changsha-Zhuzhou-XiangTan, Nanchang-Jiujiang and Shijiazhuang-Taiyuan) located in central and western regions.

5. Conclusions

Based on the growth pole theory, regional interaction theory, and energy space theory, this paper puts forward the spatial field model and defines and elaborates this evaluation method. "Spatial field" is an abstract concept that can be used to define the spatial potential difference resulting from the process of the growth pole of a regional economy driving the economic development of periphery areas through transportation and communication corridors. The proposed model is characterised by the application of a nodality index which indicates both the control ability of central cities (as growth poles) in aggregating and spreading resources and factors of production, and regional accessibility—the convenience of transportation and communication infrastructure. The spatial field model makes possible the realisation of comprehensive evaluations of regional inequality with respect to current features and potential regional socioeconomic development. In addition, through the model multi-scale analysis of regional inequality can be realised, as the results are calculated based on a pixel-based raster. Moreover, the dasymetric method can yield more reasonable and precise regional inequality estimates and graphically represent the geographic distribution of various production factors, through intuitive and highly visual presentations.

The spatial field model was applied in this study in order to evaluate the spatiotemporal pattern of regional development inequality across the country and the scope of intraregional inequality between the inland eastern, central, and western regions as well as the north and the south. The evaluation addressed the period between 2000 and 2012, and analysis was undertaken at three geographical scales. The results indicated that China has made great achievements in socioeconomic development since the turn of the new millennium. Meanwhile, along with rapid economic growth, the process of economic development across the country has also become increasingly uneven; the eastern region is still far ahead of the central and western regions and the south is more developed than the north. Coastal areas have experienced more uneven development than the interior; at the same time, the interior parts of the south was characterised by greater unevenness than the interior parts of the north. In the past decades, a number of regional development strategies have been formulated and implemented that have improved the socioeconomic performance of China's regions. However, those same strategies and policies also pose the risk of significantly increasing and widening inter- and intra-region inequality. In essence, pursuing balanced and coordinated development is a long-term process that requires further research and solutions.

This study attempts to evaluate regional inequality in China based on a new method—"the spatial field model". Whilst a number of valuable implications have been drawn from the study, a number of gaps in the research remain. One potential extension to be addressed through future research would be to attach greater importance to the construction of the spatial field as a measure. Both the nodality index and the measure of regional accessibility used in this study to define the spatial field should be improved in future applications of the model. Another potential extension lies in setting the size of the grid cells to approximately 50 m × 50 m (*i.e.*, about the width of a road) so as to obtain a more precise inequality evaluation result for a smaller region (e.g., city/county). Undoubtedly, it is necessary to strengthen computational facilities, as the evaluation process would be rather computationally complex and time-consuming at such a fine scale. Moreover, the extended model could be applied to the identification of urban agglomerations (metropolitan areas) and to the evaluation of inequality in other dimensions of regional development—e.g., production, consumption, education, and healthcare, as well as living environments.

Acknowledgments

The authors are indebted to the National Natural Science Foundation of China (Grant No. 41401192 and No. 41271174), the Humanities and Social Sciences Foundation of the Ministry of Education (No. 14YJCZH106), and the Key Program of the National Natural Science Foundation of China (Grant No. 71433008). The insightful and constructive comments of editors and three anonymous reviewers are greatly appreciated.

Author Contributions

Shasha Lu and Xingliang Guan contributed to the development of the idea and participated in all phases. Danlin Yu, Yu Deng and Linjie Zhou helped perform the analysis with constructive discussions and improve the figures and manuscript. All authors have read and approved the final manuscript.

Conflicts of Interest

The authors declare no conflict of interest.

References

1. Sutherland, D.; Yao, S.J. Income inequality in China over 30 years of reforms. *Camb. J. Reg. Econ. Soc.* **2011**, *4*, 91–105.
2. Li, G.; Fang, C. Analyzing the multi-mechanism of regional inequality in China. *Ann. Reg. Sci.* **2014**, *52*, 155–182.
3. Yang, D.T. What has caused regional inequality in China? *China Econ. Rev.* **2002,** *13*, 331–334.
4. Deng, X.Z.; Huang, J.K.; Rozelle, S.; Uchida, E. Economic growth and the expansion of urban land in China. *Urban Stud.* **2010**, *47*, 813–843.
5. Zhang, W. Rethinking Regional Disparity in China. *Econ. Plan.* **2001**, *34*, 113–138.
6. Kanbur, R.; Zhang, X. Which regional inequality? The evolution of rural-urban and inland-coastal inequality in China from 1983 to 1995. *J. Comp. Econ.* **1999**, *27*, 686–701.
7. Fan, C.C.; Sun, M. Regional inequality in China, 1978–2006. *Eurasian Geogr. Econ.* **2008**, *49*, 1–18.
8. Fan, S.; Kanburb, R.; Zhang, X. China's regional disparities: experience and policy. *Rev. Dev. Financ.* **2011**, *1*, 47–56.
9. Li, Y.H. Resource flows and the decomposition of regional inequality in the Beijing-Tianjin-Hebei Metropolitan Region, 1990–2004. *Growth Chang.* **2012**, *43*, 335–357.
10. Li, X.J.; Qiao, J.J. County level economic disparities of China in the 1990s. *Acta Geogr. Sin.* **2001**, *56*, 136–145 (in Chinese).
11. Chen, J.; Fleisher, B.M. Regional income inequality and economic growth in China. *J. Comp. Econ.* **1996**, *22*, 141–164.
12. Liu, H. Changing regional rural inequality in China 1980–2002. *Area.* **2006**, *38*, 377–389.
13. Yu, D.L.; Wei, Y.H.D. Analyzing regional inequality in post-Mao China in a GIS environment. *Eurasian Geogr. Econ.* **2003**, *44*, 514–534.
14. Ye, X.; Wei, Y.H.D. Geospatial analysis of regional development in China: The case of Zhejiang Province and the Wenzhou model. *Eurasian Geogr. Econ.* **2005**, *46*, 445–464.

15. Fleisher, B.; Li, H.; Zhao, M.Q. Human capital, economic growth, and regional inequality in China. *J. Dev. Econ.* **2010**, *92*, 215–231.
16. Lu, D.D. Objective and framework for territorial development in China. *Chin. Geogr. Sci.* **2009**, *19*, 195–202.
17. Fujita, M.; Hu, D. Regional disparity in China 1985–1994: The effects of globalization and economic liberalization. *Ann. Reg. Sci.* **2001**, *35*, 3–37.
18. Pan, F.H.; He, C.F. Regional difference in social capital and its impact on regional economic growth in China. *Chin. Geogr. Sci.* **2010**, *20*, 442–449.
19. Ho, C.; Li, D. Rising regional inequality in China: Policy regimes and structural changes. *Pap. Reg. Sci.* **2008**, *87*, 245–259.
20. Lu, M.; Wang, E. Forging ahead and falling behind: Changing regional inequalities in post-reform China. *Growth Chang.* **2002**, *33*, 42–71.
21. Lyons, T.P. Intraprovincial disparities in post-Mao China: A multidimensional analysis of Fujian Province. *J. Dev. Areas* **1997**, *32*, 1–28.
22. Li, Y.; Wei, Y.H.D. The spatial-temporal hierarchy of regional inequality of China. *Appl. Geogr.* **2010**, *30*, 303–316.
23. Bradfield, M. *Regional Economics: Analysis and Policies in Canada*. McGraw-Hill Ryerson Ltd.: Toronto, ON, Canada, **1988**.
24. Christaller, W. *How I Discovered the Theory of Central Places: A Report about the Origin of Central Places*; Oxford University Press: Oxford, UK, **1972**.
25. Perroux, F. Note on the concept of growth poles. In *Regional Economics: Theory and Practice*; Mckee, D.L.; Dean, R.D.; Leahy, W.H. Eds.; Free Press: New York, NY, USA, **1970**; pp. 93–104.
26. Barkley, D.L.; Henry, M.S.; Bao, S. Identifying "spread" versus "backwash" effects in regional economic areas: A density functions approach. *Land Econ.* **1996**, *72,* 336–357.
27. Grafton, D.J. Small-Scale growth centres in remote rural regions: The case of Alpine Switzerland. *Appl. Geogr.* **1984**, *4,* 29–46.
28. Richardson, H.W. Growth pole spillovers: The dynamics of backwash and spread. *Reg. Stud.* **2007**, *41*, S27–S35.
29. Krugman, P. Increasing returns and economic geography. *J. Polit. Econ.* **1991**, *99*, 483–499.
30. Castell, M. *The Rise of Network Society*. Social Science Academic Press: Beijing, China, 2000.
31. Scott, A.J. *Global City-Regions: Trends, Theory, Policy*. Oxford University Press: Oxford, UK, 2002.
32. Short, J.R. *Global Metropolitan: Globalizing Cities in a Capitalist World*. Routledge: London, UK, 2004.
33. Guan, X.L.; Fang, C.L.; Luo, K. Regional economic development disparity of China: An application of spatial field. *Scientia Geographica Sinica*. **2012**, *32*, 1055–1065.
34. Haynes, K.E.; Fotheringham, A.S. *Gravity and Spatial Interaction Models*; Sage Publications: Beverly Hills, CA, USA, 1984.
35. Wang, F.H.; Guldmann, J.M. Simulating urban population density with a gravity-based model. *Socio-Econ. Plan. Sci.* **1996**, *30,* 245–256.
36. Batten, D.F.; Boyce, D.E. Spatial interaction, transportation, and interregional commodity flow models. *Handb. Reg.Urban Econ.* **1987**, *1*, 357–406.

37. Chen, Y.G. The distance-decay function of geographical gravity model: Power law or exponential law? *Chaos Solitons Fractals*. **2015**, *77*, 174–189.
38. Carrothers, G.A.P. An historical bedew of the gravity and potential concepts of human interaction. *J. Am. Inst. Plan.* **1956**, *22*, 94–102.
39. Preston, R E. Two centrality models. *Yearb. Assoc. Pac. Coast Geogr.* **1970**, *32*, 59–78.
40. Song, S.F.; Kevin, H.L. Urbanization and city size distribution in China. *Urban Stud.* **2002**, *39*, 2317–2327.
41. Fang, C.L.; Guan, X.L.; Lu, S.S.; Zhou, M.; Deng, Y. Input-Output efficiency of urban agglomerations in China: An application of Data Envelopment Analysis (DEA). *Urban Stud.* **2013**, *50*, 2766–2790.
42. Goodall, B. *Dictionary of Human Geography*; Penguin: London, UK, 1987.
43. Wan, N.; Zhan, F.B.; Zou, B. A relative spatial access assessment approach for analyzing potential spatial access to colorectal cancer services in Texas. *Appl. Geogr.* **2012**, *32*, 291–299.
44. Wang, Z.B.; Xu, J.G.; Fang, C.L.; Xu, L.; Qi, Y. The study on county accessibility in China: Characteristics and effects on population agglomeration. *J. Geogr. Sci.* **2011**, *21*, 18–34.
45. Wang, L.; Deng, Y.; Liu, S.H.; Wang, J.H. Research on urban spheres of influence based on improved field model in central China. *J. Geogr. Sci.* **2011**, *21*, 489–502.
46. Chan, K W. The Chinese hukou system at 50. *Eurasian Geogr. Econ.* **2009**, *50*, 197–221.
47. Zhou, Y.X.; Ma, L.J.C. China's urban population statistics: A critical evaluation. *Eurasian Geogr. Econ.* **2005**, *46*, 272–289.
48. Wang, Y.; Fang, C.L.; Xiu, C.L.; Liu, D.Q. A new approach to measurement of regional inequality in particular directions. *Chin. Geogr. Sci.* **2012**, *22,* 705–717.
49. Deichmann, U.; Balk, D.; Yetman, G. Transforming Population Data for Interdisciplinary Usages: From Census to Grid. **2001**. Available online: http://sedac.ciesin.org/gpw-v2/GPWdocumentation.pdf (accessed on 12 July 2015)
50. Mennis, J. Generating surface models of population using dasymetric mapping. *Prof. Geogr.* **2003**, *55*, 31–42.
51. Langford, M.; Unwin, D.J. Generating and mapping population density surfaces within a geographical information system. *Cartogr. J.* **1994**, *31*, 21–26.
52. Wei, Y.H.D. Regional inequality in China. *Prog. Hum. Geogr.* **1999**, *23*, 49–59.
53. Poon, J.P.H.; Shang, Q.Y. Regional wage inequality in China, 1996–2010. *Eurasian Geogr. Econ.* **2012**, *53*, 338–355.

9

Metropolises in the Twittersphere: An Informetric Investigation of Informational Flows and Networks

Thorsten Förster * and Agnes Mainka †

Department of Information Science, Heinrich-Heine-University Düsseldorf, Universitätsstraße 1, 40225 Düsseldorf, Germany; E-Mail: agnes.mainka@hhu.de

† The author contributed equally to this work.

* Author to whom correspondence should be addressed; E-Mail: t.foerster@hhu.de

Academic Editors: Jochen Schiewe and Wolfgang Kainz

Abstract: Information flows on social media platforms are able to show trends and user interests as well as connections between users. In this paper, we present a method how to analyze city related networks on the social media platform Twitter based on the user content. Forty million tweets have been downloaded *via* Twitter's REST API (application programming interface) and Twitter's Streaming API. The investigation focuses on two aspects: firstly, trend detection has been done to analyze 31 informational world cities, according the user activity, popularity of shared websites and topics defined by hashtags. Secondly, a hint of how connected informational cities are to each other is given by creating a clustered network based on the number of connections between different city pairs. Tokyo, New York City, London and Paris clearly lead the ranking of the most active cities if compared by the total number of tweets. The investigation shows that Twitter is very frequently used to share content from other services like Instagram or YouTube. The most popular topics in tweets reveal great differences between the cities. In conclusion, the investigation shows that social media services like Twitter also can be a mirror of the society they are used in and bring to light information flows of connected cities in a global network. The presented method can be applied in further research to analyze information flows regarding specific topics and/or geographical locations.

Keywords: social media; microblogging; Twitter; informational city; big data; network analysis

1. Introduction

Social media has become a major issue in today's society. It changed everyday life for many people as well as the internet itself. Online platforms like Facebook or Twitter have become an important aspect in information science because new opportunities of producing and sharing information arise. Social media is not only a regional phenomenon but influences the way of communication in the world and society. According to Castells [1] we refer to the "network society" which is predominantly represented in so called "Informational Cities" [2–4].

"The new spatial logic, characteristic of the Informational City, is determined by the preeminence of the space of flows over the space of places. By space of flow, I refer to the system of exchanges of information, capital, and power that structures the basic processes of societies, economies and states between different localities, regardless of localization" ((p. 136), [5]).

Furthermore, flows of information often play a role when it comes to world city analyses [6]. World cities overcome distances by their connections. Those connections can be physical, like airports which offer flights from and to all mayor destinations in the world, or digital, like emails or telephone correspondence. Referring to Stock [3], today we may find "Informational Cities" which are global player in a world city hierarchy. Those cities are hubs in a global network [2] regarding their space of flows. According to Mainka *et al.* [7], we will call those cities "Informational World Cities". Their definition of "Informational World Cities" and the 31 cities they identified form the foundation of this paper.

The question arises of how to measure flows of information. On the city level, we could measure the communication that is occurring between actors located in different cities. The flow of information through electronic communication between employees in firms is not an available resource for researchers [8], but, through the mass communication on social media channels, we are able to analyze connections between users located in cities or spreading information about cities. We will not separate the users in hierarchical levels, e.g., celebrities and ordinary users, but deem the connections between them as vertical. To analyze vertical connections, we will use a scientrometric approach. Here, the flow of information is measured e.g., by scientific publications and their reputation or co-authorship [9]. Those flows can also be measured on country or city level. For example, Haustein *et al.* [10] studied the citation and collaboration rates between researchers in the Asia-Pacific region on the country level. Thus, it is possible to see from where the information is coming, e.g., an information flow from China to Indonesia is given when a researcher from Indonesia cites a research paper which was written by a Chinese researcher. Furthermore, the closeness of both researchers is given if both are cited by a third researcher. We will adapt the idea of this study in regard to an informetric investigation of microblogging connectivity of 31 informational world cities on Twitter. Regarding this, an information flow on Twitter is given on a metadata level, (1) when a user located in city A replies to a tweet from a user located in city B, or (2) the content of a tweet is referring to more than one city. The use of metric methods from scientometry in other environments, like the web, is referred to as webometrics or altmetrics [11]. The broader term that combines all these measuring methods is still informetric [12]. Thus, we measure whether information flows through cities exist and if cities are connected in the Twittersphere. The Twittersphere includes all postings (*i.e.*, tweets) made on Twitter and therefore just refers to users of this microblog. Hence, these are the limitations of this analysis. Twitter is not equally spread in all cities in the world. Furthermore, Twitter is blocked in China [13] and therefore just visitors or citizens who know how to bypass this barrier are included in this analysis.

Launched in 2006 with a total number of about a billion registered users today and 241 million monthly active users that created 300 billion tweets until October 3rd 2013, sending 500 million new tweets every day [14], Twitter is among the biggest social networks and microblogging services available. Twitter is an example of a microblogging platform that is based on the production of real-time updates [15]. Every post is restricted to 140 characters. Users can be addressed by user mentions (e.g., @ladygaga) and tweets can be indexed into categories by using hashtags (e.g., #tweetsandthecity). In addition, publicly-traded companies are highlighted by using the dollar sign right before their stock ticker symbol (e.g., $SI for Siemens). Hence, we are dealing with rapidly changing huge data sets that cannot be analyzed with classical data processing methods, so-called Big Data. Thus, Twitter is a great target of scientific investigations and mining approaches regarding many different topics like the usage by sport clubs [16,17], political sentiment analyses [18], spreading and recommendation of news [19,20] natural disasters and events [21–23] to only mention a few. Weller *et al.* [24] provide a great collection of recent important work in their publication "Twitter and Society" by bringing the work of the leading international Twitter researchers together in one publication. It shows the importance and diversity of Twitter research and reinforces our engagement to investigate Informational World Cities on Twitter.

Investigations that examine cities or locations on Twitter in general have already been done for different aims. Heverin and Zach [25] investigated the Twitter activities in 60 U.S. cities with populations greater than 300,000 of which 30 police departments have active Twitter accounts. They observed that most of the tweets posted by these accounts contain crime or incident information. Topic detection could also be used to identify the geo-location of social media users. Han and Cook [26] (p. 452) assume that some topics like "Piccadilly" and "tube" are more often used in tweets by persons located in London than in another city.

Weidemann [27] showed that only 6% of all Twitter users share their geo-location. An investigation on general Twitter activities in London, Paris and New York City has already been done by Adnan & Longley [28]. They compared names, probable ethnicities and genders of Twitter users in these cities. According to their findings, the majority of twitter users are male and English speaking, In addition, users in certain areas are more active than others. For example, Queens shows a rather low Twitter activity in comparison to Manhattan.

Mossberger, Wu and Crawford [29] created an "Interactivity Score" for the 75 largest U.S. cities that determines the interactivity between governments and citizens in social networks, such as Twitter. They reach the conclusion that the use of Twitter by governments in these cities increased from 25% (2009) to 87% (2011). In their ranking, New York City, Los Angeles and Chicago occupy places in the top 10 of the most interactive cities. San Francisco (#18) and Boston (#26) are lower ranked. Mainka *et al.* [30] did a similar study for the 31 informational world cities. They come to the conclusion that Twitter is the most actively used social media platform for goverment and library use. This confirms our suspicion that Twitter activities of these cities are worth being investigated. Thus, our investigation of 31 cities in the Twittersphere will analyze the following research questions:

(1) How many Twitter users do the analyzed cities have?
(2) Which websites are predominant in information flows of the analyzed cities in the Twittersphere?
(3) Which content is predominant in information flows of the analyzed cities in the Twittersphere?
(4) Is it possible to measure the connectedness between cities *via* Twitter content analyses?

These four questions build on each other. Firstly, we want to know how active users in the 31 cities are. To be more precise, we want to know how many tweets are produced in each of the cities to obtain an indication of how relevant the more profound results are. Answering questions 2 and 3 gives as a good overview of what content is produced in each city. With these results, we can finally try to find content-based explanations for the connectedness between the cities. As a result, the first three questions contribute to the analysis of the results of question 4. The flow of information will be measured by geo located metadata and content analysis. Finally, the findings of this investigation show a short term investigation and represent a small excerpt of the Twittersphere. To our knowledge, it is one of the first attempts to measure information flows between cities using the microblog Twitter with an informetric approach.

2. Collecting City-Related Content on Twitter

Since Twitter provides diverse parameters that can be used to find tweets, it was possible to define two criteria that a tweet has to meet to be related to a city [31].

(1) The city name is mentioned in the tweet as a hashtag or as a regular term.
(2) The tweet is tagged with a geo-location that is located within the city territory.

For this purpose, Twitter offers two different APIs (application programming interfaces): the REST API and the Streaming API. The former one results from multiple resources like users, timelines, friends and followers. To find tweets matching user queries, we use the Search resource—Search API—respectively. Tweets tagged by a specified geo-location, without taking the content into account, can be detected by using the Streaming API. For searching tweets, we use the city names of the 31 informational world cities in diverse spellings. In order to retrieve a corpus as exhaustive as possible, all cities are searched in the respective national language and in the nine most spoken languages using the Latin alphabet: Spanish, English, Portuguese, German, French, Italian, Polish, Romanian and Dutch [32]. To keep the error rate as low as possible, the queries that contain non-Latin characters are split into two or more queries. For example the queries for Beijing are split as follows:

(1) Beijing OR #Beijing OR Běijīng OR #Běijīng OR Pechino OR #Pechino OR Pekin OR #Pekin OR Peking OR #Peking OR Pekín OR #Pekín OR Pequim OR #Pequim OR Pékin OR #Pékin
(2) 北京 OR #北京

For those queries that do not contain any non-Latin characters, it is sufficient to just use only one single query.

The restriction to Latin alphabet-based languages has been chosen to read and translate languages more easily. Considering that in countries without Latin alphabet-based languages, many tweets are in

English anyway—according to Semiocast [33] about 40% of all tweets are written in English—the amount of tweets that are not collected even if they contain one of the cities names should be within acceptable limits. In addition, some popular abbreviations like #la for Los Angeles and #nyc for New York City are included in queries, if a well-known abbreviation exists. We decided to leave out terms that are strongly related to cities like "Times Square" (New York) or "Piccadilly" (London) to receive a first good in impression of how cities are represented in tweets and if the city names are a sufficient indicator for measuring the frequency of city-related tweets.

When using Twitter's Streaming API to find tweets tagged within a specific area, a bounding box has to be specified. A bounding box is characterized by its southwestern and northeastern corners which in turn are defined as longitude, latitude pairs. As a consequence, a rectangle is being formed that defines the area to be monitored. The sizes of the rectangles are based on the official sizes of the city areas if an official declaration is provided. Otherwise, we used the Google Maps definition of the city area that can be achieved by using the Google Maps API v3 (The website displays bounding boxes using Google Maps API v3 like at: http://www.mapdevelopcrs.com/gcocode_bounding_box.php).

2.1. The Database and the Collecting Process

The tweets were collected between 2nd December 2013 and 16th December 2013. As a result, our dataset contains two weeks of Twitter data which seems to be a short time frame. Our purpose is to get indications of how to analyze these datasets to create the optimal method for determining characteristics of information flow and connectedness between informational world cities. In addition, since a tweet only "lives" for about one hour, two weeks can give a good impression about the flows happening on Twitter. In order to handle the different conditions of Twitter's REST and Streaming API two different bots (automatic programs to gather data) have been created using the scripting language PHP (version 5.4.4): the Search Bot and the Stream Bot. Both bots are working with a queue of queries, which is stored in a MySQL (version 5.6) database. The difference between the Stream Bot and the Search Bot is that the Search Bot periodically fetches the next query from the database, sends the HTTP request to the REST API, and updates the query to be placed at the end of the query queue. The Stream Bot instead fetches the queries only once when it is initialized and keeps a persistent HTTP connection to the Streaming API open until the bot is stopped by the user. After either of the two bots retrieves new tweet data from one of the two APIs, the JSON (JavaScript Object Notation)-response is stored into the cache table. A permanently running process checks this table for newly inserted data and delegates it to the scripts that parse and normalize the data to the format defined by the database model and finally import the data into it. This approach can be considered as a modified version of the simple model used in the architecture for Twitter data collection described by Oussalah, Bhat, Challis and Schnier [34], which only uses Twitter's Streaming API to collect tweets. The full process is visualized in Figure 1.

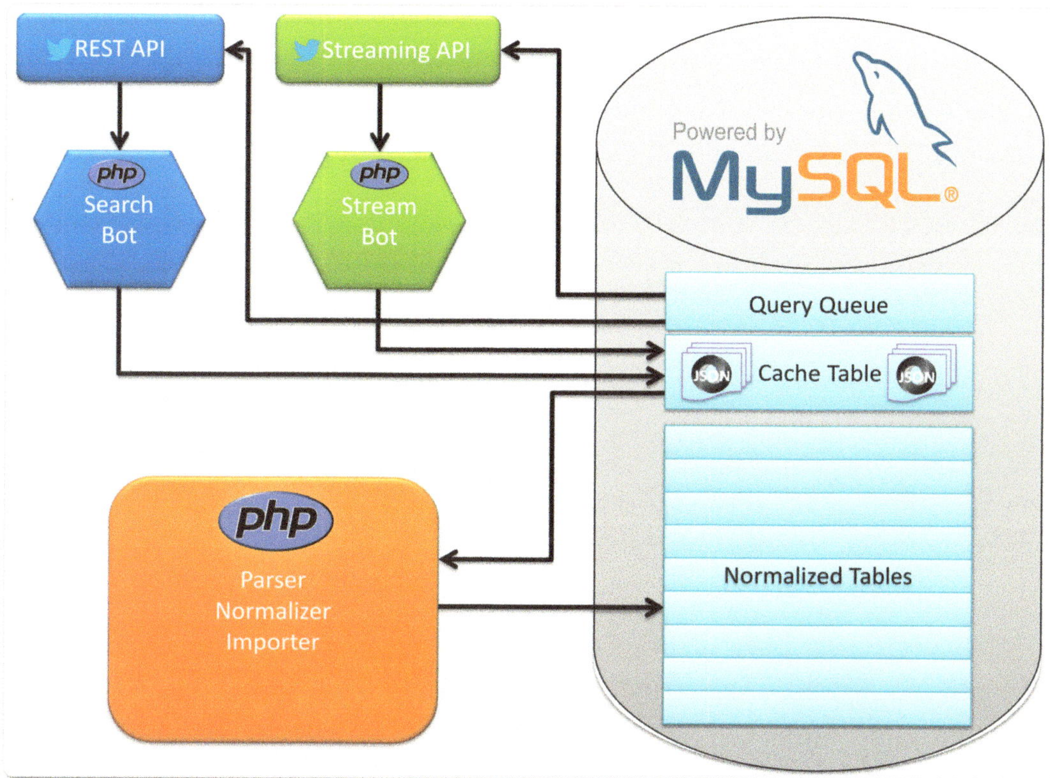

Figure 1. The collecting process.

2.2. The Dataset

The database contains 40 million tweets that were found using the different queries for the two APIs mentioned before and therefore has duplicates. For example, when using Twitter's REST API, a tweet is found *via* the search term "New York" if it contains the string "New York City". If this tweet with the unique id provided by Twitter does not exist in the database, a new record is stored and the query that matches will be connected with it. If the tweet is found again because it additionally contains #paris, only the connection between the query for Paris and the tweet will be stored in the database and the record, containing the tweet data, will not be stored twice. In addition, it is possible that the previous tweet has geo-information assigned and its coordinates are located within the rectangle of Berlin. Thus, this tweet is found a third time but this time *via* the Streaming API and a new connection between tweet and query will be stored. Hence, there is a difference between counting how often a tweet was found using any of the APIs and counting the unique tweets stored in the database. Figure 2 shows the number of tweets the corpus contains segmented into two categories with three subcategories each. The subcategory REST API represents all tweets that are only found using Twitter's REST API. If our example tweet was found using the term "New York" and a second time using the term "Paris," it would fall into this category. As a result, it increases the number of unique tweets found only *via* the REST API by one and the number of all tweets containing duplicates by two because it has been found using the queries for New York City and Paris. However, since our example tweet has been found twice using the REST API and once using the Streaming API, we have only one unique tweet but three hits for the category of all tweets containing duplicates that have been found in both APIs. As Figure 2 shows, there is a difference of 1,019,548 tweets between unique and duplicate tweets. The values of the unique and duplicate tweets found by the Streaming API are identical, because a tweet cannot be tagged with two different geo-locations. Comparing the two query methods used to create the corpus, it is obvious that more tweets were found using search terms than using geo-locations, which is due to users not specifying their locations. About 68% of all (not unique) tweets were found using the REST API, 31% using the Streaming API and 1% using both.

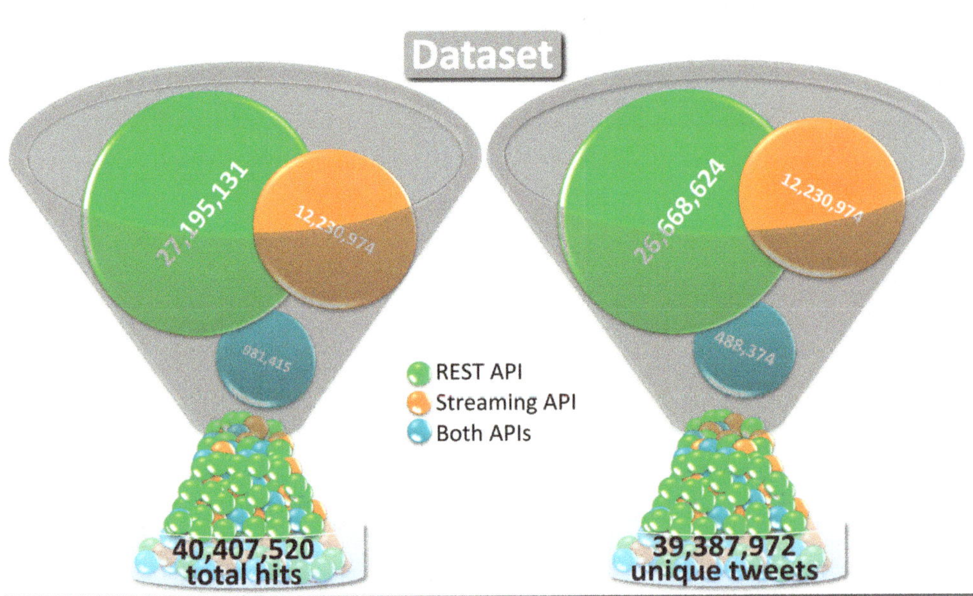

Figure 2. The dataset.

2.3. Topic Detection

The largest part of the analysis has been realized using MySQL's features like full-text search or aggregate functions. In addition, automated PHP (server-side scripting language) scripts that have been written for our purpose did the work when it comes to MySQL's limits. For content analyses, MySQL's text analyzing methods are not sufficient. Thus, the tweet data additionally has been indexed into an Apache Lucene Index that was designed to handle Big Data and provides a full-text search engine with more suitable features. Each tweet has been indexed as one single document with the following data and metadata:

- Tweet ID,
- Text,
- Date of creation,
- Related cities (Geo or Query),
- ID of the including conversation if existing,
- Is Retweet,
- Is Spam,
- Favorite Count,
- Retweet Count,
- Language provided by Twitter.

Therefore, it is possible to query the corpus in all imaginable combinations of these attributes. It is now possible to formulate queries like "get all tweets that are related to Munich and have more than 1000 favorites" as an example.

For each city, we have now created two lists of the top 100 used hashtags in our corpus. We used the simple tweet count as the value for the weighting and totalized the numbers to get an ordered list of the most assigned topics to hashtags and therefore to the tweets in the dataset. The topics have been manually defined through content analysis and all hashtags added using a specified coding scheme. Again, we used two hashtag lists for determining the topics represented by hashtags. One list is for the geo-tagged tweets and the other for the tweets found through search terms. Unlike the term lists, the lists of hashtags each consist of the 25 most used hashtags. As an example, the hashtag #music and the hashtag #song are assigned to the topic "Photos, Movies and Music". The firstly mentioned hashtag has been found in 500 tweets and the secondly mentioned hashtag in 350 tweets. As a result, we get a total tweet count of 850 tweets for the topic "Photos, Movies and Music".

2.4. Spam Detection

We used a pared-down form of spam detection algorithm with some manual aspects to abolish most of the spam tweets. Firstly, we detected the top 100 users in the whole dataset classified by tweet count. Secondly, for all these users, a request to the users/show resource of Twitter's REST API has been done. If the API response contains the information, that the account has been suspended by Twitter, all tweets from this account have been marked as spam. Thirdly, we manually checked all remaining accounts whether they produce tweets with relevant content. Accounts that only produce retweets for a certain hashtag or topic like the "Toronto Retweeter" (@toronto_rt) or the "Vancouver Retweeter" (@vancouver_rt) are classified as spammers because they do not produce relevant content for this analysis. Most of these accounts contain the info "I am a bot" in their account description. Finally, some of the previously identified spam accounts produce tweets that can be identified as spam by their content. For a user, it is obvious that the tweet is a spam tweet just by reading the tweet text. For an example tweet of this kind, see Figure 3.

To identify other similar tweets and mark them as spam we queried the Lucene index (http://lucene.apache.org/) for all spam tweets and all unique tokens the result set of the documents contains. All terms that are not potential spam words have been filtered out manually with a resulting list of 580 potential spam words. To find additional spam tweets, we queried the database for tweets that contain at least two of these spam words because a tweet that contains the term mature is not necessarily a spam tweet. A tweet that contains mature and porn is most likely to be a spam tweet. To keep the false positive rate as low as possible, we checked all found tweets manually and removed the ones that cannot be marked as spam. The selection of tweets took place by keeping the following questions in mind:

- Does the tweet come from a real user?
- Does the tweet not only contain hashtags?
- Aren't there any other tweets with the same content (*i.e.*, created by retweet bots, *etc.*) in the dataset?

If one of the questions about a tweet can be answered with "no", in most cases the tweet can be marked as a spam tweet. These spam tweets are filtered out in all investigations where they may distort the results.

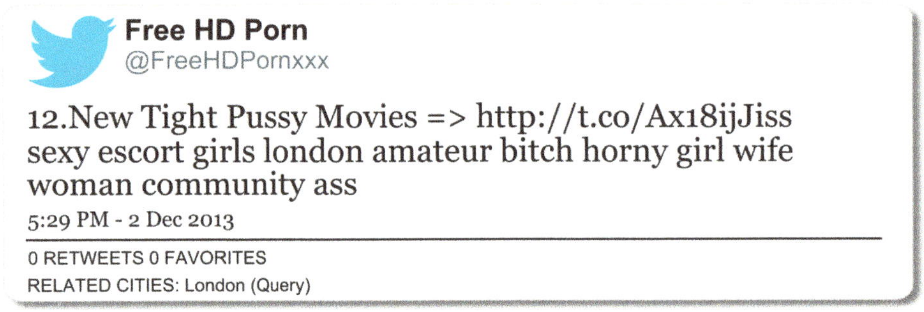

Figure 3. Example of a spam tweet.

3. Results and Discussion

Table 1 shows the number of tweets found for each city, either through a query of terms and Twitter's REST API or through a provided geo-location and Twitter's Streaming API ranked according to the total amount of Tweets. It is obvious that we have a clear leader: Tokyo. With almost 4.5 million tweets found through search terms, Tokyo absconds from the other cities with a notable gap from second place, which is held by New York City with about three million tweets. It is striking that from Paris to Barcelona another significant gap occurs and from that point on, the number of tweets shrinks constantly. To summarize, the four cities amount to 45% of tweets found using search terms for 31 cities. This phenomenon could be explained by the fact that these four cities show a high rate of urban tourism [31]. New York City (over 10 million visitors), London (over 15 million visitors), and Paris (over 8 million visitors) belong to the cities in our set which had the most international visitors in 2011 [35]. Tokyo had only about 2.7 million international visitors, but it has had always a strong domestic tourism (430 million domestic visitors in 2008).

Taking the numbers into account that represent the tweet counts per city and have been found through geo-locations, we can identify the two frontrunners Paris and São Paulo with an almost equal number of tweets (about 1.8 million). New York City and London follow as the next pair (both about 1.3 million tweets) after a gap of about 470 thousand tweets less than São Paulo and then the third pair of cities, Los Angeles and Kuala Lumpur, with about 1.4–1.7 million tweets, represent the last group that is prominent before the number of tweets shrinks constantly. As a relatively logical result, we can locate the three Chinese cities—Beijing, Shanghai and Shenzhen—which rank 29–31, if we take a look at tweets that have their origin in either of these cities. This is a result of the restricted availability of Twitter in China and the fact that there exists an extremely widespread Twitter-clone, called Weibo [36].

In our previous quantitative analysis [31], we already quoted some influencing factors that can explain the immense differences between the numbers we calculated from our dataset. For example, we observe a large number of tweets found through geo-locations for São Paulo and Kuala Lumpur, but a rather small amount of tweets that mention the name of the two cities. The German cities Berlin, Frankfurt and Munich show a contrary picture. For these cities, we found rather few geo-tagged tweets. We assume that the high level of awareness of data privacy protection in Germany [37] can be a factor for the small amount of geo-tagged tweets. Thus, German users do not share their location the same way users in Brazil or Malaysia do. We also found out that the penetration rate of smartphones, which obviously influences the amount of geo-tagged tweets due to the built-in GPS modules, seems to have an impact on the number. Rowinski [38] counts 70 million smartphone users in Brazil and only 32 million in Germany. France, however, also has only 33 million smartphone users, which leads to the assumption that this factor cannot be generalized.

Table 1. Tweets per city ranked by the total amount of collected tweets between 2nd December 2013 and 16th December 2013.

No.	City	Query	Location
1	Tokyo	4,465,123	865,492
2	New York City	3,055,425	1,323,259
3	London	2,724,022	1,316,485
4	Paris	1,971,704	1,814,664
5	Barcelona	1,451,510	114,892
6	Milan	1,378,095	92,402
7	Chicago	1,371,202	632,030
8	Dubai	1,210,888	186,166
9	Boston	964,364	253,953
10	Los Angeles	857,534	1,178,099
11	Sydney	808,115	126,103
12	Singapore	761,102	315,672
13	Toronto	724,243	237,821
14	São Paulo	522,466	1,793,413
15	Melbourne	482,487	59,677
16	Amsterdam	479,377	69,303
17	Munich	467,448	20,288
18	Hong Kong	400,388	61,769
19	San Francisco	399,487	520,670
20	Berlin	390,895	55,305
21	Seoul	366,721	233,164
22	Beijing	286,691	11,503
23	Vancouver	278,951	40,380
24	Montréal	246,757	89,092
25	Shanghai	225,418	11,127
26	Vienna	217,948	25,945
27	Kuala Lumpur	152,464	1,140,118
28	Stockholm	92,843	90,359
29	Frankfurt am Main	80,964	12,996
30	Helsinki	32,579	19,301
31	Shenzhen	14,768	1155

3.1. Twitter as a Sharing Service for Other Social Media Services

The extraction of the top links shared in tweets related to informational world cities have bring into light that other social media services lead the ranking in almost every city. Figure 4 shows the five most common services that are shared. The values are inverted. Rank one gain 100 points, rank two 90 points, rank three 80 points and so on. If a website does not make it to the top 10 of the most shared links in tweets related to a specific city, it ends up with zero points.

The biggest winner of this calculation is Instagram. Almost all rankings of the most shared websites result in a score of 70–100 points for Instagram. Merely Munich is an exception here. Instagram comes to a score of 40 points in tweets related to Munich only. A similar phenomenon can be observed when looking at the values for the second ranked website Foursquare that does not even make it to the top 10 of Munich-related. However, Munich has this in common with Barcelona. People in these cities obviously are not very interested in sharing their current location *via* Foursquare or they just did not connect their Foursquare account to their Twitter account. Users from Tokyo, Frankfurt, Kuala Lumpur and São Paulo seem to use this feature much more frequently, though, as Foursquare is the top scorer for these cities. YouTube and Facebook show almost a similar importance when looking at tweets linking to other websites. YouTube comes to a total score of 2040 and Facebook to a score of 1980. Facebook scores a hundred points in London and YouTube in Seoul scores hundred points, as well. In all the other cities, the two services can be located in the midfield. With a score of 400, Tumblr also belongs to the top five of social media services shared *via* Twitter. All other services, like Pinterest, Flickr or LinkedIn, are just of marginal interest in this investigation and therefore not included in our visualization. The Chinese microblogging service Weibo, for example, only matters in the Chinese cities but does not make it to the top ten in any of the cities, though.

3.2. The Main Topics in Tweets

As we are not only interested in quantitative results, we also want to follow a more qualitative approach by determining the main topics in tweets. Which content is predominant in information flows between cities in the Twittersphere? The analyzed data shows that mentioning other cities or geographical locations is very popular in tweets that are related to informational world cities. Since we previously identified Foursquare as one of the most shared websites in our dataset, and sharing locations is the business model of this service, this is not surprising.

Twitter is also frequently used for job offers, either by job agencies or the companies themselves. As a result, the topic category "Economy, Industry, Finance & Jobs" is even more present then "Events & Dates" which we also expected to be one of the top topics due to the event-driven nature a microblogging service like Twitter is. Because of cities like Barcelona or Munich with their famous soccer teams, but also cities like Chicago or Los Angeles with their famous American Football, Basketball or Baseball teams, "Sports" can also be found among the most popular topics. This also impacts the number of conversations between users from different informational world cities, which we will show in the next chapter.

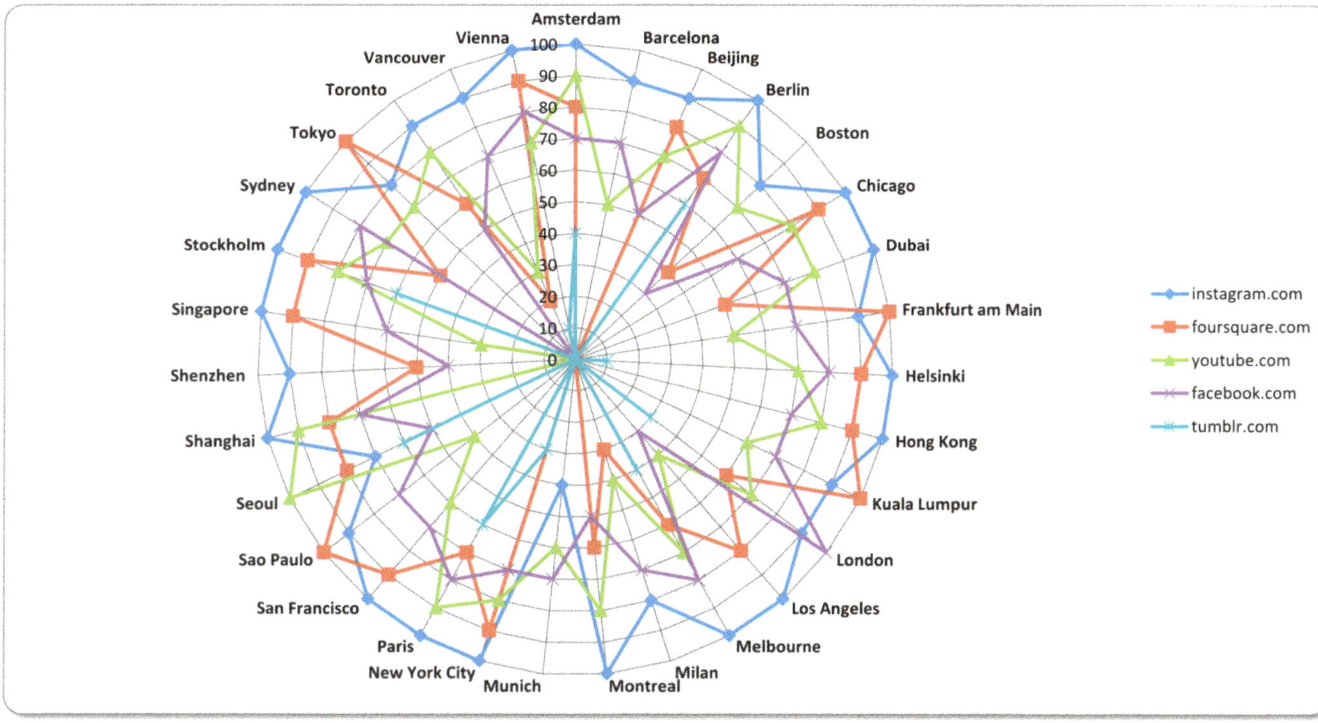

Figure 4. The top five shared services on Twitter.

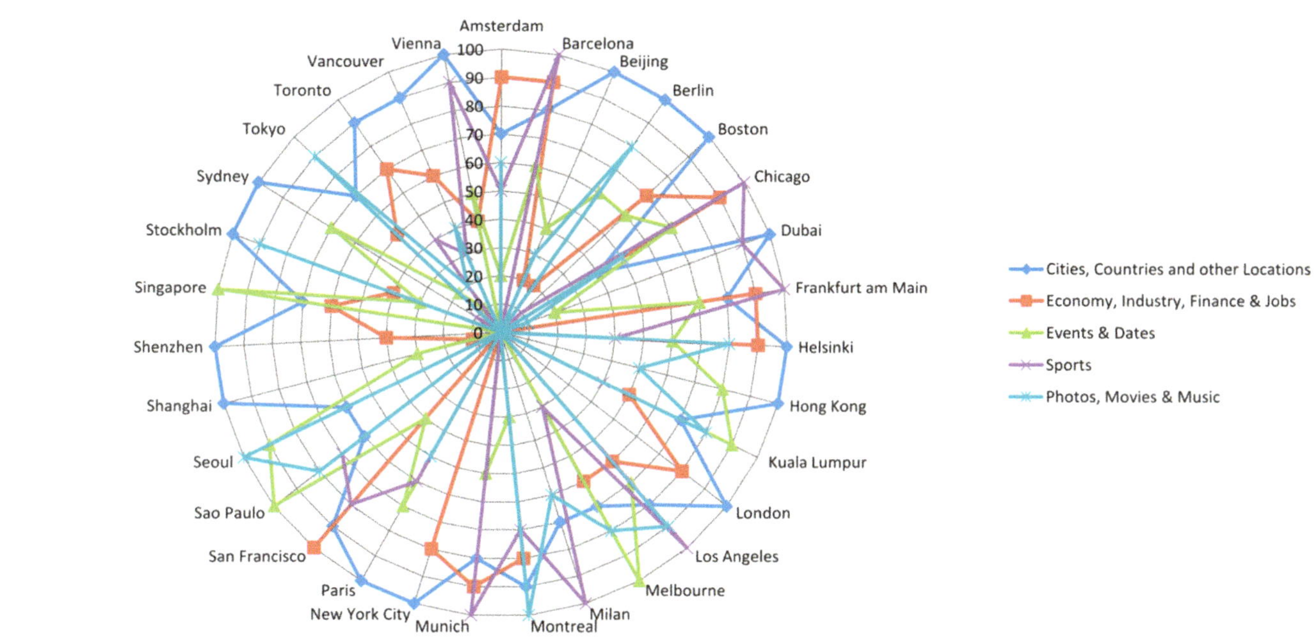

Figure 5. The most important topics in tweets I.

Due to the possibility to share multimedia content or links to multimedia content, there is also a remarkably huge number of tweets in our dataset that we could assign to the category "Photos, Movies and Music". In addition, many radio stations of the cities have their own account and share tweets about their currently playing music (#nowplaying).

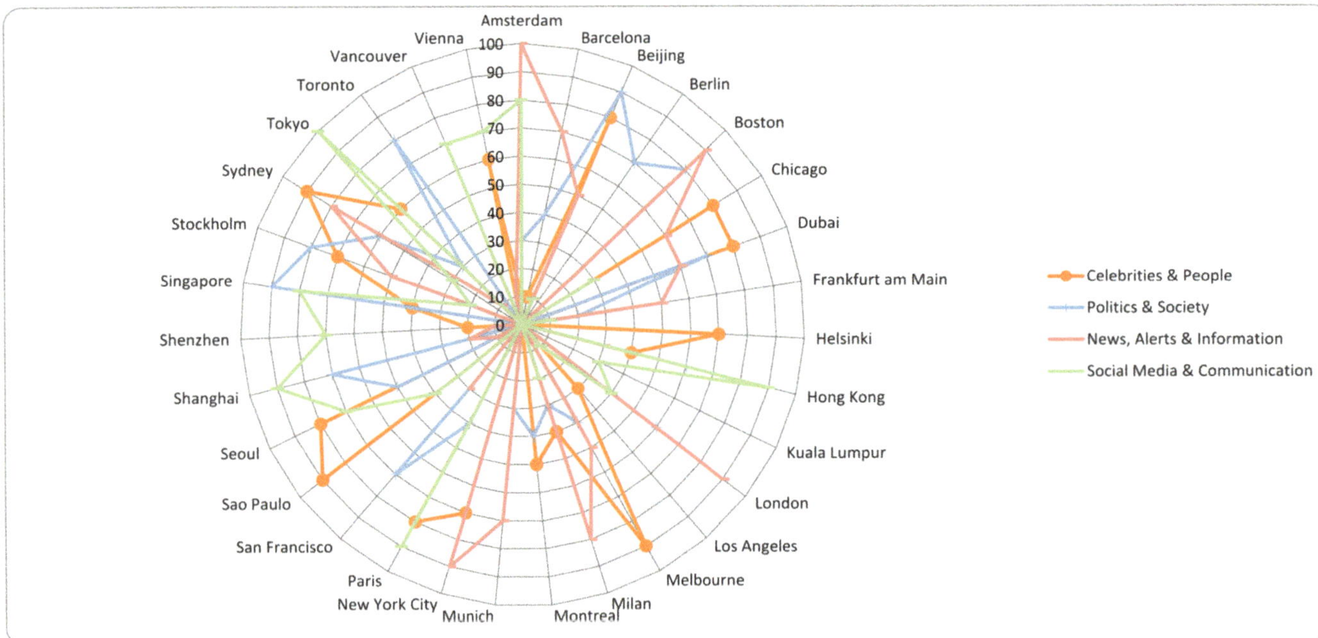

Figure 6. The most important topics in tweets II.

In addition to the already mentioned topics, Twitter is also used to talk about celebrities, share news from online news services like the websites of newspapers or to talk about politics. During our collecting period, a political affair in Toronto regarding Mayor Rob Ford occurred. A video has been revealed that shows him smoking crack [39]. As a result, "Politics and Society" was one of the most important topic categories during that time.

The last category to mention we call "Social Media and Communcation". It refers to all tweets that contain content that is commonly used within Twitter to communicate or participate in certain events. An example would be the hashtag #ff or #followfriday or also the hashtag #tbt ("Throwback Thursday"). The nine most important topics are shown in Figures 5 and 6 ranked by our inverted calculation on a scale from 0 to 100.

3.3. The Closeness between Informational World Cities

An interesting aspect of informational world city research is the determination of the closeness of the cities to each other. The closeness can be analyzed through connections in a network of hubs. The data provided by Twitter has the potential of extracting conversations by a conversation id within the metadata. Thus, we extracted all conversations in our dataset and analyzed the cities that either are mentioned or tagged with geo codes. The cities form the nodes in this network. As a result, we built a clustered graph containing all cities connected through the number of conversations. For example, 1438 conversations can be assigned to Chicago and New York at the same time so this number shows us the degree of "connectedness" between the two cities. By using the OpenOrd algorithm in Gephi, an open tool for the visualization of network graphs (gephi.github.io), we now can calculate a graph that clusters all our cities based on the number of conversations they have with each other (Figure 7). The type size indicates the importance of a node that represents a city. The more conversations a city can be assigned to in total, the more important the node in the graph gets. Hence, the investigated network is just a snapshot which represents a given timespan. The relations we see here, may be influenced through special events, e.g., the formula one takes place in a city, which gains worldwide attention and also would affect conversations on Twitter and accordingly result in a bigger node of this city.

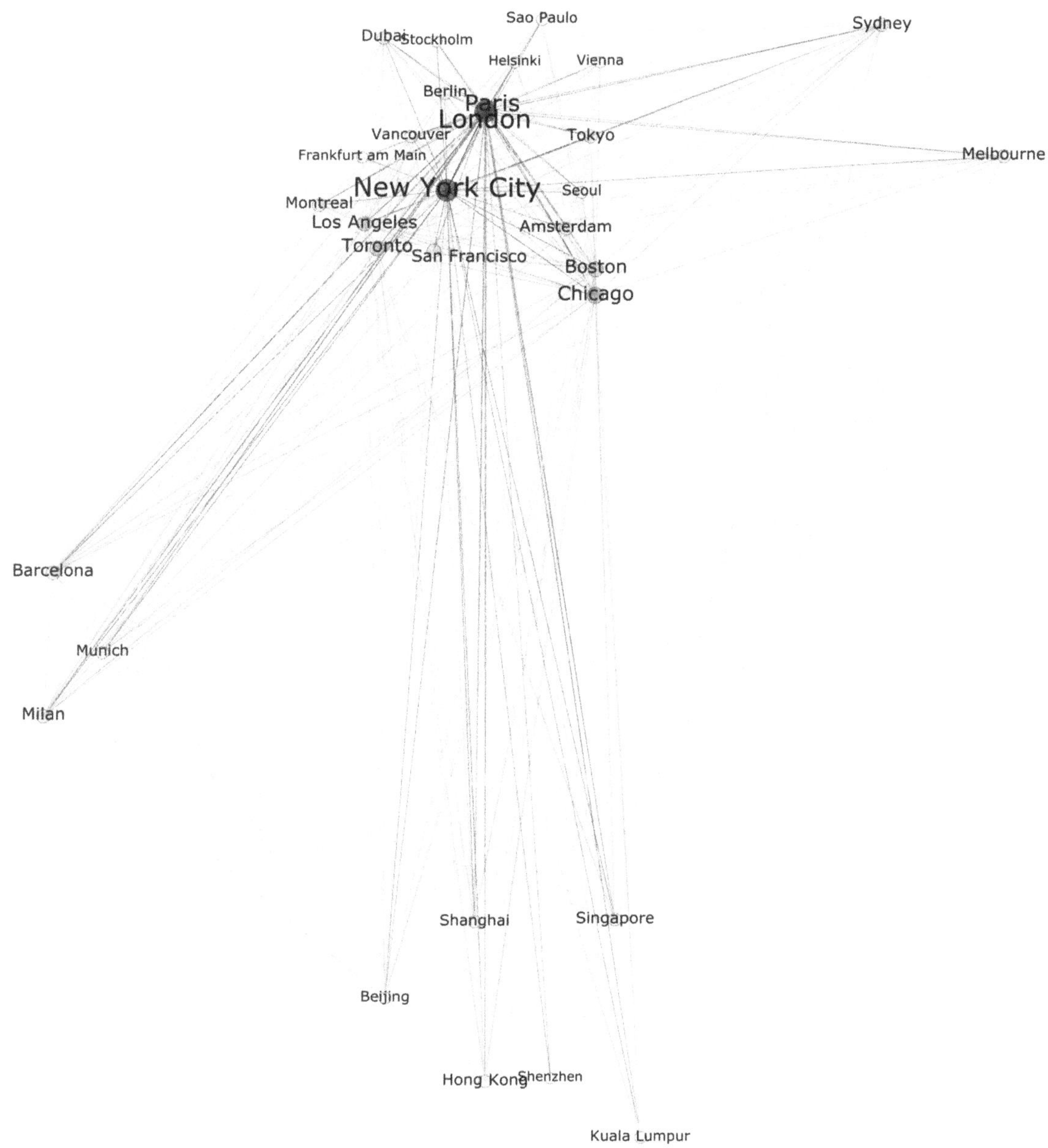

Figure 7. The closeness between cities according user conversations.

The most obvious phenomenon in the graph is the distribution of four clusters of different sizes. We find an Australian cluster consisting of Melbourne and Sydney and a South East Asian cluster consisting of Beijing, Shanghai, Singapore, Hong Kong, Shenzhen and Kuala Lumpur. Besides these two clusters that mainly include cities that in the real world are geographically close as well, two other clusters popped up that do not fulfill this characteristic. The cluster that contains Barcelona, Munich, and Milan, we call the "soccer cluster". As the top topics in tweets related to each of these cities in our dataset are clearly dominated by sports or more precisely soccer, it therefore can be assumed that the content of the tweets has also an influence of where the users come from that communicate through direct conversations *via* Twitter or which cities they talk about. By analyzing the conversations assigned to the three cities we found out that most of them are dealing with the top soccer teams which are originated in these cities: FC Barcelona, AC Milan and FC Bayern Munich. One could think, that this occurs because the sport club names include the city name but other clusters are not dominant, e.g., the University of Amsterdam does not build a cluster with other universities. As a result, common interests of people who live in informational world cities seem to lead to more inter-city communication. Figure 8 shows an example conversation related to all three cities.

The last cluster shown in the graph is also the biggest. It contains all Northern and Southern American cities that are part of our investigation and except for Barcelona, Milan and Munich all European cities as well. With Tokyo, we can also find the only city from South East Asia that is close to the other cities in this cluster but not close to the other cities from its own region. As a result, Tokyo seems to be more strongly connected to other informational world cities than other cities from South East Asia.

The biggest cluster also contains the biggest nodes represented by Paris, London and New York. As shown in the figure, the nodes of Paris and London almost lie on top of each other. This phenomenon results from the fact that, with a value of 1833, these cities showed the highest number of common conversations in our dataset. New York City, however, shows the highest amount of conversations, with other cities in total coming to a number of 9230 conversations that are connected to at least one of the other thirty cities as well. By way of comparison, London comes to a number of 8300 conversations and Paris comes to 6335 conversations. Without any surprise, Shenzhen represents the smallest node in the graph (79 conversations in total).

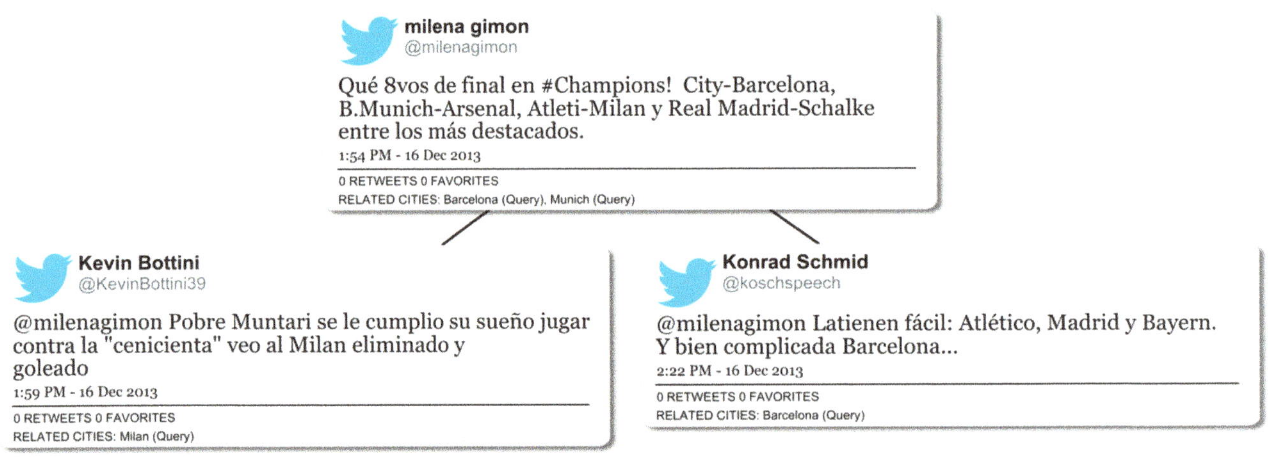

Figure 8. Example conversation related to Milan, Barcelona and Munich.

4. Informational World Cities in the Twittersphere

Our results strongly contribute to current informational world city research. If we want to know how "informational" a city is, we have to analyze social media behavior of users that produce city-related content. We have seen some hints of how connected some of the cities are to each other by analyzing conversations. There is a connection based on content (e.g., soccer) and a connection based on the geographical location. This gives us a hint that cities which are "close" to each other on Twitter are also close to each other due to other factors. As a result, we could assume that social media services can be a mirror of the society they are used in. Neighbors talk with and about neighbors and soccer fans talk with other soccer fans, no matter of where they are located.

The most important topics in city-related tweets can give us a hint of the quality of the content produced at social media platforms in and about informational world cities. Long-term investigations can show if Twitter is mainly used to (automatically) share locations or if it is also used to discuss topics of higher quality. Similar to the topics, we can also use links to make assumptions about content-quality. The domination of other social media services in the top list of shared websites suggests that there's also a huge amount of content that is only a reference to other content and services.

Our analysis underlays a few limitations. The first limitation to mention is the Twitter APIs and their technical restrictions. For our investigation, we had to rely on the data Twitter provides users with no access to the Twitter Firehose. Studies showed that users can receive from 1% up to 40% of public tweets from the Streaming API [40]. Furthermore, server or database failures may lead to temporary short-term data loss, although we did not track any of these issues during our collecting period.

Due to the strong multilingualism in our dataset represented by 59 languages, we received from Twitter's language recognition algorithm, including uncommon languages like Cherokee, we were confronted with several challenges. There exist some writing systems (e.g., Chinese) which do not necessarily separate words with whitespaces and thereby have to be analyzed using dictionaries which we did not have access to. Since we were not able to translate all languages in the corpus, we had to rely on automatic translation tools (*i.e.*, Google Translate, Bing Translator) or several online dictionaries (*i.e.*, dict.cc, leo.org, Wiktionary) to identify topics in tweets. Thus, topic-detection in this investigation is best-effort and may underlie a few inaccuracies like most of the automatic or semi-automatic investigations on corpora which are based on natural language.

Our results show that there are still problems concerning homonymy, although we did filter the severest problems (*i.e.*, "Paris Hilton") out. It is difficult to filter tweets that mention a city that is named similar to a city we are looking for and, additionally, are not geo-tagged. If a tweet contains, for example, the text "I'm in Sydney", how do we know whether the user means the Australian city or Sydney, MT, USA? It seems that we have to accept a certain percentage of false positives in the dataset while searching tweets *via* Twitter's REST API. At least, the search for geo-locations provides only tweets from the city we are looking for. However, other homonymy problems can be solved manually. For example, tweets that are found due to a user mention contains the city's name can easily be filtered by manually checking all tweets, though it will take some time in a dataset as big as the one we used for our investigation.

5. Conclusions and Future Work

Our analysis is a great step forward in the research of informational world cities presented in social media. We have been given many indications of how an Informational World City can be classified regarding Twitter activities. Indeed, research on this field does not end here. As we have seen, several limitations underlay this investigation that, if possible, must be overcome in future investigations. For this, it is necessary to create and collect heuristics that eliminate most of the problems with homonymy. In addition, a fully developed spam filter algorithm has to be created or used to avoid the vast majority of spam tweets. The results of the given article can only be confirmed by following investigation with new datasets. For that purpose, another period of time can be used to collect tweets. It can be interesting to analyze a dataset that contains tweets produced in a whole month or even a year and make the data comparable to our two-week-dataset.

Apart from the further research on the aspects, we investigated in the given paper, other aspects are also in our interest. A sentiment analysis of the tweets and the topics of the tweets can provide further conclusions about user behavior in informational world cities. In addition, Twitter provides other key figures like follower counts, retweets or favorite counts and others. These key figures can also be used to classify the quality of the content produced that can be related to either one of the cities.

Altogether we showed, using Twitter as an example, that social media research cannot be neglected when investigating the prototypical cities of the knowledge society, since social media has been established as an important part of it.

Acknowledgements

We would like to thank the anonymous reviewers and the editors of Geo-Information Fostering Innovative Solutions for Smart Cities for very useful comments.

Author Contributions

All authors contributed equally to this work.

Conflicts of Interest

The authors declare no conflict of interest.

References

1. Castells, M. *The Rise of the Network Society: The Information Age, Economy, Society, and Culture*, 2nd ed.; John Wiley & Sons Publisher: Chichester, UK, 2010.
2. Castells, M. *The Informational City: Information Technology, Economic Restructuring, and The Urban-Regional Process*; Basil Blackwell: Oxford, UK, 1989.
3. Stock, W.G. Informational cities: Analysis and construction of cities in the knowledge society. *J. Am. Soc. Inf. Sci. Technol.* **2011**, *62*, 963–986.
4. Yigitcanlar, T. Informational city. In *Encyclopedia of Urban Studies*; Hutchinson, R., Ed.; Sage: New York, NY, USA, 2010; Volume 1, pp. 392–395.
5. Castells, M. Cities, the information society and the global economy. In *The Global Cities Reader*, 2nd ed.; Neil, B., Roger, K., Eds.; Routledge: London, UK/New York, NY, USA, 2006; pp. 135–136.
6. Friedmann, J. Where we stand: A decade of world city research. In *World Cities in a World-System*; Paul, L.K., Peter, J.T., Eds.; Cambridge University Press: Cambridge, UK, 1995; pp. 21–47.
7. Mainka, A.; Hartmann, S.; Orszullok, L.; Peters, I.; Stallmann, A.; Stock, W.G. Public libraries in the knowledge society: Core services of libraries in Informational World Cities. *Libri* **2013**, *63*, 295–319.
8. Taylor, P.J. The interlocking network model. In *International Handbook of Globalization and World Cities*; Derudder, B., Hoyler, M., Taylor, P.J., Witlox, F., Eds.; Edward Elgar Publishing: Cheltenham, UK/Northampton, UK, 2012; pp. 51–63.
9. Haustein, S. *Multidimensional Journal Evaluation: Analyzing Scientific Periodicals beyond the Impact Factor*; Walter de Gruyter: Berlin, Germany, 2012.

10. Haustein, S.; Tunger, D.; Heinrichs, G.; Baelz, G. Reasons for and developments in international scientific collaboration: Does an Asia-Pacific research area exist from a bibliometric point of view? *Scientometrics* **2010**, *86*, 727–746.
11. Prime, J. Altmetrics. In *Beyond Bibliometrics: Harnessing Multidimensional Indicators of Scholarly Impact*; Cronin, B., Sugimoto, C.R., Eds.; MIT Press: London, UK, 2014; pp. 263–287.
12. Tague-Sutcliffe, J. An introduction to informetrics. *Inf. Process. Manag.* **1992**, *28*, 1–4.
13. Zhang, Q.; DiFranzo, D.; Hendler, J.D. Social networking on the World Wide Web. In *Encyclopedia of Social Network Analysis and Mining*; Alhajj, R., Rokne, J., Eds.; Springer: New York, NY, USA, 2014; pp. 1879–1892.
14. By The Numbers: 138 Amazing Twitter Statistics. Available online: http://expandedramblings.com/index.php/march-2013-by-the-numbers-a-few-amazing-twitter-stats/#.UzKoTfnWUqJ (accessed on 12 December 2014).
15. Kietzmann, J.H.; Hermkens, K.; McCarthy, I.P.; Silvestre, B.S. Social media? Get serious! Understanding the functional building blocks of social media. *Bus. Horiz.* **2011**, *54*, 241–251.
16. Weller, K.; Bruns, A. Das Spiel dauert 140 Zeichen-Wie deutsche Fußballvereine Twitter für Marketing und Fan-Kommunikation entdecken. In Proceedings of des 8. Hildesheimer Evaluierungs- und Retrievalworkshop, Hildesheim, Germany, 25–16 April 2013.
17. Hutchins, B. The acceleration of media sport culture: Twitter, telepresence and online messaging. *Inf. Commun. Soc.* **2011**, *14*, 237–257.
18. Tumasjan, A.; Sprenger, T.O.; Sandner, P.G.; Welpe, I.M. Predicting elections with Twitter: What 140 characters reveal about political sentiment. *ICWSM* **2010**, *10*, 178–185.
19. Lerman, K.; Ghosh, R. Information contagion: An empirical study of the spread of news on Digg and Twitter social networks. *ICWSM* **2010**, *10*, 90–97.
20. Phelan, O.; McCarthy, K.; Smyth, B. Using twitter to recommend real-time topical news. In Proceedings of the Third ACM Conference on Recommender Systems, New York, NY, USA, 23–25 October 2009.
21. Sakaki, T.; Okazaki, M.; Matsuo, Y. Earthquake shakes Twitter users: Real-time event detection by social sensors. In Proceedings of the 19th International Conference on World Wide Web, Raleigh, NC, USA, 26–30 April 2010.
22. Acar, A.; Muraki, Y. Twitter for crisis communication: Lessons learned from Japan's tsunami disaster. *Int. J. Web Based Communit.* **2011**, *7*, 392–402.
23. Vieweg, S.; Hughes, A.L.; Starbird, K.; Palen, L. Microblogging during two natural hazards events: What Twitter may contribute to situational awareness. In Proceedings of the SIGCHI Conference on Human Factors in Computing Systems, Paris, France, 27 April–2 May 2013.
24. Weller, K.; Bruns, A.; Burgess, J.; Mahrt, M.; Puschmann, C. *Twitter and Society*; Peter Lang: New York, NY, USA, 2014.
25. Heverin, T.; Zach, L. Twitter for city police department information sharing. In Proceedings of the American Society for Information Science and Technology, Pittsburgh, PA, USA, October 22–27 2010.
26. Han, B.; Cook, P. Text-based Twitter user geolocation prediction. *J. Artif. Intell. Res.* **2014**, *49*, 451–500.
27. Weidemann, C. Social media location intelligence: The next privacy battle—An ArcGIS add-in and analysis of geospatial data collected from Twitter.com. *Int. J. Geoinform.* **2013**, *9*, 21–27.

28. Adnan, M.; Longley, P. *Analysis of Twitter Usage in London, Paris, and New York City*; AGILE: Leuven, Belgium, 2013.
29. Mossberger, K.; Wu, Y.; Crawford, J. Connecting citizens and local governments? Social media and interactivity in major US cities. *Gov. Inf. Q.* **2013**, *30*, 351–358.
30. Mainka, A.; Hartmann, S.; Stock, W.G.; Peters, I. Government and social media: A case study of 31 informational world cities. In Proceedings of the 47th Hawaii International Conference on System Sciences, Waikoloa, HI, USA, 6–9 January 2014.
31. Förster, T.; Lamerz, L.; Mainka, A.; Peters, I. The Tweet and the city: Comparing Twitter activities in informational world cities. In Proceedings of the 2014 Conference: Informationsqualität und Wissensgenerierung, Frankfurt am Main, Germany, 8–9 May 2014.
32. Lewis, M.P. *Ethnologue: Languages of the World*, 16th ed.; SIL International: Dallas, TX, USA, 2009.
33. Arabic Highest Growth on Twitter. Available online: http://semiocast.com/publications/2011_11_24_Arabic_highest_growth_on_Twitter (accessed on 23 August 2013).
34. Oussalah, M.; Bhat, F.; Challis, K.; Schnier, T. A software architecture for Twitter collection, search and geolocation services. *Knowl.-Based Syst.* **2013**, *37*, 105–120.
35. Top 100 Cities Destination Ranking. Available online: http://blog.euromonitor.com/2013/01/top-100-cities-destination-ranking.html (accessed on 23 August 2013).
36. Eichstädt, B.; Wei, C. Im social web der mitte. In *Social Media*; Leinemann, R., Ed.; Springer: Berlin, Germany, 2013; pp. 107–110.
37. Unisys Security Index. Available online: http://www.unisyssecurityindex.com/usi/germany (accessed on 23 August 2013).
38. Think Mobile is Big Now? Here's Proof That It's Just Getting Started. Available online: http://readwrite.com/2013/05/29/huge-potential-only-15-of-global-internet-traffic-is-mobile#awesm=~ofKwdMIxymYk5E (accessed on 29 May 2013).
39. Rob Ford: "Yes, I have smoked crack cocaine". Available online: http://www.thestar.com/opinion/editorials/2013/08/04/toronto_mayor_rob_ford_remains_shadowed_by_crack_video_scandal_editorial.html (accessed on 18 April 2014).
40. Twitter Firehose *vs.* Twitter API: What's the Difference and Why Should You Care. Available online: http://www.brightplanet.com/2013/06/twitter-firehose-vs-twitter-api-whats-the-difference-and-why-should-you-care/ (accessed on 2 March 2013).

10

Exploratory Method for Spatio-Temporal Feature Extraction and Clustering: An Integrated Multi-Scale Framework

Wen Luo [1], Zhao-Yuan Yu [1,2,3,*], Sheng-Jun Xiao [1], A-Xing Zhu [1,2,3] and Lin-Wang Yuan [1,2,3]

[1] Key Lab of Virtual Geographic Environment, Ministry of Education, Nanjing Normal University, Nanjing 210023, China; E-Mails: luow1987@163.com (W.L.); xiaoshengjun@126.com (S.X.); axing@njnu.edu.cn (A.Z.); yuanlinwang@njnu.edu.cn (L.Y.)

[2] State Key Laboratory Cultivation Base of Geographical Environment Evolution (Jiangsu Province), Nanjing 210023, China

[3] Jiangsu Center for Collaborative Innovation in Geographical Information Resource Development and Application, Nanjing 210023, China

* Author to whom correspondence should be addressed; E-Mail: yuzhaoyuan@njnu.edu.cn

Academic Editor: Wolfgang Kainz

Abstract: This paper presents an integrated framework for exploratory multi-scale spatio-temporal feature extraction and clustering of spatio-temporal data. The framework combines the multi-scale spatio-temporal decomposition, feature identification, feature enhancing and clustering in a unified process. The original data are firstly reorganized as multi-signal time series, and then decomposed by the multi-signal wavelet. Exploratory data analysis methods, such as histograms, are used for feature identification and enhancing. The spatio-temporal evolution process of the multi-scale features can then be tracked by the feature clusters based on the data adaptive Fuzzy C-Means Cluster. The approach was tested with the global 0.25° satellite altimeter data over a period of 21 years from 1993 to 2013. The tracking of the multi-scale spatio-temporal evolution characteristics of the 1997–98 strong El Niño were used as validation. The results show that our method can clearly reveal and track the spatio-temporal distribution and evolution of complex geographical phenomena. Our approach is efficient for global scale data analysis, and can be used to explore the multi-scale pattern of spatio-temporal processes.

Keywords: clustering; scale; tracking; oceanography; classification; data mining

1. Introduction

Most geographical phenomena have different characteristics at different scales in spatio-temporal domain. Various influences caused by many different factors simultaneously acting across scales lead the data on the phenomena to contain a mixture of multi-scale signals. The mixture of the multi-scale characteristics affects the data characteristics in both spatial and temporal domains. In the spatial domain, the characteristics of a given phenomenon should be represented with certain spatial range or resolution to support the complex remote sensing analysis including the object extraction, image segmentation, change detection, object tracking, and geostatistical analysis, *etc.* [1,2]. In the temporal domain, the multi-scale characteristics of the data are acting as periodical or quasi-periodical signal at frequency ranging from hours to several decades. The effects of the multi-scale characteristics have attracted much attention, have been discussed in many different disciplines, *etc.* [3–6].

Remote sensing data have been quickly accumulated continuously. More and more long-term remote sensing sequences (e.g., satellite and aerial images, SAR, *etc.* [7]), which cross the time span of more than several decades, are logged and used in various fields. These long-term data observations provide valuable information for detailed revelation of the evolution of the geographical processes [8]. Along with the wide application of multi-temporal remote sensing data in many different fields, researchers have also proposed several different kinds of advanced analysis methods for feature extraction, segmentation, *etc.* [9].

There have already been some techniques proposed to solve these problems. Multidimensional data modeling and databases [10,11] and Array Databases (e.g., rasdaman [12], SciDB [13]) have largely improved the performance of storing and querying of the massive multidimensional spatio-temporal data. MOLAP (Multidimensional On Line Analysis Process), commonly used in data warehouse, is rapidly improved for multidimensional data queries and aggregation [14]. Technologies, such as wavelets and hierarchical clustering of data are introduced to MOLAP to partly support the computation [15,16]. However, these analyses require accessing directly the most detailed data since no prior knowledge for the query exists. The spatio-temporal unified approaches (e.g., STIM [17], Clifford Algebra/Tensor approaches [18,19]) are also proposed for the unified analysis of such huge amount of spatio-temporal data. However, these approaches are still in an early stage and too complex to be used directly in the operational environment.

Despite the large progresses achieved in the developing of multi-temporal remote sensing analysis methods, two critical problems still exist: (1) Most of these methods treat spatial and temporal dimensions separately. Several approaches can do multi-scale analysis in either spatial or temporal domain (e.g., EMD, wavelet, and parallel adaptive kernel smoothing) [20–22]. Few methods can support for the spatio-temporal domain analysis in the spatio-temporal unified way [23,24]. With the separation of spatial and temporal signal in data analysis the spatio-temporal data can only be analyzed either spatially or temporally, which will lead to information missing, pattern inconsistency, and synchronizing difficulties. (2) Rarely methods integrate the multi-scale information during the exploratory data analysis. Although there are several exploratory data analysis methods, which can well reveal certain characteristics (e.g., value distributions, spectrum characteristics, and feature variations), developed, large percent of these methods are only applied to the overall data rather than the features at each scale. It is still complicated to reveal the multi-scale characteristics from different perspectives and perform the exploratory multi-scale spatio-temporal feature extraction.

Feature clustering, one of the most typical and commonly used exploratory analysis methods for multi-temporal remote sensing data, is largely restricted by the two problems mentioned above. Integrating multi-scale information in feature clustering can produce rich information about the spatio-temporal process and increase the feature clustering performance for feature extraction and process investigation. For example, various signals, including ocean currents and tides, high frequency variations (e.g., Madden-Julian Oscillation (MJO)), inter-decade variations (e.g., El Niño-Southern Oscillation (ENSO)), *etc.* can be separated from the global ocean data according to their multi-scale characteristics differences [25,26]. With the multi-scale cluster in the spatio-temporal domain, the spatio-temporal evolution characteristics and processes of the above various signals can be revealed.

A fine performance multi-scale spatio-temporal feature clustering method should satisfy the following demands: (1) Since most geographical processes change continuously in the spatio-temporal domain, the analysis should be performed in the spatio-temporal domain with the consideration of the data continuity. (2) The feature clustering should be applied to components of different scales, which have different data distributions and different noise levels. Thus the feature clustering is better to be data adaptive and flexible. (3) Since the data volume of the original multi-temporal remote sensing data is usually large, and the multi-scale decomposition increases the data volume significantly, the feature clustering should be efficient for real world operational analysis. To achieve all the above requirements, new approaches should be developed to perform the multi-scale spatio-temporal feature clustering analysis.

This paper presents an integrated approach to perform the multi-scale spatio-temporal exploratory feature clustering. The paper is organized as follows: Section 2 presents the overall idea of the proposed method. Section 3 provides the details of the method. The case study with the satellite altimeter data is shown in Section 4. The technical and geographical effectiveness of the method is evaluated in Section 5. Discussions and conclusions are presented in Sections 6 and 7, respectively.

2. Overall Idea

The overall idea of the exploratory multi-scale spatio-temporal feature clustering of multi-temporal remote sensing data is depicted in Figure 1.

From the unified spatio-temporal perspective, the original multi-temporal remote sensing data can be considered as a unified spatio-temporal data cube. In this data cube, each location (pixels with same spatial coordinates) is a sampling with a time series indicating the temporal fluctuations through the temporal dimensions. Due to the lack of spatio-temporal unified multi-scale decomposition methods, the spatio-temporal multi-scale decomposition can only be performed in the spatial domain or the temporal domain. Therefore, the original spatio-temporal data, which are considered as a unified spatio-temporal data cube, should be unfolded as a multi-channel time series or spatial series. Under this notion, the entire multi-temporal remote sensing data over an area can be perceived as a set of time series with each location being a channel of time series. Therefore, the whole spatio-temporal data can be seen as a single multi-signal time series that is continuously observed at a synchronized temporal sampling rate. The spatial distributions of the data are inherited in the channel-to-channel relations.

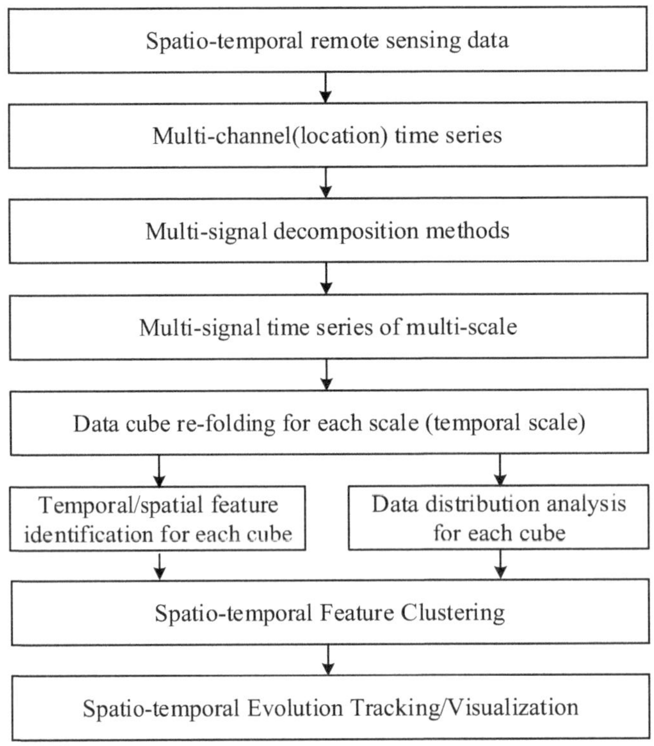

Figure 1. The overall framework of the exploratory multi-scale spatio-temporal feature clustering.

Multivariate multi-scale analysis methods, such as multi-signal wavelet, can be applied to the multi-signal time series to extract the multi-scale variation patterns. The multi-signal wavelet decomposition estimates the common multi-scale variation pattern of all channels synchronically using the full information of the spatio-temporal data. Both global and the local spatio-temporal variations at different spatio-temporal scales can be extracted synchronically [27]. With the multi-signal wavelet decomposition, the original multi-temporal remote sensing data can be decomposed into several multi-scale components. Each decomposed component, which is a multi-signal time series that indicates a certain feature of the multi-scale spatio-temporal variations, can be re-folded to a data cube with the same size of original data. Therefore, the connotative signals that only affect at certain significant scales can be extracted and revealed.

To enhance the feature extraction performance and filter out possible uncertainties or pseudo-variations caused by scale mixture or noise, classical exploratory data analysis methods, such as principal components analysis, power spectrum analysis and histograms, can be used to characterize the spatial/temporal features (e.g., spatial shapes or temporal indexes) and the data value distributions of the multi-scale variations. This can be achieved by comparing the features extracted by principal components analysis and power spectrum analysis with existing geographical indexes or spatial patterns to reveal geographical meanings of the multi-scale components. To identify the spatio-temporal distribution characteristics of the geographical phenomena, feature cluster methods such as K-Means, Fuzzy C-Means Cluster (FCM), *etc.*, can be applied to the spatial/temporal histograms to achieve the spatio-temporal feature clustering. To achieve the data adaptive clustering for each decomposed multi-scale component, data indexes, such as entropy, can be used to develop the data adaptive optimized cluster number selection procedure. To improve the computational performance, the histograms can be used for feature clustering. The histogram-based cluster of the decomposed components helps to enhance the features of the data, which makes the data more separable. Since the data value distributions in the histogram bins are interlinked with different regions of data in the spatio-temporal domain, the original spatio-temporal data can be clustered by separating the histogram with the value distribution corresponding to the desired regions. With the clustered features, the evolution of the features can be tracked and visualized crossing different time stamps.

3. Methods

According to the overall framework, several methods are assembled and steadily integrated in a processing workflow (Figure 2). The original spatio-temporal data are firstly reorganized as a multi-signal time series, which is then decomposed into several feature series cube and a noise cube by the multi-signal wavelet decomposition. The empirical orthogonal function analysis (EOF) and power spectrum analysis are applied to extract the spatio-temporal and frequency characteristics of each decomposed component. The integer approximation, which stretches the data, is used to enhance the data feature and reduce the computational cost. By reordering the data of integer approximation of each feature series cube in an ascend order and splitting them into several data bins, the histogram approximations can be constructed. Then the Fuzzy C-Means Cluster (FCM), with adaptively selection of cluster numbers, is applied to the stretched feature cubes. The segmentations of histograms are then reversely mapped to the spatio-temporal domain to achieve the spatio-temporal segmentation of original multi-temporal remote sensing data.

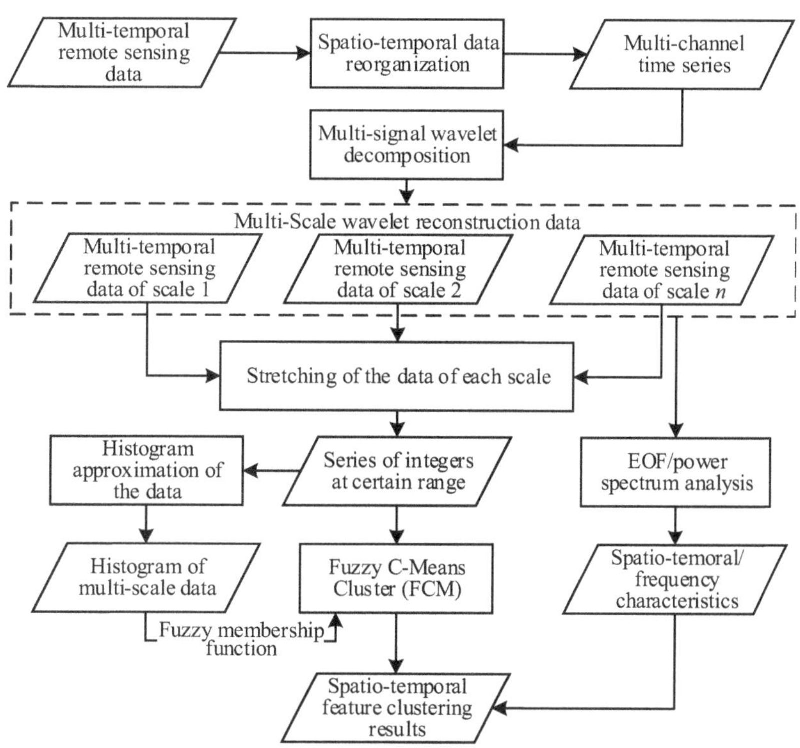

Figure 2. The processing workflow.

3.1. Spatio-Temporal Data Reorganization

In the beginning, we organize the multi-temporal remote sensing data as a unified multidimensional data cube [28]. The unfolding is then applied to the data cube to form the multi-signal time series. Since there are usually more than one coordinates in the spatial domain, unfolding the original data cube into a set of time series with each spatial coordinates a time series (*i.e.*, spatial unfolding) will make the data much simpler. Therefore, we reorganize original data as unified spatio-temporal cube and then apply spatial unfolding to form the multi-signal time series. Although, the unfolding process of the original spatio-temporal data into the multi-channel signal appears to separate the spatial and temporal dimensions, the pattern extraction is done through the spatio-temporal distribution of the data and the data synchronization problems are naturally solved during the analysis.

3.2. Multi-Signal Wavelet Decomposition

Multi-signal wavelet decomposition is applied to the reorganized multi-signal time series to retrieval the multi-scale spatio-temporal variations. Assuming $X(t) = \{x_1(t), x_2(t), \ldots, x_M(t)\}$ is the M dimensional multi-signal time series unfolded from the multi-temporal remote sensing data, the multi-signal wavelet decomposition of $X(t)$ can be formulated as follows:

$$X(t) = \sum_k F_k(t) + \varepsilon(t) \tag{1}$$

where $F_k(t)$ is the deterministic feature series signal that has certain spatio-temporal scale characteristics, $\varepsilon(t)$ is the spatial correlated residuals, which can be seen as noise, and k is the level of features extracted.

Let $h(t) = \sum_n h(n)\, h(Mt - n) \in L^2(R)$ be the scale function. $M - 1$ wavelet functions (notated as $j^{(i)}(t)$) can be constructed. For each of the wavelet function that meet:

$$j^{(i)}(t) = \sum_n g^{(i)}(n)h(Mt - n)\,, i = 1,2,\cdots, M - 1 \tag{2}$$

where $g^{(i)}$ is the coefficient function of the wavelet filter of wavelet function $j^{(i)}(t)$.

Rapid algorithm [29], which is constructed by parameterization of the mother wavelet and decomposition of the signal in the feature space, is used here for efficient decomposition. For each channel of the multi-signal, irrelevant elements involved in the signals are eliminated by a dyadic decimation, which reduces the original resolution into half of its length. This procedure is recursively applied on approximation coefficients to produce increasingly smoother versions of the original signals. Thus, we have the following decomposition:

$$\begin{cases} A_{j+1}^{1}X = M^{1/2} \sum\limits_k h(Mn\text{-}k) A_j^k X \\ D_{j+1,n}^{(i)}X = M^{1/2} \sum\limits_k g^{(i)}(Mn\text{-}k) A_j^k X \end{cases} \tag{3}$$

where $A_{j+1}^{1}X$ is the approximation coefficient of the first scale, $A_j^k X$ is the approximation coefficients of the k-th scale, $D_{j+1,n}^{(i)}X$ is the detailed coefficient of the $(k + 1)$-th scale, and $g^{(i)}$ is the high pass filter of the scale function. Since the original data X are a signal that have total M channels, all the approximation coefficients A_j^k is a $M \times p_k$ matrix, where p_k is the same length as the DWT approximation coefficients at the *k*-th scale.

The reconstruction of the multi-signal wavelet decomposition can be adopted in wavelet synthesis by up-sampling and inverse filtering of the decomposed coefficients. That is:

$$A_{j,n}^k X = M^{-1/2} \sum_k h(k - Mn) A_{j+1}^k X + M^{-1/2} \sum_k g^{(i)}(k - Mn) D_{j+1}^{(i)} X \quad (4)$$

According to Equation (4), the approximation component at level j can be reconstructed with the approximation and detail components at level $j + 1$.

Since all the time series are decomposed at the same time, they are at the same time scales. As the decomposed components have the same size of the original data, exploring methods, such as principle components, histograms and power spectrums, can be direct applied to identify the characteristics and geographical meaning of these spatio-temporal variations at each scale.

3.3. The Stretching of the Data

To enhance the feature pattern of each decomposed component and reduce the computational cost, the data are stretched to the range that best represents its features. A simple mechanism, integer approximation, which projects float point data into a series of integers at certain range, is used. Given a multi-signal time series $X(t) \in (a, b)$, which has a value range (a, b), *i.e.*, $X(t) \in (a, b)$, where a and b are float values. The mapping of the multi-signal time series $X(t)$ to the integer approximation in the range of (m, n) can be formulated as follows:

$$f: \mathbb{R}_{R(X(t))} \mapsto \mathbb{R}^+_{(m,n)} \quad (5)$$

where $\mathbb{R}$ is the real number set, $R(X(t))$ is the value range of the series $X(t)$, and the $\mathbb{Z}^+_{(m,n)}$ is the positive integer set with the value range of (m, n). A typical solution of the mapping is:

$$x_i' = \{m + (n-m) \frac{x_i\text{-Min}(X)}{\text{Max}(X)\text{-Min}(X)}\} \quad (6)$$

where (x) is the function which aims at finding the closest integer of real value x.

For each decomposed feature series cube, we can apply Equation (6) to project the whole data from the original value range (a, b) into the new value range (m, n) with their spatial and temporal distribution rarely changed. Since both the boundary value of the data range (m, n) and the level value i are all customizable, they can approximate the original signal at any given precision.

3.4. Histogram Approximation of the Data

Histogram is used to summarize the value distribution of the integer-approximated data to reduce the data value and data dimensions, which also helps to enhance the features of the data.

Let $f \in L^1(\mathbb{R}^n)$ is the integer approximation of multi-signal time series $X(t)$, the cumulative histogram operator H is defined by:

$$H(f)(y) = \mu\{x: f(x) \leq y\} \quad (7)$$

where μ is the Lebesgue measure. *H(f)* is the real function that represents the cumulative histogram. With the integer approximation of the multi-signal series, the histogram m_i counts the number of observations that fall into each of the bins meets the conditions: $n = \sum_{i=1}^{k} m_i$, where n is the total number of observations and k is the total number of bins. The data value distributions in the histogram bins are interlinked with different regions of data in the spatio-temporal domain, clustering the histogram with the value distribution corresponding to the desired regions can result in a spatio-temporal cluster of original data according to their value distributions. Since the data volume of the histogram are much less than the original data, the cluster of the histogram is much more efficient than that of the original data.

For each decomposed feature series, we can calculate one histogram by recognizing the data cube as a one dimensional data vector. In the histogram, the data are organized in an ascending order, so the clustering is very simple. The multi-signal wavelet decomposition, which can be seen as a denoising procedure, largely reduces sudden changes in the data. Therefore, the data samples that have close spatial distances are highly possible to be classified into the same bins. This will further lead to large potential to keep the spatial continuity of the clustering. The feature cluster performance is customizable by controlling the number of bins, which make the feature cluster flexible to achieve the balance between performance and the approximation level.

3.5. Adaptive FCM on the Histogram for Feature Clustering

Original spatio-temporal clustering measures the distances between pixels in the spatio-temporal domain. This kind of spatio-temporal domain is computational intensive and time consuming. The histogram, which is a reordered index about the value distribution of the original spatio-temporal data, abstracts the value features into different bins. Since each bin of the histogram relates to certain pixels in the spatio-temporal domain, the clustering of the histogram bins is much simpler and can better indicate the value distribution of original data in the spatio-temporal domain. In our approach, the histogram used can be seen as a pre-process, which gathers the pixels into fewer groups according to their value data. Then the similarities between different bins are then clustered. In this way, not only the features are enhanced but also the computation can be largely reduced. Since for each bin, there are bidirectional mapping functions to map the data from the value domain to the spatio-temporal domain, the clustering between bins of the histogram in deed performs the clustering in the spatio-temporal domain. Therefore, the clustering is definitely a spatio-temporal domain clustering, which can be used to track the spatio-temporal evolution of the features.

To reduce the shape changes between the boundaries to achieve more smooth separation in the spatio-temporal domain, the Fuzzy C-Means Cluster (FCM), which has soft thresholds for data separation, is used. Here, FCM, which clusters the data sampler to two or more clusters according to the fuzzy membership function, is applied to the histogram. The computation of the FCM is an iterative method that moves the cluster centers and assigns the data points to each cluster centers recursively to optimize the data segmentation according to a certain object function [30]. The objective function [31] we have used is:

$$J_m = \sum_{i=1}^{N}\sum_{j=1}^{C} u_{ij}^{m} \parallel x_i - c_j \parallel^2, 1 \leq m \leq \infty \tag{8}$$

where m is a real number greater than 1, u_{ij} is the degree of membership of x_i in the cluster j, x_i is the i-th of d-dimensional histogram of the observed data (original data or the decomposed feature series), c_j is the d-dimension center of the cluster, which is initially randomly selected and recursively updated. $||*||$ is any norm expressing the similarity between any histogram data and the center.

The determination of the cluster numbers is a critical but also important issue for the feature clustering. Since the clustering is applied to the histogram, the difference between histogram bins (*i.e.*, the value distribution) are much smaller than the difference between the original data, lots of traditional cluster number determination methods are not efficient for such task. So the fuzzy membership between different histogram bins, which measures the relative variability and similarity between different value ranges, is used to index the affection of the cluster numbers to the whole data segmentation and their statistical distributions. The Classification Entropy (CE) [32], which measures the fuzziness of the cluster partition, is used to determine the optimized segmentation cluster number adaptively according to the data distribution. The definition of the CE is:

$$CE(c) = -\frac{1}{N}\sum_{i=1}^{c}\sum_{j=1}^{N} u_{ij}\log(u_{ij}) \tag{9}$$

where u_{ij} is the membership of data point j in cluster i. According to the definition of the *CE*, the larger *CE* indicates the large fuzziness between different clusters. When the cluster number is growing, the value of *CE* is going smaller, which suggests a more disorderly data distribution. Therefore, according to the change characteristics of the *CE* against to the cluster numbers, the data segmentation structures against value distributions can be identified. The change point of the *CE* suggests the significant change of the cluster distribution *i.e.*, change of the characteristics. Thus, the cluster number located at the significant change point of the *CE* can be selected as the optimal cluster number of each decomposed component.

4. Case Studies

4.1. Research Data and Experiment Configuration

The Monthly Global Delayed Time Mean Sea Level Anomaly (MSLA) data produced by SSAlto/Duacs, AVISO [33] were used as the experiment data. The data were produced by combining the satellites of T/P, Jason-1, Jason-2 and Envisat. The spatial resolution of the data is 0.25° and the time period of the data is from January 1993 to December 2013. The total data are 1440 × 721 × 252. The data covers most of the global ocean containing the area between 81°S–81°N, the land part and the ocean in the polar area are represented as missing values. The original data in the form of several individual NetCDF files is read recursively into the MATLAB 2011b environment on an Inspur NP 3560 server with two Intel Xeon E5645 (2.4G) processors and 48 GB DDR-3 ECC Memory. The operating system is Windows Server 2008 R2. All the test codes are written in the same software and hardware environment. The Multivariate ENSO Index (MEI) is used as an index of ENSO event [34].

The MSLA data are firstly represented as multi-signal time series, with the time series at each pixel as a channel. Then the multi-signal matrix is decomposed with the multi-signal wavelet with the nearly symmetric wavelet "Sym4" as mother wavelet. Symlet wavelet family (SymN) is a reversion of the commonly used Daubechies wavelet. It is an orthogonal and nearly symmetric wavelet with support width 2N-1 and vanishing moments N. Due to its nearly symmetric properties, it can reduce the phase shifts during the wavelet reconstruction [35]. Similar to the individual DWT analysis, the decomposition level in this work was restricted to the greatest integer contained in $\log_2$ (n), where "n" is the length of the time series. According to the length of the data, 6-level decomposition is applied to the whole data set.

With the decomposed multi-signal data, the minimum of the data is subtracted to keep all the data positive. According to Equations (5) and (6), the integer approximation of the data is applied. The data range parameters (m, n) are selected according to the decomposition value distribution of the data. To best distinguish the difference of all the decomposed components, all components are re-normalized to the value that ranging from 0–160. With the integer approximation, the histogram distribution of each decomposed component is calculated and the histogram-based FCM algorithm is then applied to each decomposed feature series. The cluster with different number of clusters is simulated and the classification entropy is computed for each simulation to determine the optimal cluster number of each decomposed component. The final data cluster at each scale is computed with the optimized cluster number for each decomposed component. The computation time and the memory occupations changing are logged and analyzed according to the cluster numbers. Finally, the distribution and the evolution of the cluster result are compared with typical indexes such as ENSO, *etc.* The evolution of the spatio-temporal pattern is also compared with existing literature.

4.2. Multi-Scale Decomposition Result

The reconstruction of the multi-scale components of the satellite altimeter data extracted from the multi-signal wavelet decomposition is depicted in Figure 3. Since the full representation of all the data is impossible, we only provide several time slices. From the spatial dimension, the A6 components are much smoother than the original data. This suggests the multi-signal wavelet decomposition not only is multi-scale filter in the temporal domain, but also affects the spatial domain. Although the information of the generalized spatial scale affection of the multi-scale decomposition is still limited, in our cases, the larger temporal scales will also result in a smoother spatial distribution.

To reveal the detailed spatio-temporal distribution of the decomposed components at different scales, we applied the empirical orthogonal function analysis (EOF) to each decomposed component. The EOF method can extract the significant and common pattern of the spatio-temporal data. The variance contribution and the principal components (PCs), which suggest the consistency of the components across the spatial distribution and the significant temporal change of certain patterns, are extracted from the EOF (Table 1). To further understand the temporal patterns of each decomposed component, the power spectrum (PSD) of each extracted PC is also calculated. The peak of the PSD suggests the significant period of the PC. The PCs and the PSD of the PCs are depicted in Figure 4. To reveal the detailed frequency characteristics of the PCs, the highest peak of PSD and its associated frequency and period are extracted and represented in Table 2.

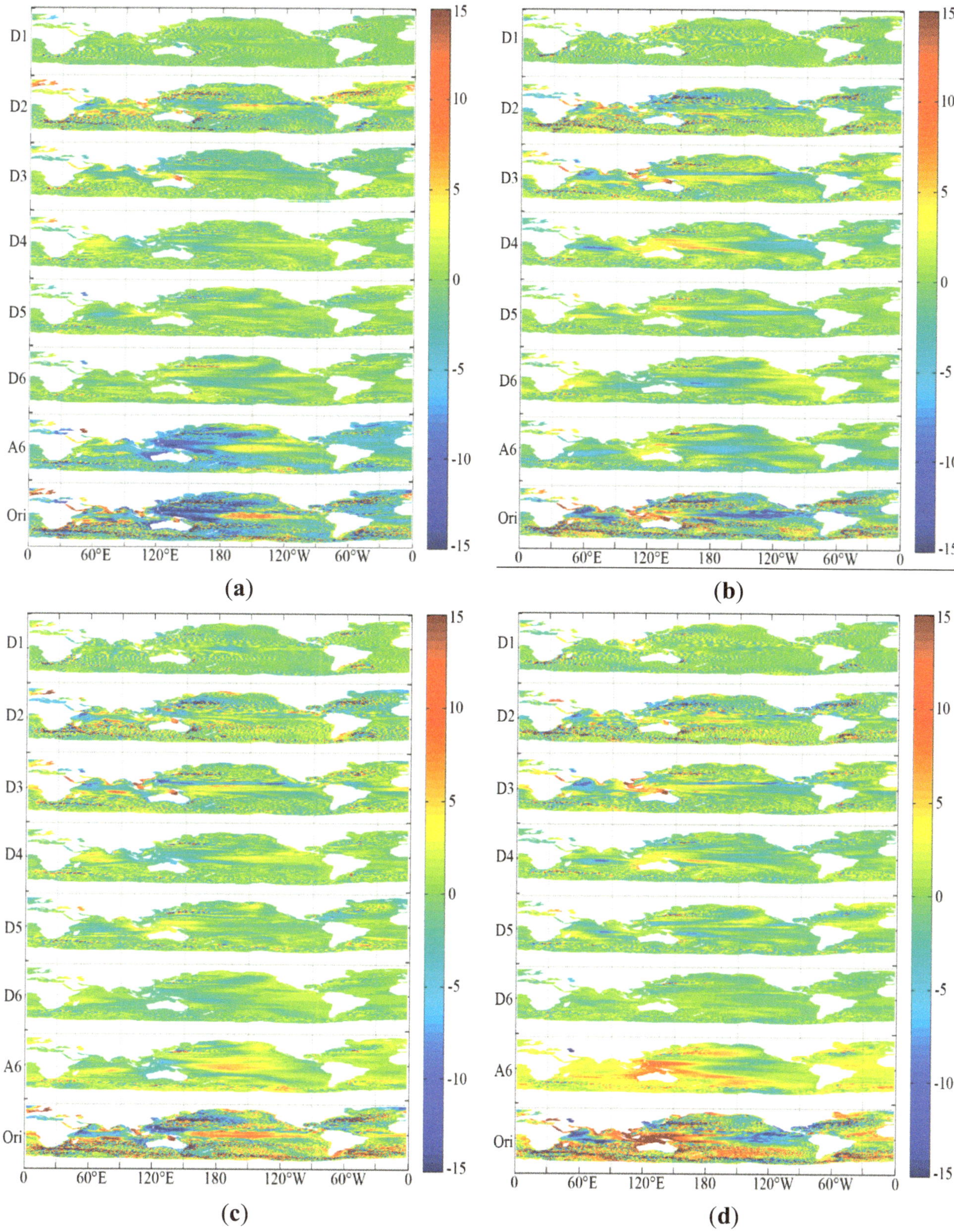

Figure 3. The multi-signal wavelet decomposition results. Four time slices are selected from the whole spatio-temporal data cubes of original data and reconstructed decomposed components ((**a**) January 1993; (**b**) January 1999; (**c**) January 2005; (**d**) January 2011). The color bar in each subgraph indicates anomaly range of the data. The multi-signal wavelet decomposition not only filters the data in the temporal domain, but also affects the spatial domain. Larger temporal scales will also result in a smoother spatial distribution.

Table 1. Explanation of the Variance from the EOF of the Decomposed Result (%).

	PC1	PC2	PC3	PC4	PC5	PC6	PC7
Oridata	31.2	27	16.2	10.5	5.5	4.9	4.3
A6	63.7	16.2	10.7	9.1	0	0	0
D6	31.7	23.7	22.9	19.8	1.7	0	0
D5	31.5	15.4	13	11.5	10.5	9.3	8.4
D4	31.2	17.8	11.8	11.2	10.4	8.8	8.4
D3	52.4	27.8	5.2	4.2	3.5	3.4	3.2
D2	32	22.8	9.8	9.6	8.8	8.5	8.3
D1	16.4	15.6	14.7	13.6	13.3	13	13

Table 2. Peak of PSD and corresponding frequency.

	PC1	PC2	PC3	PC4	PC5	PC6	PC7
Oridata	0.0833 (12)/ $1.01 \times 10^{9*}$	0.0079 (126)/ 4.45×10^7	0.0833 (12)/ 4.49×10^8	0.0833 (12)/ 5.33×10^7	0.0079 (126)/ 7.47×10^7	0.0278 (36)/ 2.62×10^7	0.0079 (126)/ 7.79×10^7
A6	0.0079 (126)/ 1.67×10^7	0.0079 (126)/ 1.45×10^8	0.004 (252#)/ 1.27×10^8	0.0079 (126)/ 4.03×10^7	0.0119 (84)/ 1.73×10^5	0.0119 (84)/ 1.25×10^3	0.0159 (63)/ 2.26×10^2
D6	0.0079 (126)/ 9.34×10^7	0.0159 (63)/ 4.84×10^7	0.0119 (84)/ 4.52×10^7	0.0119 (84)/ 6.18×10^7	0.004 (252#)/ 3.24×10^6	0.0198 (50)/ 1.65×10^4	0.004 (252#)/ 1.74×10^2
D5	0.0198 (50)/ 1.06×10^8	0.0278 (36)/ 5.35×10^7	0.0198 (50)/ 5.73×10^7	0.0238 (42)/ 3.12×10^7	0.0159 (63)/ 1.59×10^7	0.0278 (36)/ 2.61×10^7	0.0238 (42)/ 2.35×10^7
D4	0.0357 (28)/ 8.45×10^7	0.0437 (23)/ 5.17×10^7	0.0556 (18)/ 2.02×10^7	0.0397 (25)/ 2.42×10^7	0.0476 (21)/ 2.07×10^7	0.0357 (28)/ 2.71×10^7	0.0437 (23)/ 1.55×10^7
D3	0.0833 (12)/ 7.18×10^8	0.0833 (12)/ 3.90×10^8	0.0714 (14)/ 1.85×10^7	0.0635 (16)/ 1.08×10^7	0.0595 (17)/ 1.22×10^7	0.0714 (14)/ 1.34×10^7	0.0754 (13)/ 1.52×10^7
D2	0.1667 (6)/ 1.24×10^8	0.1667 (6)/ 8.81×10^7	0.1429 (7)/ 5.47×10^6	0.1389 (7)/ 1.12×10^7	0.1389 (7)/ 8.33×10^6	0.127 (8)/ 6.87×10^6	0.1349 (7)/ 4.71×10^6
D1	0.25 (4)/ 3.11×10^6	0.25 (4)/ 1.44×10^7	0.254 (4)/ 2.87×10^6	0.2619 (4)/ 2.89×10^6	0.2579 (4)/ 2.26×10^6	0.3135 (3)/ 1.41×10^6	0.2659 (4)/ 2.75×10^6

XX (YY)/ZZ in the table means principle frequency (principle period)/peak PSD value. The unit of the frequency and principle period is Hz and month, respectively. # in the table means principle frequency (principle period) is not trustable due to the period is larger than the half of the data length. For these three components, no other peaks of the PSD can be founded. This is caused by the limitation of the base theory of FFT. Since all the three components are not the first principle components, their impacts on the frequency characteristics of original data is small.

From Figure 4, we can find that the PCs of the original data, which are not easy to separate, are very similar. The PSD suggests that most of the PCs of original data have strong annual cycle. Most of the differences between different PCs are the phase difference caused by the spatial distribution lag. However, the PCs and the PSD of the PCs suggest the decomposed components are much more separable. The dominant periods of the PSD of each component are significantly different. The major period of the components from A6 to D6 is changing from inter-decades to bi-month. In addition, the PSD of each EOF at each scale can also be classified. This indicates that the patterns of the spatial different are also clearly extracted. The multi-scale decomposition can clearly improve the pattern classification of the data.

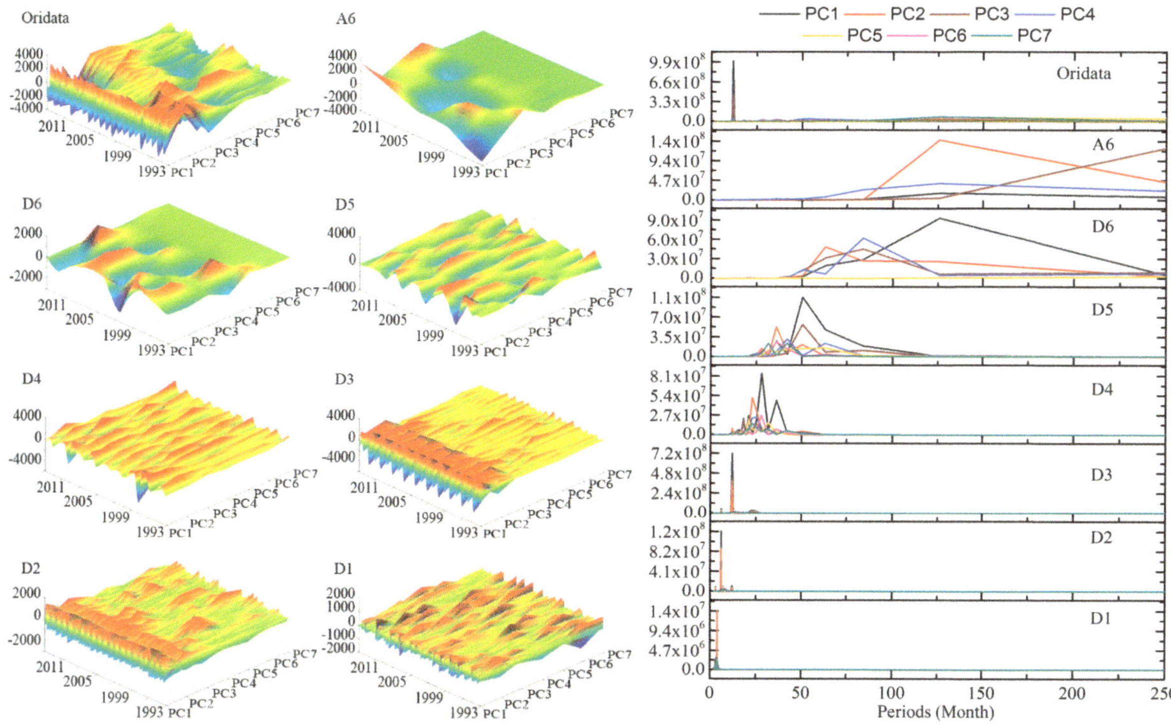

Figure 4. The principal components of original data and each decomposed components and their power spectrums. The 3D plot of the principal components of original data and each decomposed components (**left**); and the power spectrums of each decompose data of each principal component of original data and each decomposed component (**right**). The frequency is changed into period for better illustration.

From the explanation of the variance from the EOF of the decomposed result, the A6 and D6 components are much simpler than the other components, which have fewer principal components. This suggests large-scale variations are more consistent and smooth. Although, the PSD for the time series that are longer than half periods of the whole data length are not very accurate, the combining investigation from the PCs also suggest that both A6 and D6 are inter-decade variations. The A6 components have a period of ten years, which can be seen as the long-term trend of the data. Meanwhile, the D6 components have a quasi-period of 5–7 years. The strong spatial variation in the Equatorial Pacific and Indian Ocean suggests D6 may have close relations to the significant ENSO variations. The D5 and D4 components are also inter-decade variations, which have spatial distribution around the global oceans. The spatial distribution also suggests these two components may also have close relations with the ENSO events. D3 component, which has very high variance contribution of the first PC, has a significant annual cycle. Since the yearly period is the most common and significant period all around the global ocean, the significance of the yearly cycle also proves the consistency of the EOF. The D2 and D1 components have periods of half-a-year and two to four months, respectively. Some irregular high frequency noise is also contained in these components.

From the above discussion, the multi-signal wavelet decomposition can well extract the multi-scale temporal variation from the original multi-temporal remote sensing data. The decomposition is performed in the time domain but also affects the spatial distribution. The decomposed components have clear temporal variation with significant temporal period. By the multi-scale decomposition, the features of original data are much more separable. Thus, it will be very useful and helpful for feature extraction and data segmentation at different scale levels.

4.3. Feature Clustering Result

The integer approximation and the histogram representation of each decomposed component are performed for the feature clustering. The histogram of the original data and the decomposed multi-scale components are shown in Figure 5. The original histogram only has a wide band peak, because various spatial features are blended together. For this case, it is not very easy to classify different patterns directly from the original histogram. The histogram of the decomposed components is much more separable. Each histogram of the decomposed component has narrow bands and clear single peaks. The values of A6, D6 and D5 are smaller than those of D4–D1 component. This corresponds to the situation where the variation amplitude of the inter-decade cycles is lower than that of the inner-decade cycles in most areas. The segmentation based on the histograms of all the decomposed components is much more structurally clear than segmentation based on the histogram of original data. The cluster distribution of the histogram bins also suggests that the data are more significant and separable than the original data.

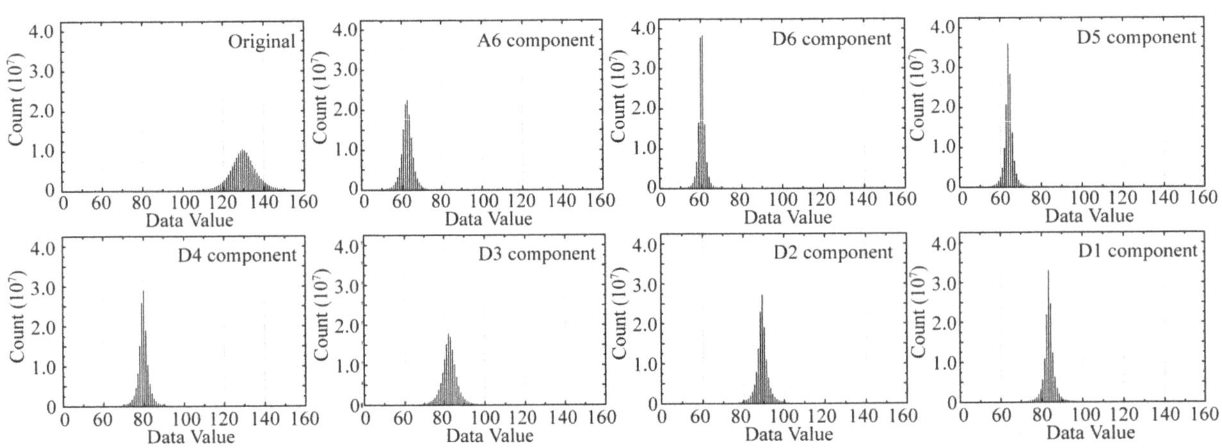

Figure 5. The histogram of the original data and the decomposed components.

Before the final feature clustering at each decomposed component, the optimized cluster number of each decomposed component is calculated from the simulation of the CE index changing according to the cluster number. The simulations are constructed by changing the number of the Clusters from 10 to 40. For each cluster number, the CE is calculated and logged. Figure 6 suggests how the CE changes according to the cluster number at each scale. For all decomposed components, the value of CE grows lower when the cluster number increases. The CE of the original data is significantly higher than that of the decomposed components. And the D1–D4 components have very similar CE. The CE of D5 and D6 components is also very similar, and they all have very low CE values. Unlike other components, the CE of A6 does not change smoothly, which suggests that the complexity degrees of the A6 component may be different. These suggest that the complexity of the inter-decade variations is lower than that of the inner-decade variations. This is partially because the inter-decade variations are much smoother and change more slowly and partially because fewer noises are contained in the long-term variations. We will notice that in the lower sections the D5 and D6 are mainly influenced by the ENSO event, which is an irregular phenomenon in both spatial and temporal dimensions. Thus this kind of irregularity in CE also demonstrates the characteristics of the signal.

Most of the CE at different scales is changing very slowly after the cluster number gets bigger than 30. This suggests that the clustering with the cluster number bigger than 30 will not produce lots of different information. Therefore, we can consider that the minimal number of 30 will well represent the distribution and feature of original data. Thus, we select the optimized cluster number of 30 for all the decomposed components to produce the final segmentation. Although, it is optimal to select the different optimal cluster number for each level, we select the optimized cluster number of 30 for all the decomposed components according to CE for simplicity.

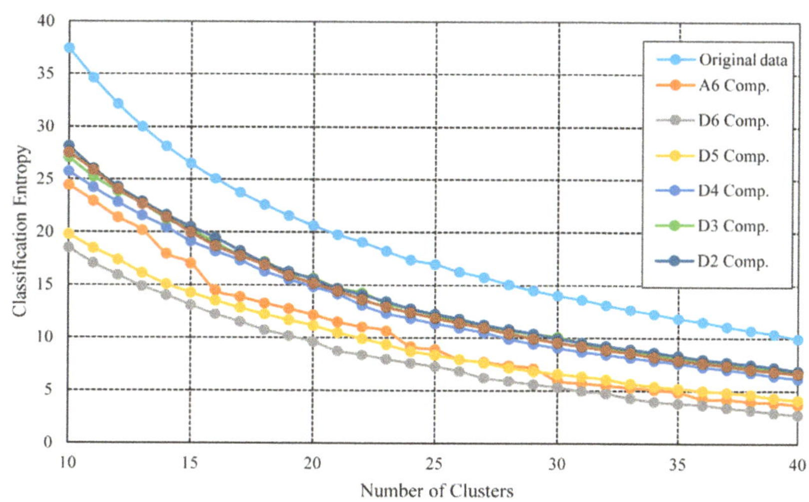

Figure 6. Determination of Cluster Numbers.

The feature clustering with the cluster number 30 is performed to each histogram of the decomposed components. The spatio-temporal distribution of the segmented result is shown in Figure 7. The cluster of the original data does not have a clear significant spatial structure. Lots of small patches can be found in the cluster result, which suggests the original segmentation cannot well separate the variations at different temporal and spatial scales. The cluster result of the decomposed components has much clearer spatial structures. The cluster result of the A6 component suggests clearly two stages before and after the year 2003. This is consistent with the trend change of the sea level height. The cluster result of the D6–D4 components is much more significant in the time dimension. The major changes are the spatial pattern changes crossing the global. Since these components mainly reveal the ENSO signal, detailed comparison of the spatio-temporal changes during ENSO events can be achieved. From the temporal cluster index changes, the evolution process, including the starting, growing, peaking, weakening and vanishing, is much clearer than the original data. The clustering of the D3–D1 components mainly changes periodically through the time which suggests that our cluster can well represent both the spatial and temporal changes of the data. All the above suggest our approach can be used as an exploring tool to find patterns in the spatio-temporal data. Unlike tensor methods, our approach can support irregular data that contain missing data at any spatial locations.

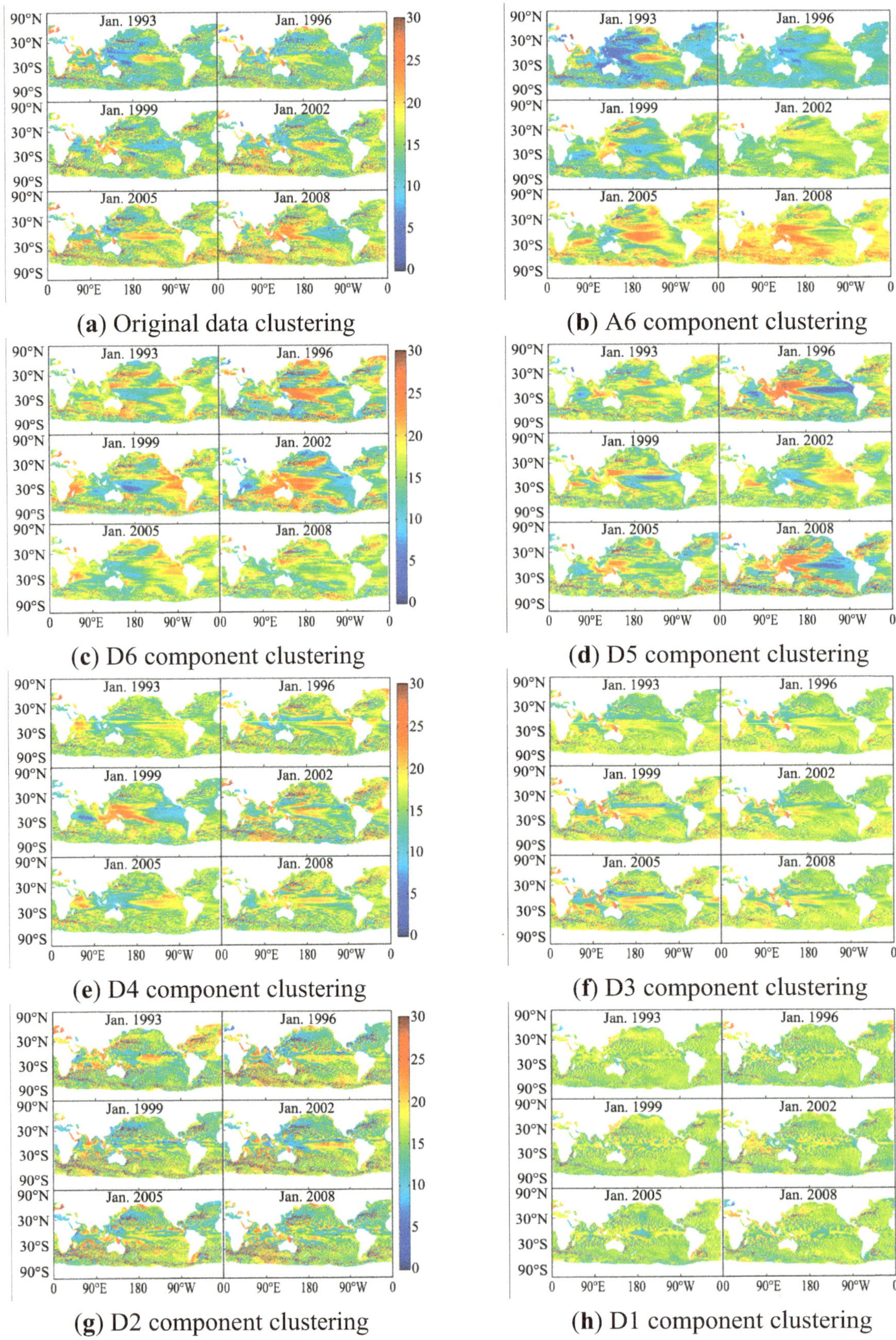

Figure 7. The cluster results of the original data and the decomposed components: (**a**) cluster result of the original data; and (**b**–**h**) cluster results of A6, D6, D5, D4, D3, D2, and D1 components, respectively. The color bar in each subgraph indicates the classification type index of the data. Due to the space limits, only six time slices are selected for demonstration.

5. Evaluations of the Tracked Spatio-Temporal Patterns

5.1. Evaluations by Tracking the 1997–1998 El Niño Evolution

The 1997–1998 El Niño is the strongest observed El Niño event that happened during the research time. The sea level anomalies have been largely affected by this El Niño event. Extensive studies have been performed concerning the pattern of sea level response to this El Niño event, especially in the tropical Pacific area [24]. Most of these response patterns are performed with certain pattern extraction methods (e.g., EOF and ICA) directly from the original data [36]. However, the ENSO signal, which is also a multi-scale phenomenon, often has quasi-periods from two to seven years, and may affect the sea level at several different scales since the interactions between ENSO events and sea level are complex [24–26]. Extracting and tracking the spatio-temporal evolution with the data cluster from the multi-scale decomposition will be helpful to reveal the possible mechanisms about how the ENSO events affects the sea level in the spatio-temporal scale domain.

The primary step of extracting the multiscale interactions between ENSO and sea level is to extract possible ENSO signal from the original SSHA data. Since there is no quantitative simple index in the spatio-temporal domain that can clearly indicate the evolution of ENSO, we use the MEI index in the temporal domain to identify the ENSO signals at different decomposed components. The MEI is compared to the first two principle components of all the decomposed feature series (Figure 8). The correlation coefficients between the principle components and the MEI index are presented in Table 3. From Figure 8, the first principle components (PC1) of D5 and D4 correspond to the MEI index well. The correlations between the PC1s of D5 and D4 and MEI are −0.534 and −0.571, respectively. Since D6–D4 components represents the sea level variations of inter-decade time scales, it has close correspondence between the ENSO scales. In the second principle components, although the principle components of the original data have high correlations between MEI, and all the decomposed components have weak correlations between MEI. However, from the Figure 8, the PC2s of D5 and D4 also have high correspondence with MEI. Other components, e.g., A6, D3, D2 and D1 only have weak correlations between ENSO signals. Therefore, we can conclude that the D5, D4 may reveal the multi-scale evolution characteristics of ENSO.

Table 3. Time and memory cost of the feature cluster procedure.

Data	Time Cost(s) *		Memory Cost(MB)	
	Our Methods	**K-Means**	**Our Methods**	**K-Means**
Original Data	122.10	742.57	330.71	958.45
A6 Component	116.24	906.87	321.84	962.61
D6 Component	115.39	921.96	337.62	966.98
D5 Component	116.93	894.73	331.56	962.59
D4 Component	116.44	866.97	335.71	966.84
D3 Component	119.57	896.69	311.54	961.58
D2 Component	116.54	829.32	332.74	989.68
D1 Component	119.44	766.87	339.84	969.84

* The time/memory cost of our methods including the time/memory cost of wavelet decomposition, histogram computation and FCM clustering.

The multi-temporal decomposed components D5 and D4 and their cluster results from the May 1997 to August 1998, are selected to validate whether the feature extracted are affected by the 1997–1998 El Niño event. To well represent the data, only the data between 40°N and 40°S are shown in Figures 9 and 10. The selected components have the inter-decade cycle, which contains the El Niño signal, and the selected time range contains all the evolution stages including the starting, growing, peaking, weakening and vanishing stage of this El Niño event. From the decomposed result, both D5 and D4 components have clear patterns that can reveal the existence of the El Niño event. However, the detailed tracking of this El Niño event is still difficult directly from the decomposed data. So the feature clustering using the FCM is further applied to extract the multiscale spatio-temporal evolution of the El Niño event.

The cluster of the decomposed components shows how the El Niño changes in the spatio-temporal domain much clearer than original components data. The D5 components suggest a significant El Niño pattern from May 1997 to February 1998 in Tropic Pacific Ocean. The tongue of the high sea level has been moving from 180° to 120°W. After March 1998, the high sea level in the Tropic Pacific is growing weaker. However, peak heights are located in the West Tropic Indian Ocean. The locations and boundaries are comparable with the existing literature [37,38]. The significance of the Indian Ocean changes in this El Niño event, which cannot be directly seen and extracted from the original and decomposed components, has already been proved in the literature [39]. The D4 components are slightly different from the D5 components. The westward high sea level starts from September 1997 and ends in April 1998. The mixture of two significant peaks of the westward trade wind at different scales is one of the causes why this El Niño becomes the strongest El Niño [40,41].

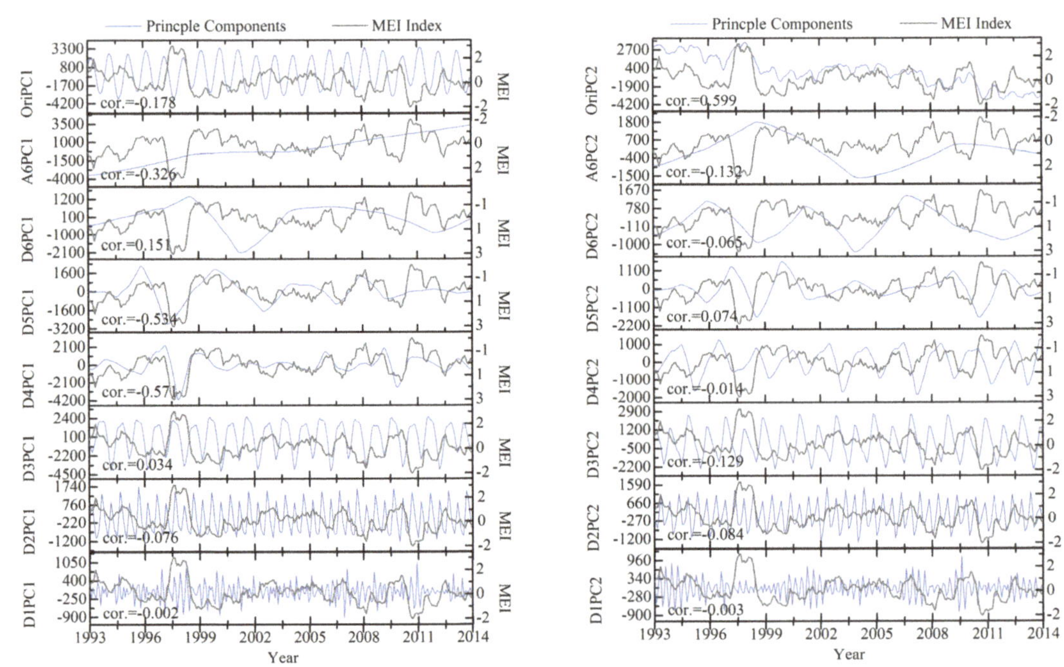

(a) **(b)**

Figure 8. The comparison between the principle components of the decomposed components and the MEI index: (**a**) comparison between the first principle components of the decomposed components and the MEI index; and (**b**) comparison between the second principle components of the decomposed components and the MEI index.

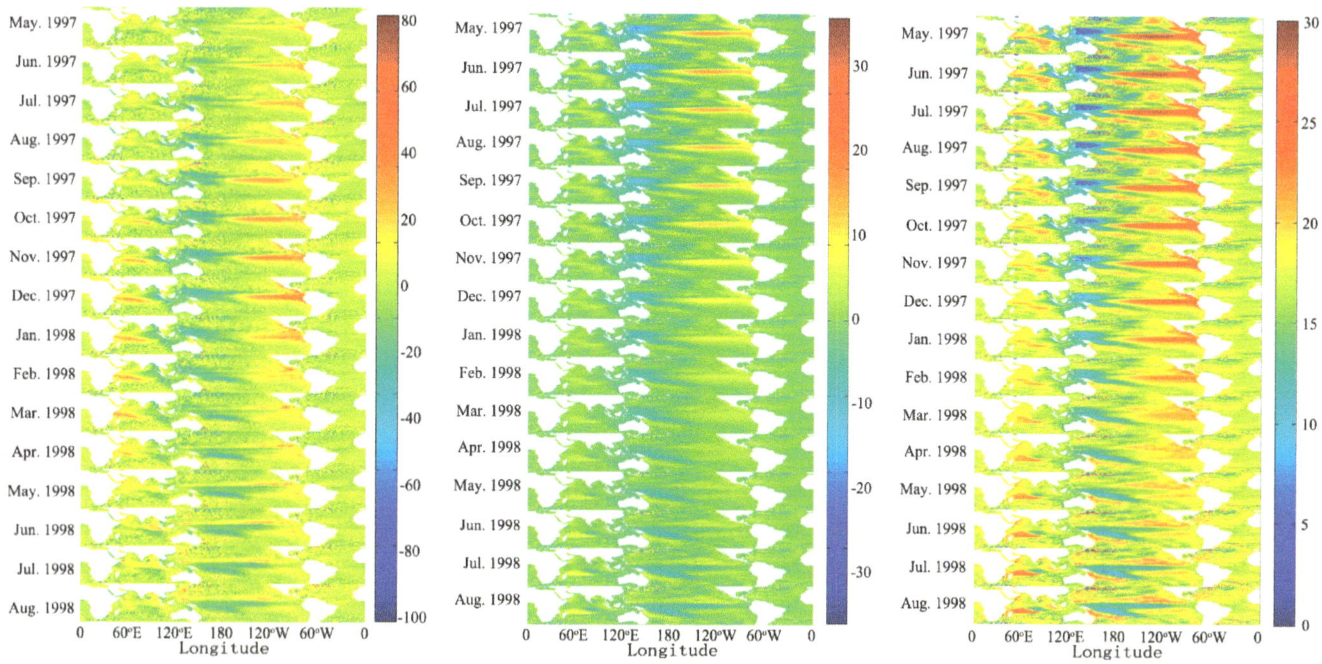

(a) (b) (c)

Figure 9. The spatio-temporal evolution of the D5 component during the 1997–1998 El Niño: (**a**) the original data; (**b**) the D5 components; and (**c**) the cluster of D5 components.

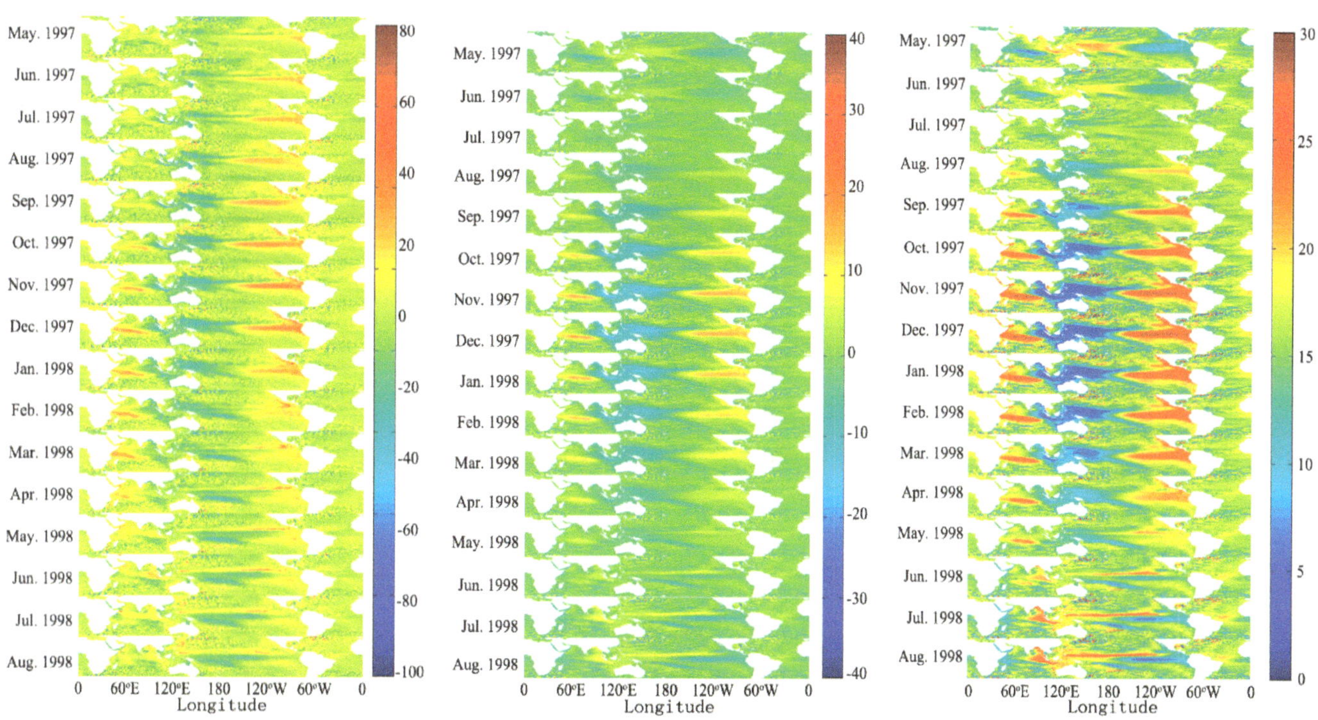

(a) (b) (c)

Figure 10. The spatio-temporal evolution of the D4 component during the 1997–1998 El Niño: (**a**) the original data; (**b**) the D4 components; and (**c**) the cluster of D4 components.

In most recent studies, there are two different El Niño events, the East Pacific El Niño (EP) and Central Pacific El Niño (CP), which may dominate the whole ENSO cycle [42]. The two different types of El Niño events can be distinguished by the starting location of the event in either East Pacific or Central Pacific.

However, the two different events are interlinked and are not easy to clearly separate. There are researches suggest the 1997 strong El Niño are caused by both two different El Niño events [43,44]. In our result, the spatial pattern difference between the D4 and D5 partly support such results. The cluster of the D5 component has its highest center in the Central Pacific while the D4 component has its center in only East Pacific. The start time of the D4 component agrees well with the time that the bloom of this El Niño event happens [45]. This suggests the two signals have different affections in spatio-temporal domain, which can be clearly revealed using the feature clustering of the spatio-temporal decomposed components at different scales. Since this pattern can not be observed clearly in both original data and the decomposed components, the cluster can reveal the multi-scale pattern from the multi-temporal remote sensing data and greatly enhance the representation. In other words, the proposed method can reveal the spatio-temporal evolution of certain kinds of patterns or events more clearly.

5.2. Computational Performance Evaluation

For the computational complexity, both theoretical computational complexity and experiment results are computed. For our method, the theoretical computational complexity is composed of the following parts: the complexity of the multi-scale decomposition, the integer representation, the histogram construction, the FCM and the cluster reverse mapping to the original data. For the multi-scale decomposition part, the theoretical complexity of the multi-signal wavelet decomposition is about $kO(n\log(n))$, where k is the number of decomposed levels; the complexity of the integer representation, histogram construction and the reverse mapping are all $O(n)$, where n is the total data count. The complexity of the FCM runs asymptotically in $O(Nc^2p)$ time, where N is the number of p-dimensional observations (*i.e.*, sum of number of histogram bins of each levels.) and c is the number of clusters [46]. Therefore, the overall complexity of the method is $kO(n\log(n))+O(n)+O(Nc^2p)$. This computational complexity is much smaller than that of the original FCM, and it is customizable according to the number of the histogram bins.

The computational efficiency and the spatial continuity of the feature cluster are evaluated in the experiments. The computational time and the memory cost of the feature cluster procedure of the original data and decomposed components are logged in Table 3. From Table 3, we can clearly see that the time cost and memory cost of different components are very similar, which suggests the computational performance of our method is stable. The computation time and the memory cost are about 1/7 and 1/3, respectively, which are less than that of classical K-Means cluster, since the computational time and memory cost are affected greatly by the histogram representation. More detailed technologies that can enhance the patterns and reduce the bins of the histogram can reduce the computational time and memory cost even more. In conclusion, our method is efficient to deal with large amount of multi-temporal remote sensing data.

6. Discussion

In our approach, the multi-signal wavelet decomposition was only used to decompose the multi-temporal remote sensing data along the time dimension to extract the temporal multi-scale variations. However, the channel-to-channel (location to location) relations at different scales during the decomposition, which also reveal the spatial interaction and distribution of the signal at certain temporal scales, are not fully studied. Revealing the detailed channel-to-channel relations during the decomposition and further integrate them in spatio-temporal feature cluster may also very helpful. In the integer representation, the adaptive selection of the bins to construct the histogram is important for data that have closed value in a very narrow band of the integer. Selecting the threshold adaptively according to the data value distribution to construct the histogram can solve this. The information or morphology criteria (e.g., entropy) can be used for the adaptive histogram construction. However, special attention should be paid to avoid the pseudo mode that may be produced by the equalizer. Since both the integer representation and the histogram representation are very fundamental representations, more advanced approximation in the time-frequency space (e.g., the representation of the data with Fourier basis or wavelet basis) may produce much better results. With the spectrum representation theory, the characteristics of the data may be extracted with much more details at a much lower computational cost.

The histogram re-organize similar data values into a single histogram bin, it interlinks the value distribution and the spatio-temporal distribution of the data. The integer representation, which is simple and efficient, not only enhances the feature pattern but also reduces the computational complexity. The soft threshold of the FCM results applied to the histogram can results in more consistency and continuity of the spatio-temporal feature cluster. In addition, the cluster is data adaptive, and a prior knowledge about numbers of objects/features and the distribution of the original data is not required. However, the cluster performance of the method can still be improved in several aspects. In the FCM, the selection of the optimal cluster numbers is still experimental and the efficiency of the selection procedure can also be optimized. In the reversed mapping procedure from the cluster of the histogram to the original data type, the constraints of spatial continuity can be integrated, which can further control the fine level of details of the cluster.

7. Conclusions

In this paper, we propose a new approach that integrates the multi-scale temporal decomposition into the feature cluster of the multi-temporal remote sensing data. The multi-signal wavelet decomposition, the integer approximation and histogram representation and FCM methods are combined to extract the segments of the significant features at different time scales. Our proposed method is efficient and stable, and can be used for extracting and tracking complex geographical phenomena such as El Niño events from the noisy multi-temporal remote sensing data. Experimental results demonstrate that our method can be used as an exploring analysis tool and a foundation to construct methods for more efficient and accurate spatio-temporal feature extraction and feature cluster of multi-temporal remote sensing or other spatio-temporal data.

Acknowledgments

This work was supported by the National Natural Science Foundation of China (Grant No. 41471319, 41231173, 41201377), NCET program (Grant No. NCET-12-0735), and the PAPD project. Support from the Vilas Associate Award, the Hammel Faculty Fellow Award, the Manasse Chair Professorship from the University of Wisconsin-Madison, and the "One-Thousand Talents" Program of China are greatly appreciated. The authors also wish to acknowledge Ming-Song Xu for his constructive comments on this paper.

Author Contributions

All four authors have contributed to the work presented in this paper. Wen Luo, Zhao-Yuan Yu and Lin-Wang Yuan formed the original idea. Wen Luo, Zhao-Yuan Yu and Sheng-Jun Xiao worked on the experiment and data analysis. Wen Luo, Zhao-Yuan Yu and A-Xing Zhu worked on the adaptive FCM algorithm. Wen Luo and Lin-Wang Yuan worked for the computation complexity analysis and evaluation. All authors worked collaboratively on writing the manuscript.

Conflicts of Interest

The authors declare no conflicts of interest.

References

1. Zhang, X.; Xiao, P.; Feng, X.; Wang, J.; Wang, Z. Hybrid region merging method for segmentation of high-resolution remote sensing images. *ISPRS J. Photogramm. Remote Sens.* **2014**, *98*, 19–28.
2. Ouma, Y.O.; Josaphat, S.S.; Tateishi, R. Multiscale remote sensing data segmentation and post-segmentation change detection based on logical modeling: Theoretical exposition and experimental results for forestland cover change analysis. *Comput. Geosci.* **2008**, *34*, 715–737.
3. Kalbermatten, M.; van de Ville, D.; Turberg, P.; Tuia, D.; Joost, S. Multiscale analysis of geomorphological and geological features in high resolution digital elevation models using the wavelet transform. *Geomorphology* **2012**, *138*, 352–363.
4. Gray, J.; Song, C. Mapping leaf area index using spatial, spectral, and temporal information from multiple sensors. *Remote Sens. Environ.* **2012**, *119*, 173–183.
5. Colditz, R.R.; Schmidt, M.; Conrad, C.; Hansen, M.C.; Dech, S. Land cover classification with coarse spatial resolution data to derive continuous and discrete maps for complex regions. *Remote Sens. Environ.* **2011**, *115*, 3264–3275.
6. Zhang, C.; Lin, H.; Chen, M.; Li, R.; Zeng, Z. Scale compatibility analysis in geographic process research: A case study of a meteorological simulation in Hong Kong. *Appl. Geogr.* **2014**, *52*, 135–143.
7. Del Ventisette, C.; Righini, G.; Moretti, S.; Casagli, N. Multitemporal landslides inventory map updating using spaceborne SAR analysis. *Int. J. Appl. Earth Observ. Geoinform.* **2014**, *30*, 238–246.
8. Corbane, C.; Jacob, F.; Raclot, D.; Albergel, J.; Andrieux, P. Multitemporal analysis of hydrological soil surface characteristics using aerial photos: A case study on a Mediterranean vineyard. *Int. J. Appl. Earth Observ. Geoinform.* **2012**, *18*, 356–367.
9. Ardila, J.P.; Bijker, W.; Tolpekin, V.A.; Stein, A. Multitemporal change detection of urban trees using localized region-based active contours in VHR images. *Remote Sens. Environ.* **2012**, *124*, 413–426.
10. Pedersen, T.B.; Jensen, C.S. Multidimensional database technology. *Computer* **2001**, *34*, 40–46
11. Jensen, C.S.; Kligys, A.; Pedersen, T.B.; Timko, I. Multidimensional data modeling for location-based services. *VLDB J.* **2004**, *13*, 1–21.
12. Baumann, P.; Dumitru, A.M.; Merticariu, V. The array database that is not a database: File based array query answering in Rasdaman. In Proceedings of the 13th International Symposium on Advances in Spatial and Temporal Databases, Munich, Germany, 21–23 August 2013.

13. Soroush, E.; Balazinska, M.; Krughoff, S.; Connolly, A. Efficient iterative processing in the SciDB parallel array engine. In Proceedings of the 27th International Conference on Scientific and Statistical Database Management (SSDBM 2015), San Diego, CA, USA, 29 June–1 July 2005.
14. Cheung, D.W.; Zhou, B.; Kao, B.; Kan, H.; Lee, S.D. Towards the building of a dense-region-based OLAP system. *Data Knowl. Eng.* **2001**, *36*, 1–27.
15. Karayannidis, N.; Sellis, T. Hierarchical clustering for OLAP: The CUBE file approach. *VLDB J.* **2008**. *17*, 621–655.
16. Smith, J.R.; Li C.; Jhingran, A. A wavelet framework for adapting data cube views for OLAP. *IEEE Trans. Knowl. Data Eng.* **2004**, *16*, 552–565.
17. Scotese, C.R.; Rees, P.M. The Spatial-Temporal Information Matrix (STIM) cube: An efficient way to store geological information. In Proceedings of the 2004 Annual Meeting of the Geological Society of America, Denver, CO, USA, 7–10 November 2004.
18. Yuan, L.; Yu, Z.; Chen, S.; Luo, W.; Wang, Y.; Lv, G. CAUSTA: Clifford algebra based unified spatial-temporal analysis. *Trans. GIS* **2010**, *14*, 59–83.
19. Yuan, L.; Yu, Z.; Luo, W.; Hu, Y.; Feng, L.; Zhu, A. A Hierarchical tensor-based approach to compressing, updating and querying geospatial data. *IEEE Trans. Knowl. Data Eng.* **2015**, *27*, 312–325.
20. Hu, Z.Q.; Liu, L.P. Applications of wavelet analysis in differential propagation phase shift data de-noising. *Adv. Atmos. Sci.* **2014**, *31*, 825–835.
21. Kaplan, N.H.; Erer, I. Bilateral filtering-based enhanced pansharpening of multispectral satellite images. *IEEE Geosci. Remote Sens. Lett.* **2014**, *11*, 1941–1945.
22. Yuan, L.; Yu, Z.; Yi, L.; Luo, W.; Chen, S. Multiscale spatial decomposition for skew-distributed data with parallel spatial kernel smoothing. *Environ. Plan. B Plan. Des.* **2014**, *41*, 613–636.
23. Xue, C.; Song, W.; Qin, L.; Dong, Q.; Wen, X. A spatiotemporal mining framework for abnormal association patterns in marine environments with a time series of remote sensing images. *Int. J. Appl. Earth Observ.* **2015**, *38*, 105–114.
24. Yu, Z.Y.; Yuan, L.W.; Lu, G.N.; Luo, W.; Xie, Z.R. Coupling characteristics of zonal and meridional sea level change revealed by satellite altimetry data and their response to ENSO events. *Chin. J. Geophys.* **2011**, *54*, 1972–1982. (in Chinese)
25. Hoell, A.; Barlow, M.; Wheeler, M.C.; Funk, C. Disruptions of el nino-southern oscillation teleconnections by the maddenjulian oscillation. *Geophys. Res. Lett.* **2014**, *41*, 998–1004.
26. Oliver, E.C.J. Intraseasonal variability of sea level and circulation in the Gulf of Thailand: The role of the Madden-Julian Oscillation. *Clim. Dyn.* **2014**, *42*, 401–416.
27. Luisier, F.; Blu, T. Sure-let multichannel image denoising: Interscale orthonormal wavelet thresholding. *IEEE Trans. Image Process.* **2008**, *17*, 482–492.
28. Yuan, L.W.; Yu, Z.Y.; Chen, S.F.; Luo, W.; Wang, Y.J.; Lu, G.N. Causta: Clifford algebra-based unified spatio-temporal analysis. *Trans. GIS* **2010**, *14*, 59–83.
29. Sindhumol, S.; Kumar, A.; Balakrishnan, K. Automated brain tissue classification by multisignal wavelet decomposition and independent component analysis. *ISRN Biomed. Imaging* **2013**, *2013*, doi:10.1155/2013/473437.

30. Chen, S.C.; Zhang, D.Q. Robust image segmentation using fcm with spatial constraints based on new kernel-induced distance measure. *IEEE Trans. Syst. Man Cybernet. B Cybernet.* **2004**, *34*, 1907–1916.
31. Cai, W.L.; Chen, S.C.; Zhang, D.Q. Fast and robust fuzzy c-means clustering algorithms incorporating local information for image segmentation. *Pattern Recognit.* **2007**, *40*, 825–838.
32. Baldi, P.; Brunak, S.; Chauvin, Y.; Andersen, C.A.F.; Nielsen, H. Assessing the accuracy of prediction algorithms for classification: An overview. *Bioinformatics* **2000**, *16*, 412–424.
33. AVISO. The Altimeter Products were Produced by Ssalto/Duacs and Distributed by AVISO, with Support from CNES. Available online: http://www.aviso.altimetry.fr/duacs/ (accessed on 3 November 2014).
34. Wolter, K.; Timlin, M.S. El niño/southern oscillation behaviour since 1871 as diagnosed in an extended multivariate enso index (mei.Ext). *Intl. J. Climatol.* **2011**, *31*, 1074–1087.
35. Percival, D.B.; Walden, A.T. *Wavelet Methods for Time Series Analysis*; Cambridge University Press: Cambridge, UK, 2000.
36. Westra, S.; Brown, C.; Lall, U.; Koch, I.; Sharma, A. Interpreting variability in global SST data using independent component analysis and principal component analysis. *Int. J. Climatol.* **2010**, *30*, 333–346.
37. Jena, B.; Sahu, S.; Avinash, K.; Swain, D. Observation of oligotrophic gyre variability in the South Indian Ocean: Environmental forcing and biological response. *Deep Sea Res. Part I Oceanogr. Res. Pap.* **2013**, *80*, 1–10.
38. Macías, D.; Franks, P.J.S.; Ohman, M.D.; Landry, M.R. Modeling the effects of coastal wind- and wind-stress curl-driven upwellings on plankton dynamics in the Southern California current system. *J. Mar. Syst.* **2012**, *94*, 107–119.
39. Lu, Q.; Zuo, J.; Li, Y.; Chen, M. Interannual sea level variability in the tropical pacific ocean from 1993 to 2006. *Glob. Planet. Change* **2013**, *107*, 70–81.
40. Chen, D.; Cane, M.A. El Niño prediction and predictability. *J. Comput. Phys.* **2008**, *227*, 3625–3640.
41. Yu, Z.; Luo, W.; Yi, L.; Hu, Y.; Yuan L. Clifford algebra-based structure filtering analysis for geophysical vector fields. *Nonlin. Process. Geophys*. **2013**, *20*, 563–570,
42. Ashok, K.; Yamagata, T. Climate change: The El Niño with a difference. *Nature* **2009**, *461*, 481–484.
43. Grinsted, A.; Moore, J.C.; Jevrejeva, S. Application of the cross wavelet transform and wavelet coherence to geophysical time series. *Nonlin. Process. Geophys.* **2004**, *11*, 561–566.
44. Liu, Q.; Liu, Z.; Pan, A. Conceptual model about the interaction between El Ninõ/Southern Oscillation and Quasi-Biennial Oscillation in far west equatorial Pacific. *Sci. China Ser. D Earth Sci.* **2006**, *49*, 889–896.
45. Wen, L.; Lin, Y.; Yu, Z.; Sun, H.; Yuan, L. Barycenter reflected equatorial Pacific sea level structure evolution and its indication of ENSO events. *Oceanologia* **2015**, *57*, 107–115.
46. Kolen, J.F.; Hutcheson, T. Reducing the time complexity of the fuzzy c-means algorithm. *IEEE Trans. Fuzzy Syst.* **2002**, *10*, 263–267.

11

A Multimedia Data Visualization Based on *Ad Hoc* Communication Networks and Its Application to Disaster Management

Youhei Kawamura [1,*], Markus Wagner [2], Hyongdoo Jang [3], Hajime Nobuhara [1], Takeshi Shibuya [1], Itaru Kitahara [1], Ashraf M Dewan [4] and Bert Veenendaal [4]

[1] Faculty of Engineering, Science and Information Systems, University of Tsukuba, Ibaraki Prefecture 305-8573, Japan; E-Mails: nobuhara@iit.tsukuba.ac.jp (H.N.); shibuya@iit.tsukuba.ac.jp (T.S.); kitahara@iit.tsukuba.ac.jp (I.K.)

[2] School of Computer Science, University of Adelaide, Adelaide, SA 5005, Australia; E-Mail: markus.wagner@adelaide.edu.au

[3] Department of Mining Engineering and Metallurgical Engineering, Curtin University, Perth, WA 6433, Australia; E-Mail: hyongdoo.jang@curtin.edu.au

[4] Department of Spatial Sciences, Curtin University, Perth, WA 6433, Australia; E-Mails: A.Dewan@curtin.edu.au (A.M.D.); B.Veenendaal@curtin.edu.au (B.V.)

* Author to whom correspondence should be addressed; E-Mail: kawamura@iit.tsukuba.ac.jp

Academic Editors: Songnian Li, Suzana Dragicevic, Xiaohua Tong and Wolfgang Kainz

---

Abstract: After massive earthquakes and other large-scale disasters, existing communication infrastructure may become unavailable and, therefore, it can be quite difficult for relief organizations to fully grasp the impact of the disaster on the affected region. Consequently, this will be the cause of delays to offer the strategic assistance, and to provide water and food, *etc.* In order to solve the problem of re-establishing communication infrastructure to allow for information gathering, we developed an *ad hoc* mobile communications network for disaster-struck areas using ZigBee. As the communication speed of ZigBee is low, we propose a problem-specific image compression method for the multimedia data visualization. By using the proposed method combined with GPS information, it is possible to quickly grasp the damage situation in the region. Through our communication experiments in Tsukuba City, Japan we confirm the effectiveness of our system as a disaster information gathering and management system.

Keywords: disaster; communications system; control of information

1. Introduction

When large-scale disasters strike, a quick damage assessment is very important. Especially in the early stages, because of an allocation limitation of available manpower, it will be a key task to rapidly develop a rescue strategy according to precise information. For example, immediately after the Great East Japan Earthquake in 2011, people could not use communication systems, such as mobile phones and the Internet, in disaster-struck areas. With this lack of communication systems, the decision-making process of aid allocation and appropriate rescue operation was delayed [1–4]. As a result, emergency provisions were not adequately delivered to some evacuation centers.

In this series of research, a disaster management system has been proposed by Ishii and Kawamura [5], which collects and manages disaster information in an integrated fashion at control centers *via ad hoc* networks in order to support rescue efforts, distributions of relief goods, and so on. Concretely, in case of the disaster, investigators or volunteers install ZigBee terminals (see Figure 1 for an example) that have been stored in the evacuation centers or at other safe storage places in areas with interrupted communications infrastructure immediately after the occurrence of a major disaster. After installation, an *ad hoc* network is constructed automatically. The locations of these terminals are determined by the radio wave propagation analysis and optimal placement method that was developed by Ishii and Kawamura [5]. Investigators who are investigating the damages with the help of PCs or PDAs can gather information and transfer these data to the control center via the created ZigBee network. Additionally, it is possible that affected members of the public join the efforts of the professional relief organizations and share their information using their own PCs or PDAs.

Figure 1. ZigBee module used in this study.

Through this long-term research project, a disaster information management system will be developed and the communication speed, its availability, and the terminal installation time between the control center and evacuation centers will be assessed. As the ZigBee communication network standard is not suitable for the transmission of large volumes of data such as image data due to its low transmission rate (256 kbps), we propose a new transmission method that is the combination of GPS information and improved PIC format compression that focuses on spatial information. In a real-world experiment, we assess the effectiveness and the compression ratios by using our proposed disaster information management system.

The structure of this article is as follows. In Chapter 2, we provide an overview of the whole proposed disaster information management system. Then, we propose our method of transmitting and receiving maps in low-speed communication systems in Chapter 3. Finally, we describe in Chapter 4 our experiments of sending and receiving data in Tsukuba City, Japan.

2. Disaster Information Management Systems

2.1. Current Status of Disaster Information Management

Currently, Multi-Channel Access (MCA) wireless communication systems and Ushahidi are put into effect as disaster information management systems after disasters.

Ushahidi (Figure 2, Watanabe *et al.* [6,7]) is the system that was used when the riots occurred during the elections in Kenya (2008). This open source web service was created to share information where such incidents occurred. Concretely, users can send information, such as "There is a riot in the area A" or "There is crime in the area B" via email, which will then appear on top of Google Maps. It has also been used to help collect situational information about military actions in the Gaza Strip (2009), Haiti (2010), after the earthquake and tsunami in Japan (2011), to monitor the elections in India and Mexico (2009), and in other cases. When the Great East Japan Earthquake occurred, the disaster information managing system named "sinsai.info", which is based on Ushahidi, was not only used to grasp the situation of the disaster area, but also to collect and share information about water supplies, official announcements, notices, news, and evacuation calls. It was quickly available and lots of people could understand the situation with the help of the online information visualization. *Via* email, posts on websites, and *via* the information sharing service Twitter it was possible to collect information continuously.

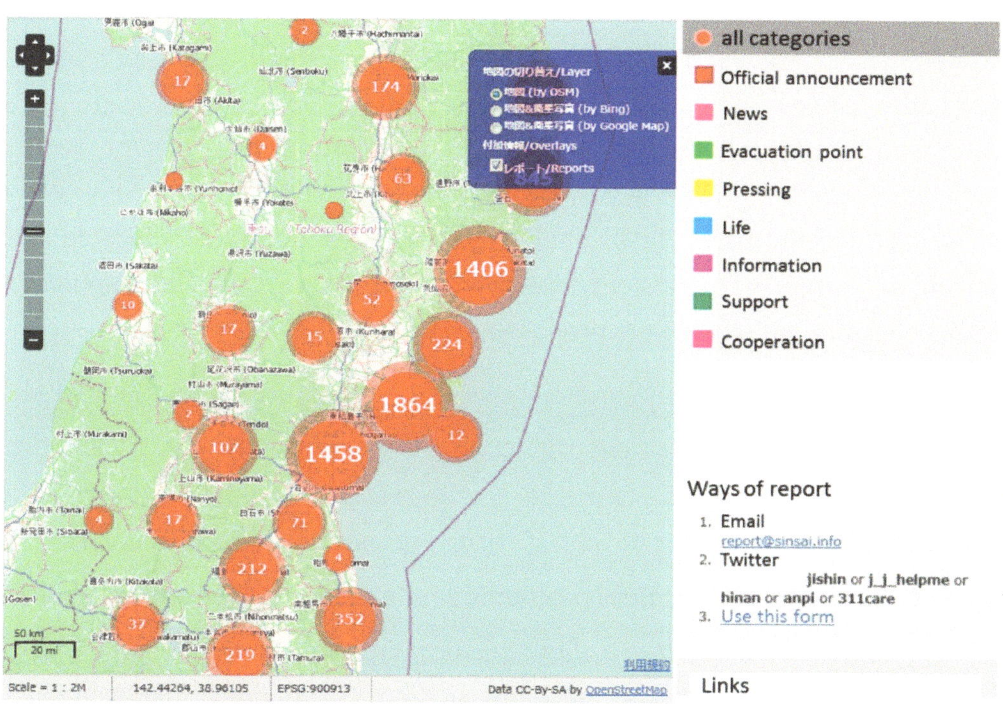

Figure 2. One example of a system based on Ushahidi.

In contrast to the web service Ushahidi, MCA (Multi-Channel Access System) is a commercial wireless communication system (Figure 3). It supports a large number of system users and can use a variety of radio channels. This system consists of a "control station" that the operating body installs and manages, and of "command" and "mobile stations" that the users install and manage in order to achieve radio communication between the units. For example, this system can be used by local governments to share information with fire-fighters and emergency personnel, even if both teams use different wireless communication technology.

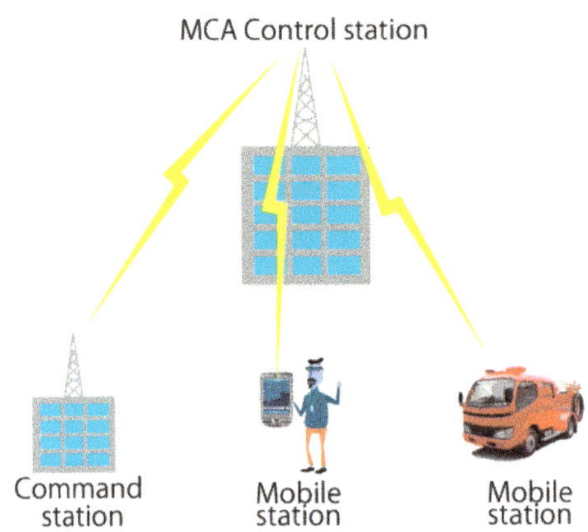

Figure 3. One example of a system based on the Multi-Channel Access System.

2.2. Overview of the Proposed Zigbee Disaster Control System

When a large-scale disaster such as the Great East Japan Earthquake occurs, the damage can be severe enough to take out MCA control stations, as well as fixed-line telephones, mobile phone stations, *etc.* Consequently, it can become difficult to use any of the two above-described disaster information management systems. We consider it an essential requirement for disaster control systems that wireless communication system can work even in such situations where hardly any previously existing infrastructure is available.

In this long-term research project, we develop a disaster information management system that uses the ZigBee communication standard (http://www.zigbee.org/). Figure 4 shows the overall flow of the proposed system. The disaster information management system is installed on PCs located in the control center and in the evacuation centers. This system contains the GIS data (*i.e.*, the geographical map data). The following is the organizational flow of the proposed system:

[Step ①] Immediately after the disaster (Corresponding to ① in Figure 4), the optimal allocation of ZigBee terminals between the control center and evacuation centers is calculated by using the PC in the control center.

[Step ②] The maps are updated with the of Step ① (Corresponding to ② in Figure 4).

[Step ③] Figure 5 illustrates a "ZigBee positioning situation". The communication terminal equipment carried by the investigators has a ZigBee communication module and a GPS module (corresponding to the red square in Figure 5); it must also display the coverage map using a PC or PDA (corresponding to ② in Figure 4). ZigBee terminals (assumed to have been stockpiled in the city hall or in evacuation centers within hours after the disaster) are placed along the way according to the computed optimal arrangement (Corresponding to ③ in Figure 4).

[Step ④] After the deployment of the ZigBee terminals, data transfers across the ZigBee network are available within the ranges that are visualized on the digital map (Corresponding to ④ of Figure 4). After this installation, this system will continue to work until the batteries run out after several months, or until the terminal are collected again.

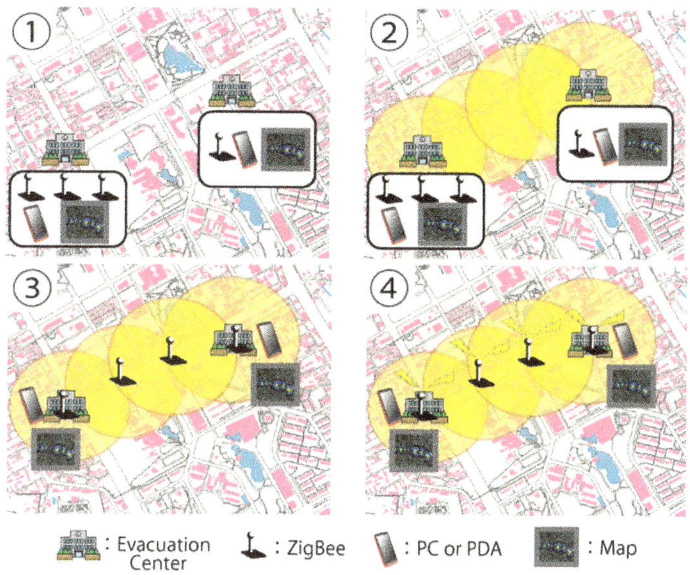

Figure 4. The process of establishing connections between centers: when the evacuation centers are established ①, the ZigBee network is planned ②, then the ZigBee nodes are located ③, and the network can be used ④.

Figure 5. Screen of the proposed system.

Disaster information is information on a variety of materials, such as water supply information, official announcements, notices, news, and location of evacuation centers. In our case, disaster information consists of GPS data and text data, which is displayed on a map as a blue square. In our system, these texts (information) are displayed on the right of the map in chronological order, and once they are selected the corresponding information is highlighted. Additionally, GPS information is highlighted as a color change from blue to green (corresponding to the green of Figure 5), and the relevant text comes up to the first line (corresponding to the yellow of Figure 5). If the PDA or PC have been newly added to this system, or that system is re-connected after a temporary cut of the connection to the network, it is possible to acquire missing information by using the "request update function".

2.2.1. ZigBee Network

ZigBee is one of the communication standards for wireless sensor networks, and it has mainly been used in instrumentation control (see, for example, Kawamura *et al.* [2]).

Table 1 is a comparison of ZigBee with other typical wireless communication standards. Due to its characteristics, ZigBee is suitable, for example, for building automation, industrial automation, and home automation. Furthermore, ZigBee is also used in automatic meter reading, in security systems, and in remote control.

Table 1. Comparison of different communication standards.

Name	ZigBee	Wi-Fi	Bluetooth
Communication distance	10–3000 m	100 m	10 m
Communication speed	250 kbps	11 Mbps	1 Mbps
Network capacity	65,536 nodes	32 nodes	7 nodes
Life-time on battery	several months	several hours	several days
Application	instrumentation control	wireless LAN	wireless accessories

ZigBee has the following four advantages for our scenario:

(1) Support of a large network
As up to 65,635 terminals ("ZigBees") can be configured to form one wireless network, and the configuration can happen independently of location and time.

(2) *Ad hoc* network
It is easy to build an *ad hoc* network, and more stable communication can be realized by providing a higher density of terminals along the intended transfer paths.

(3) Low electric power consumption
Maintaining a long-term battery-powered network is possible.

(4) Low cost
The costs of the ZigBee terminals are low, so it is possible to configure a wide area network by using a large number of them.

With the recent introduction of Bluetooth Low Energy (BLE) a competitor to ZigBee has been developed. While the energy consumption of BLE is significantly below that of Bluetooth, other features still make it unsuitable for our purposes (see for example, Mohan *et al.* [8] and Baker [9]). First, BLE is designed for small, local area networks and it is typically set up in star topologies or in one-to-one connections. ZigBee on the other hand is typically run in wider area networks and in mesh topologies that are significantly more robust when individual nodes fail. This is why we decided to build our disaster information management system based on ZigBee.

Despite these desirable advantages, ZigBee has the disadvantage of providing a relatively low communication speed. In practice this means that it is possible to send and receive data, such as text data of some kilobytes, however, it is difficult to transmit and receive multimedia data of some megabytes. To compensate for this problem, we are using a lightweight text-based communication in the present system. For the transmission of map data, we require significantly compressed data.

2.2.2. Optimized Arrangement Method and Radio Wave Propagation Analysis

It is necessary to construct the network as quickly as possible immediately after the disaster. In order to establish communication between the control center and all evacuation centers, we must decide on the installation locations of the ZigBee terminals. While deciding, we must consider the locations of woodland and water, because these influence the signal attenuation. There is also a need to respond flexibly to changes in the map information, as structures can collapse, and tsunamis and earthquakes can uproot forests. It is necessary to invest a small amount of computation time for estimating the radio wave propagation.

A radio wave propagation analysis method that was developed by Kawamura *et al.* is used in this system. This method uses a sequential tracking method that considers the transmission along the line-of-sight only (no transmission through obstacles, reflections, and diffractions) to reduce the computational complexity. The radio attenuation is estimated for different types of vegetation, and digital maps created based on GIS (Geographic Information System) to get the geographic information in the affected area (Ishii and Kawamura [5]). This method is then used by an iterated ray-launching method that greedily connects the control center with the evacuation sites; the eventual terminal locations are then determined along the paths of the simulated connections. The same algorithms are used by Moridi *et al.* [10] in a mine monitoring and communication system.

3. Transmitting and Receiving Map-Data in Low-Speed Communication

It can be expected that severe disasters change the geographical information, for example, the vegetation is changed, as are man-made structures. Nevertheless, it is necessary that members of the public and of relief organizations can communicate, even when the variations happen. In order to deal with such alterations, our system predicts the radio wave propagation and computes the optimal arrangement using the new map that is obtained and sent to each evacuation center after the disaster. Figure 6 shows the proposed map data transmission method. New map data based on satellite images can be available a few hours after a disaster. This, then, allows us to optimally arrange the terminals based on the situation at hand. Investigators, who have PCs or PDAs to collect disaster related information, need the map of the area and also a visualization of the available range of the radio communication. Therefore, an operator at an evacuation center needs to send updated geographical information (bit-mapped image, ① in Figure 6 shows in the red square the map as updated by a person who is present at the site) based on updated GPS information that is obtained via PCs and PDAs. All information will be combined for the modified map (② in Figure 6 shows the original map, ③ in Figure 6 shows the new map).

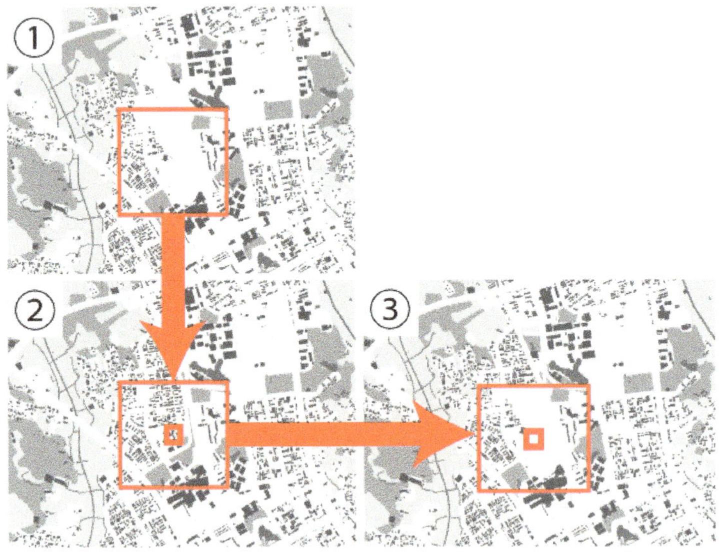

Figure 6. Map data update method.

Initially, this map image would be a bitmap and its image size would be too large to send across low-speed communication networks, such as a ZigBee network; therefore, image-compression is necessary. In this chapter, we propose a new problem-specific image-compression method.

3.1. Existing Image-Compression Methods

The PIC image format can be used for images that have continuous large areas of same colors, such as maps showing geographical information. Our method adopts PIC's run-length encoding method to longitudinal (vertical) direction and it has the characteristic of invertible transformation; for an in-depth description, see Carlson [11].

Figure 7 shows the image-compression method by PIC. Figure 7a shows the original state of the image. First of all, the picture is explored for each pixel from upper left to lower right line-by-line. If the color of an exploring pixel is same as some of the next pixel, the color information of the next one is deleted. If the color of the exploring pixel is different from some of the next pixel, the color information of next one is kept. This process is repeated until the lower right corner of the image is reached in order to find all the points where the color changes. Afterwards, all such points form "chains" from the top to the bottom of the image (Figure 7, ■ and ▲). In the actual compression step, the distances between the points where the colors change is recorded, as is the color itself, and also the direction of the chain and the offset.

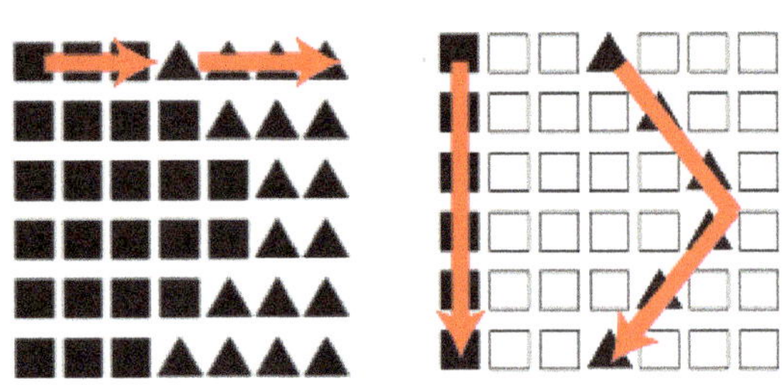

(a) **(b)**

Figure 7. PIC format. (**a**) shows the original picture, (**b**) the right outlines the compression technique.

3.2. Proposed Image-Compression Method

The proposed method is an adaption of the PIC method to improve the compression for disaster-related information management systems. Images of maps have few colors because the colors represent just a small number of geographical information categories. The correspondences of geographical information and colors are stored in a database. Since this database does not change, the control center, the evacuation centers, and the investigators can share it in advance. In the database, geographical information and colors are encoded using three bits, e.g., "000 = road, 001 = water area, *etc.*". We call this suggested compression method GIM-format (Geographical Information Image). In case further data compression is necessary, we will limit the GIM images to only the sector around the current investigators (suggested image data size is 400 × 400 pixels). We will call this method GIMS-format (GIM Sector).

3.3. Comparison of Image-Compression Methods

Figure 8 shows map images that have four different levels of detail. First, we compress each image with GIM, GIMS, PIC, GIF, PNG, and JPEG using Adobe Illustrator; then we compare the resulting data-sizes. The original image data size is 28,473 kB (3006 × 2622 pixels, bitmap format). Note that, in the case of the GIMS-format, we report the mean of nine 400 × 400 pixels cut-outs from the original image.

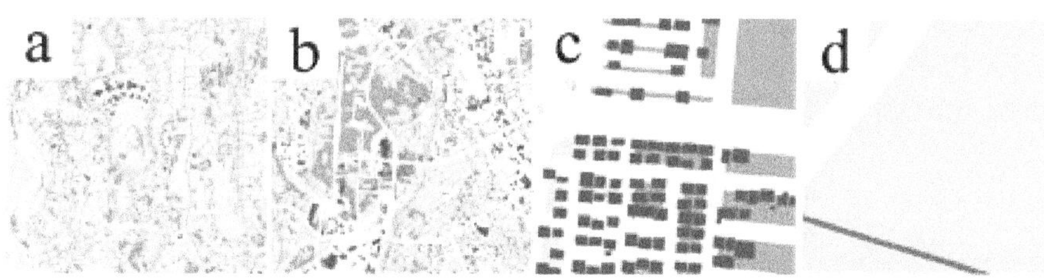

Figure 8. Image data used. The four figures (**a**–**d**) show shows map sections at four different levels of detail.

The results of the compressions are shown in Table 2. The GIM- and GIMS-format are the most effective compression methods. The resulting size of the GIM-format is approximately half of the others format images (except PIC-format) and is about 0.6 times the size of the PIC-format. In particular, the more large areas of the same colors are in the images, the more effective the compression is. PIC is more effective for images that have many small areas of same colors. This complements the effectiveness of the GIM-format for images that have many areas of same color.

Table 2. Comparison of compression methods, number are file sizes in KB.

	a	b	c	d
GIMS	3.62	3.43	0.43	0.14
GIM	252	161	21	6
PIC	355	226	26	8
GIF	567	350	79	20
PNG	615	444	66	50
JPG	980	824	268	187

It is shown that all images can be compressed to 4k bytes or less by cutting out the important sections; thus, dividing the images makes it possible to transmit only necessary map information via the low-speed ZigBee network.

4. Evaluation Experiment for the Proposed System

4.1. Outline of the Evaluation Experiment

We conducted three experiments to evaluate the proposed disaster control system.

(1) Following the system flow, the time is measured for the calculation of the optimal locations of four ZigBee terminals.
(2) The experimenters transmit and receive disaster related information at 20 different spots. On each spot, the communication speed and the communication quality were measured.
(3) Then, the experimenters obtain new map information about their vicinity from a control center using GPS information and update it. After that, the new range of the radio communication is calculated and visualized.

The investigated area for the installation of the ZigBee terminals is 800 × 500 meters around Kasuga 4-chome of Tsukuba City, Japan. Figure 9 shows the calculated optimal arrangement of ZigBee terminals. In the first experiment, four ZigBee terminals were placed; the time was measured while an experimenter placed the terminals.

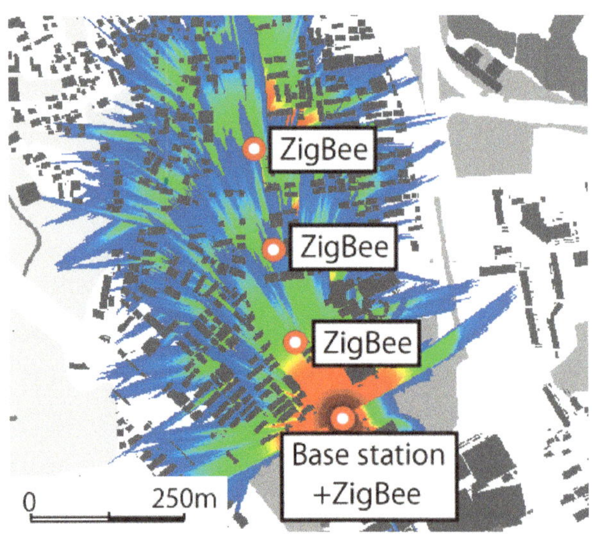

Figure 9. Figure of placement plan.

In the second experiment, the data transmitted and received consists of GPS data and "a communication experiment spot + the number of spot". The data was transmitted from a PC at an evacuation center to a PC on a control center through the ZigBee network. Additionally, the time required to send text data was measured using synchronized clocks on both PCs. By comparing the time-stamps of the messages with the receiving PC's time, we could calculate the transmission times. In addition, we consider a special (and likely) scenario in this second experiment: in one of the 20 measurements, an investigator leaves the area with ZigBee network coverage while investigating (either accidentally or on purpose) and then returns back into the ZigBee covered area. For this scenario, we have implemented a "request update function" to fetch data from the evacuation center was not received due to the missing signal.

In the third experiment, the map image only around an investigator (experimenter) was cut out and the map data was compressed at the evacuation center. The compressed image was sent to the control center and uncompressed. This process was conducted five times at different spots. As in Section 3, the original image size was 3006 × 2622 pixels, while the images were cut out into the size of 400 × 400 pixels for the GIMS-format.

4.2. Results of the Evaluation Experiment

In this section, we describe the results of our study in Tsukuba City, Japan. In this study, we measured the setup time needed, we measured the availability and speed of our ZigBee network, and we updated the geographical information locally and centrally.

4.2.1. Installation and Preparation Time

The experiment's starting time was 13:40 and system installation and preparation was completed at 14:12. In these roughly 30 min four ZigBee terminals (including the control center) were placed according to optimal arrangement calculated with a PC in an evacuation center. Based on results of a previous experiment in this area, it was assumed that one person can cover (investigate) 0.4 km^2 area in 30 min. We can use these numbers to estimate the resources needed in a real large-scale disaster: if the residential area of Tsukuba City (241.07 km^2) would be struck by a disaster, only 100 people (with 24 ZigBees each) would be needed to establish complete coverage with our proposed system within three hours.

4.2.2. Communication Availability

Figure 10 shows the results of transmitting and receiving the multimedia data at 20 spots. 19 spots were located inside of coverage of ZigBee communication and one was located "out of range" as described above. At 13 spots (Figure 10, numbers on white background) out of the 19 spots, data reception was possible. At the six others spots (Figure 10, numbers on red background), communication failed. There are buildings or elevated areas between these spots (Figure 10, numbers 8, 10, 12, 14, 16). The reasons of these failures are probably diffractions of the radio waves. As reflection, diffraction, and permeation of radio waves were not considered when the radio wave attenuation was estimated, the estimation error will be large in case there are a lot of buildings at that moment. Data obtained by an experimenter at spot 15 (Figure 10, yellow square) was stored when it was observed, and it was sent after the experimenter came inside the coverage of the ZigBee network. Subsequently, the stored data was received at the control center. This was a successful proof of the concept that an investigator can collect data while being out of the communication range and then share this data when coming into the communication range again.

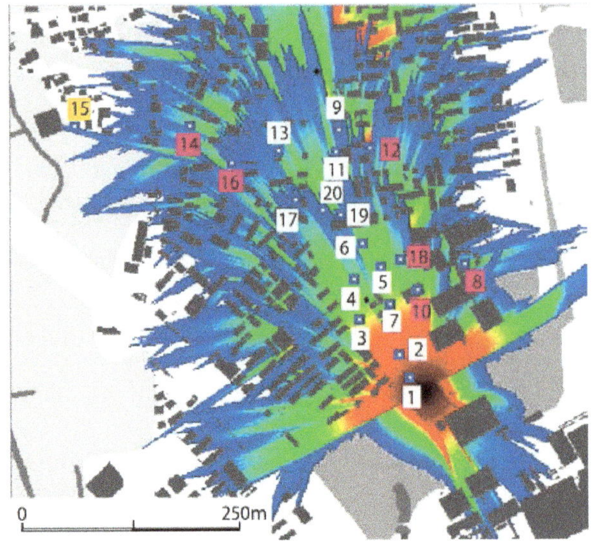

Figure 10. Experimental results of communication situation.

4.2.3. Communication Speed

Figure 11 shows the communication speed at each spot. Communication speeds were measured while the data was transmitted from an evacuation center to a control center at each spots. The images were transmitted and received at 300 ms intervals to prevent one image from splitting into two data sets. Many results concentrated around 300 ms regardless of the distance between the measured spot and the control center, and regardless of how many terminals the data went through. Interestingly, at three spots (Figure 11, numbers 5, 11, 17), results are around 700 ms. This means that first transmission or recipient failed, so the data was transmitted a second time. Thus, the resulting communication time increased to approximately 300 × 2 (ms).

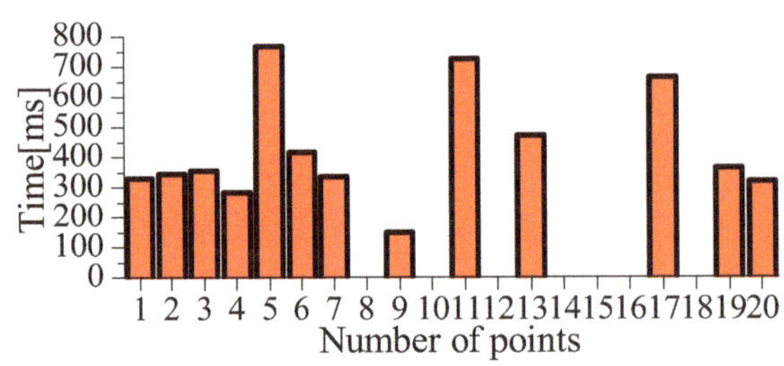

Figure 11. Experimental results of the communication speed tests.

4.2.4. Sending and Receiving Geographical Data Updates

Figure 12 shows the process of a geographical information update (image update) with a PC in an evacuation center. ① in Figure 12 is sent by an investigator to the control center after a disaster, and ② in Figure 12 is a stored image obtained before a disaster at an evacuation center. ③ in Figure 12 in the evacuation center is update from ② to ① only around the investigators GPS location. ④ in Figure 12 shows the result of prediction of radio wave propagation based on the old pre-disaster map and ⑤ in Figure12 shows the result of it based on the updated post-disaster map. Using only the geographical information around an investigator significantly reduces the size of the transmitted data; moreover, the investigators know where to go to communicate based on the updated coverage prediction, so they can easily receive and transmit disaster-related information.

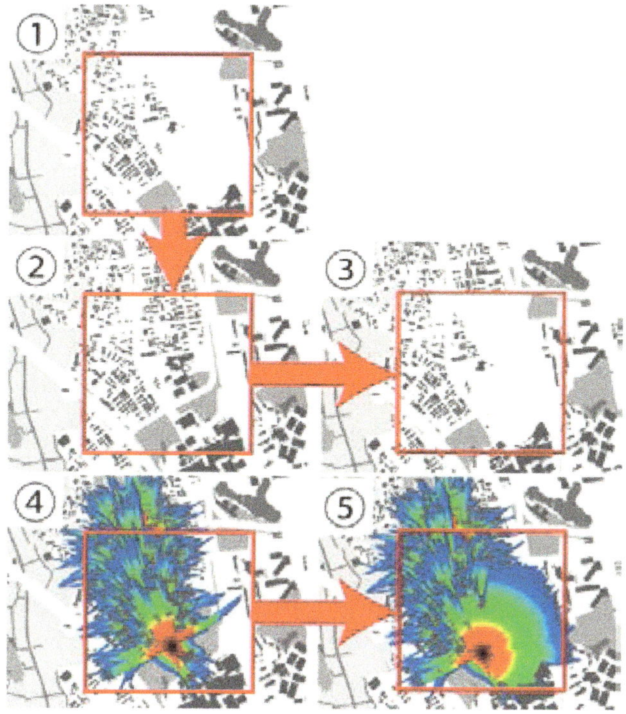

Figure 12. Experimental results of sending and receiving geographical data updates.

5. Conclusions

In this article, we described and evaluated a ZigBee wireless communication system that can be used to quickly gather information and to coordinate efforts when existing communication networks are become unavailable.

We proposed the GIM-format for geographical information and compared it with other image-compression methods. The size of GIM-format compressed images is approximately half of those generated by the other formats (except for the PIC-format). In particular, when the images have many large areas of identical colors (as it is the case of geographical data), the compression is very effective. In addition, we showed that the file size can be reduced further by limiting the image dimensions to the area of interest around the investigator's GPS location.

Through extensive experiments in Tsukuba City, we confirmed that our proposed disaster communication system can be used to gather disaster-related information and to send that data to the control center, which can then update the map for all investigators.

In addition, the experiment of transmitting and receiving map-data proved that the method is effective for the low-speed communication network of ZigBee terminals.

To conclude, the proposed disaster control system can be an alternative communication infrastructure when existing infrastructure is damaged by a major disaster. The system is easy to use even for laypeople, which enables the affected people to assist the professional organizations in their emergency relief efforts.

Acknowledgments

The authors would like to thank the reviewers and the editors for their comprehensive feedback, which helped to improve the quality of the article.

Author Contributions

All authors collaborated on this article in different ways that range from the inception of the idea, to the implementation of the tool, the conduct of the study, and the editing of the article. All authors contributed enough to warrant their coauthorship.

Conflicts of Interest

The authors declare no conflict of interest.

References

1. Sugit, H.; Fukuta, T.; Tamura, T.; Yokoi, T.; Hara, T.; Kashima, T.; Azuhata, T.; Shibasaki, B.; Yagi, Y. IISEE-NET: Information network for earthquake disaster mitigation of developing countries. In *Bulletin of the International Institute of Seismology and Earthquake Engineering*; Building Research Institute: Ibaraki, Japan, 2003; pp. 137–143.
2. Kawamura, Y.; Dewan, A.; Veenendaal, B.; Shibuya, T.; Hayashi, M.; Kitahara, I.; Nobuhara, H.; Ishii, K. Using GIS to develop a mobile communications network for disaster-damaged areas. *Int. J. Digit. Earth*, **2014**, *7*, 279–293.
3. Kawamura, Y.; Ishii, K.; Jang, H.; Wagner, M.; Nobuhara, H.; Dewan, A.; Veenendaal, B.; Kitahara, I. Analysis of radio wave propagation in an urban environment and its application to initial disaster response support. *J. Dis. Res.* **2015**, *10*, 655–666.
4. Nakahata, Y.; Kawamura, Y. Development of the landslide observation system using ZigBee. In Proceedings of the 2010 Society of Instrument and Control Engineers Annual Conference (SICE), Taipei, Taiwan, 18–21 August 2010; pp. 1191–1194.
5. Ishii, K.; Kawamura, Y. Analysis of radio wave propagation in an urban environment and its application to initial disaster response support. In Proceedings of the 2010 Society of Instrument and Control Engineers Annual Conference (SICE), Taipei, Taiwan, 18–21 August 2010; pp. 1195–1198.
6. Watanabe, N.; Uno, K. Utilization of sensor Web and spatial information platform as a supportive measure for field work monitoring. *Rep. Inst. Sci. Tech. Res.* **2011**, *23*, 62–66.
7. Watanabe, K.; Ichikawa, T. A study of the information exchanging system for a person requiring care at the time of earthquake disaster. *SIG Tech. Rep.* **2006**, *92*, 25–29.
8. Mohan, M.; George, N.; Davis, D. ZigBee technology for data communication—A comparative study with other wireless technologies. *Int. J. Adv. Res. Comput. Sci.* **2014**, *5*, 261–265.
9. Baker, N. ZigBee and Bluetooth strengths and weaknesses for industrial applications. *Comput. Contr. Engin. J.* **2005**, *16*, 20–25.
10. Moridi, M.A.; Kawamura, Y.; Sharifzadeh, M.; Chanda, E.K.; Wagner, M.; Jang, H.; Okawa, H. Development of underground mine monitoring and communication system integrated ZigBee and GIS. *Int. J. Min. Sci. Technol.* **2015**, in press.

11. Carlson, W.E. A Survey of Computer Graphics Image Encoding and Storage Formats. *Comput. Graph.* **1991**, *25*, 67–75.

12

Developing a Relative Ranking of Social Vulnerability of Governorates of Yemen to Humanitarian Crisis

Joseph Kandeh [1,*] and Lalit Kumar [2]

[1] United Nations Office for the Coordination of Humanitarian Affairs (UN OCHA), Sana'a, Yemen

[2] School of Environmental and Rural Science, University of New England, Armidale NSW 2351, Australia; E-Mail: lkumar@une.edu.au

* Author to whom correspondence should be addressed; E-Mail: kandehj@un.org

Academic Editors: Christoph Aubrecht and Wolfgang Kainz

Abstract: The social vulnerability of the Yemeni population to humanitarian emergencies is not evenly distributed between the governorates. Some governorates may be more susceptible to the impacts than others, based on the circumstances of the people residing within them. In this paper, we present a methodology for assessing social vulnerability of governorates of Yemen to humanitarian emergencies using a Geographic Information Systems approach. We develop a spatial index of social vulnerability from an initial list of 80 variables that were reduced to 12 factors through Principal Component Analysis. Our findings show that the differences in social vulnerability between governorates are primarily driven by 12 factors, of which education, lack of basic services in health, water and sanitation, electricity, housing quality, poverty, limited livelihood opportunities, and internal and external displacement are the major determinants. The results show that the factors that contribute to social vulnerability are different for each governorate, underpinning the need for context-specific vulnerability reduction approaches. The most vulnerable governorates are characterized by conflicts, armed clashes and violence. The geographic variability in social vulnerability further underpins the need for different mitigation, humanitarian response and recovery actions. The use of Geographic Information Systems approach has contributed to our understanding of the geographies of vulnerability to humanitarian emergencies in Yemen.

Keywords: risk; vulnerability; social vulnerability; natural disasters; humanitarian emergencies; humanitarian crisis

1. Introduction

Yemen is one of the countries described by the United Nations (UN) as a complex humanitarian crisis. There are about 15 million people out of a total population of 24 million requiring some form of humanitarian assistance. According to the 2015 Yemen Humanitarian Needs Overview, 10.6 million Yemenis are food insecure [1] and 13 million are without adequate safe drinking water or sanitation [2]. One million children under the age of five years are malnourished, while over one million are either internally displaced, primarily migrants from the Horn of Africa, or Yemenis returning from the Kingdom of Saudi Arabia [3].

The threats of conflicts, acts of terrorism, insecurity, access constraints, natural disasters such as flooding, landslides, pandemics and many more undermine the efforts of humanitarian assistance. In addition, Yemen's humanitarian crisis continues to be driven by under-development, weak governance, environmental stress, demographic pressure, dwindling natural resources, extreme poverty and continued political instability. At the beginning of 2015, there was a significant increase in conflict in Yemen with the Al Houthis, a group belonging to the Shia form of Islam, expanding their control across many governorates in the country. The expansion of the Al Houthis met fierce resistance from the local tribes and elements of the Al Qaeda of the Arab Peninsular (AQAP) and its affiliates. On 26 March 2015, a Saudi-led coalition began airstrikes targeting Al Houthi positions and military installations. At the end of March, all international staff of humanitarian actors in Yemen, including the UN, evacuated all their staff out of the country. The fighting has also escalated in the south between the Al Houthis and the Popular Committees. The World Health Organization reports that over 2000 people have been killed with over 8000 injured since 19 March 2015. The crisis has generated over a million internally displaced persons [4].

At the beginning of April 2015, the UN launched an appeal of US$274 million to reach 7.5 million people with life-saving assistance and protection covering a three-month period. The appeal identified priority needs in the areas of health, water and sanitation, protection, emergency shelter, non-food items and food aid. At the beginning of July 2015, the United Nations declared the Yemen humanitarian emergency as Level 3, a category reserved for the most severe, large-scale and complex crises. The Yemen humanitarian emergency has further increased the vulnerability of the Yemeni people. The crisis has revealed that some places are more vulnerable to the humanitarian crisis than others, with the demographic characteristics of the Yemeni people also contributing to this vulnerability [5,6].

1.1. Vulnerability to Humanitarian Crisis

In this paper, we broadly define humanitarian emergency drawing from Byrne and Baden [7] six pronged classification, which includes: (1) emergencies of natural rapid onset (e.g., earthquakes, cyclones); (2) emergencies of technological rapid onset (e.g., fuel, chemical and nuclear accidents); (3) emergencies of slow onset (triggered by natural disasters such as drought and pest attacks);

(4) permanent emergencies (e.g., structural poverty); (5) emergencies of mass population displacements; and (6) "complex" emergencies (generally associated with civil war and armed conflicts). The Humanitarian Coalition defined humanitarian emergency as an event or series of events that represents a critical threat to the health, safety, security or well-being of a community or other large group of people, usually over a wide area [8]. Most of the humanitarian crisis in the world are caused by natural disasters but quite a substantial number have also been caused by civil war and armed conflicts [9].Yemen's humanitarian crisis is caused by poverty, extensive violence and loss of lives, massive displacement of populations, widespread damage to property and the economy, war and insecurity. In this paper, humanitarian crisis and humanitarian emergencies are used interchangeably.

The concept of vulnerability is highly contested and is often used with differing connotations [10–12]. While there are many different definitions of vulnerability [12,13], in this paper, we define vulnerability as being a function of both the physical environment and of the socio-economic and political context which increase the susceptibility of a community during humanitarian emergencies.

Evidence from recent humanitarian crisis, such as in Libya, Syria, Iraq, Central African Republic and Yemen, suggest that vulnerability to humanitarian emergencies vary between populations, age groups, households, communities, and the location in general [2,3,14–16].

A lot has been done to understand the circumstances and needs of people in order to tailor appropriate responses during humanitarian crisis [9] using tools such as the Multi-cluster/sector Initial Rapid Assessment (MIRA), Assessment Capacities Project (ACAPS), risk and early warning initiatives such as the drought early warning system in Karamoja, Uganda, and the World Food Programme's Comprehensive Food Security and Vulnerability Assessment. However, less has been done to understand vulnerability to humanitarian crisis, specifically the potential impact of humanitarian crisis on people and places. Given the devastating impact of the Yemen humanitarian crisis, humanitarian agencies need understanding of who and what places are vulnerable to the crisis to enable them to prepare context-specific and effective responses for individuals based on their vulnerability and the vulnerability of the governorates in which they live.

The current approaches to measuring vulnerability broadly falls into the categories of the indicator approach using both quantitative [6,17] and qualitative, engineering approaches focusing on loss estimations with an emphasis on the hazard component, and the broader aspects of vulnerability such as trying to capture institutional aspects [11]. By far, the use of indicators is the most used in the literature. Numerous indices to estimate various vulnerabilities exist, with different conceptual frameworks and methodologies, underpinning the lack of a consensus [18]. Cutter, Boruff and Shirley [6] used the hazards of place vulnerability model to create an index of social vulnerability to environmental hazards, calling it the Social Vulnerability Index (SoVI). The SoVI approach has been applied at various spatial and temporal scales [17,19–22] in the areas of climate change, mortality, morbidity, heat waves, flood disasters and agriculture [23–26].

Use of indices to assess vulnerability in the field of humanitarian assistance, refugee camp situation, humanitarian funding, and humanitarian crisis also exist, with most available at the national scale. These include the Global Vulnerability Assessment (GVCA), and the Forgotten Crisis Assessment (FCA) of the European Union, the Humanitarian Response Index [27], the IDP Vulnerability Assessment and Profiling (IVAP) [28] , Global Hunger Index [29], and the Global Conflict Risk Index [30]. At the beginning of 2015, the United Nations and European Commission released the results of its Index for Risk Management (INFORM) [31]. INFORM is a tool for predicting where humanitarian crisis will occur based on a series of indicators that measure hazards and exposure, vulnerability, and capacity across 191 countries. Most of the current tools for assessing vulnerability to humanitarian crisis operate at a national or regional scale. National scale assessments of vulnerability and its consequent rankings are of strategic relevance but provide limited operational value in addressing local scale vulnerability reduction.

The development of a geospatial vulnerability assessment that can identify places and characteristics of populations with the greatest risk to humanitarian crisis at sub-national scale is scant in the literature. Despite the years of conflict, studies of Yemen's vulnerability to humanitarian crisis are limited. Empirical evidence of the underlying factors that generate vulnerability to humanitarian crisis in Yemen is lacking. Developing an index of social vulnerability to humanitarian emergencies will provide a useful tool for comparing spatial variability in social vulnerability. Such an index can also be linked to physical vulnerability in calculating the overall vulnerability of a place to humanitarian emergency.

Measuring vulnerability of governorates in Yemen is a pre-requisite to any attempt in reducing the impact of humanitarian crisis. The goal of this study is to develop a relative ranking of social vulnerability of governorates in Yemen to humanitarian emergencies, and determine the underlying factors that contribute to the increase or decrease of it. Specific objectives include: (i) developing a methodology that uses a Geographic Information Systems (GIS) that maps social vulnerability; and (ii) determining the geographies of social vulnerability of governorates in Yemen to humanitarian emergencies.

1.2. Framework for Assessing Social Vulnerability to Humanitarian Crisis

Vulnerability to humanitarian crisis generally means the potential for human suffering. The overarching goal of responding to humanitarian crisis is to alleviate human suffering. In this paper, the risk of humanitarian emergencies is the sum of crisis exposure, vulnerability and capacity measures (see Figure 1). With low capacity and high vulnerability, the risk of humanitarian emergencies will be high. However, with high vulnerability, even with a high capacity, the risk of humanitarian emergency is high even with a low magnitude crisis exposure. The overall vulnerability is viewed as a combination of physical and socioeconomic characteristics to provide a measure of the overall vulnerability of the community, which Cutter [12] termed the place vulnerability [12,32]. The focus of this paper is assessing social vulnerability of governorates of Yemen to humanitarian emergencies. Adapting Cutter's definition [6], we define social vulnerability to humanitarian crisis as a "measure of both the sensitivity of a population to events and conditions that trigger humanitarian crisis and its ability to respond to and recover from the impacts of such emergencies and crisis".

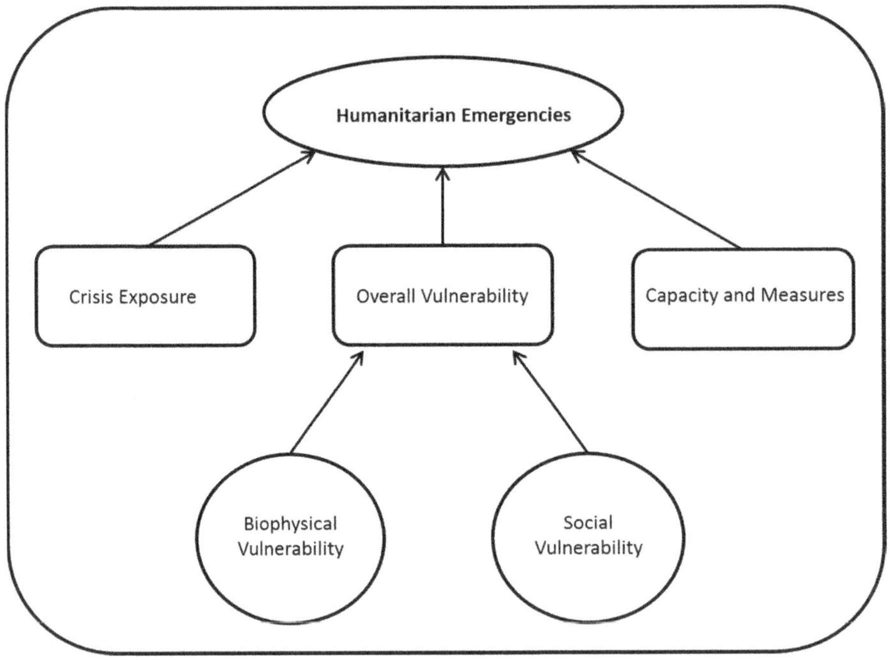

Figure 1. Humanitarian Crisis Framework.

1.3. Vulnerability to Humanitarian Crisis Factors

Empirical evidence suggests that different groups of people often experience different degrees of injury, loss and disruption from the same or similar event and also manifest differential ability to cope and recover [33,34]. Different groups of people are affected by crises in different ways based on their conditions and the conditions of the places in which they live [6]. Some will have more resources, human and material, to deal with the crisis, whilst others will not have the required resources. A number of vulnerability factors to humanitarian crisis exists in the literature [27,29,30]. In addition, the most commonly cited vulnerable groups and socio-economic and demographic characteristics of vulnerability can be found in the literature [6,35–39].

Evidence from Yemen and other countries experiencing severe humanitarian situation suggests that the most vulnerable groups and factors influencing vulnerability include age (the under-five, the elderly above 60 years, particularly the frail and impoverished), children under five years, particularly those that are malnourished, malnourished lactating mothers, large families, households that are food insecure, women, female headed households, widows and divorced, socio-economic status, households below the poverty line and, in the case of Yemen, including the Social Welfare Fund Beneficiaries, ethnic minorities such as the Mohamedsheen, and people displaced, particularly those from the Horn of Africa, internal displacement, and Yemenis deported from the Kingdom of Saudi Arabia [2,3,14–16]. Significant also are special needs populations such as the disabled, infirmed, and stranded migrant workers looking for opportunities to go to the Kingdom of Saudi Arabia and other Gulf countries [2,14–16]. In humanitarian crisis, such as in South Sudan, women and girls are extremely vulnerable as they are increasingly in danger of sexual and gender-based violence [9]. In refugee and IDP camps resulting from humanitarian crisis, women and children are over-represented [40].

Despite the fact that the identified vulnerable groups are often the most affected during a humanitarian crisis, it is the intersection of the various groupings that further increases vulnerability. Being poor, female-headed household with malnourished children below five will be disproportionately affected by a humanitarian crisis. Consensus exists that other factors that also influence vulnerability include insufficient access to resources, information, knowledge and technology, limited access to political power and representation, beliefs and customs, and building stock; all of which will, in turn, be influenced by the socio-economic and demographic factors [6,34]. Remittances have been cited as playing a vital role in supporting families affected by humanitarian crisis [40,41]. According to the World Bank, 62% of the remittance into Yemen in 2014 originated from the Kingdom of Saudi Arabia. However, at the beginning of 2014, due to changes in the labor laws in the Kingdom of Saudi Arabia, a large number of Yemenis were deported to Yemen thereby significantly reducing the amount remitted and further adding to the vulnerability of families that depended on migrant remittance.

2. Methods

According to the Yemen Central Statistics [42], Yemen is bordered by Oman on the northeast, Saudi Arabia in the north, by the Red Sea in the west and the Arabian Sea in the south (Figure 2). Yemen has an area of 527,970 km^2 (203,850 sq mi), including the islands of Perim at the southern end of the Red Sea and Socotra at the entrance to the Gulf of Aden. Yemen is a mountainous country with only the coastal plains being the lowest-lying areas. The elevation ranges from sea level to about 3666 m (12,028 ft) with an average elevation of about 2000 m (6600 ft). The southwestern tip of Yemen links up with the Bab el-Mandeb Strait, which links with the Horn of Africa. It is this close proximity with the Horn of Africa that contributes to thousands of African migrants from Eritrea, Somalia, Ethiopia and Djibouti crossing into Yemen with the hope of reaching the gulf countries. Yemen generally has a hot and extremely dry climate except in the western highlands, where the climate is more moderate. Yemen can be divided geographically into five major areas: the Mountainous area, the Coastal area, the Plateau area, the Desert (Al-Ruba Al-Khali) area, and the Yemeni islands. In terms of administrative division, Yemen is composed of four Regions at the national level (northern, southern, western and eastern), 22 Governorates, 333 Districts and over 37,000 Settlements. In the context of Yemen, governorate scale is where most of the decisions relating to delivery of humanitarian assistance occurs.

Due to armed conflicts and insecurity in the past years, up-to-date data on needs and on prevailing conditions are not available. Most of the up-to-date data on socio-economic variables are scarce. The last census was conducted in 2004, while the recent Yemen National Demographic Health Survey, which was completed in 2014, has still not been made public. For this research, we used the 2012 governorate level socioeconomic and demographic data from the Yemen Central Statistics Office to construct an index of social vulnerability.

Using the literature, appropriate indicators were selected to represent social vulnerability of governorates of Yemen [6,26,30,43]. The methodology used is shown in Figure 3. An initial list of 80 variables was selected and data for these variables were gathered for 22 governorates. Most of the data was gathered from the 2012 Yemen Statistical Yearbook [44] with the rest obtained from the Yemen Humanitarian Programme Cycle, particularly the 2015 Humanitarian Needs Overview [3,16].

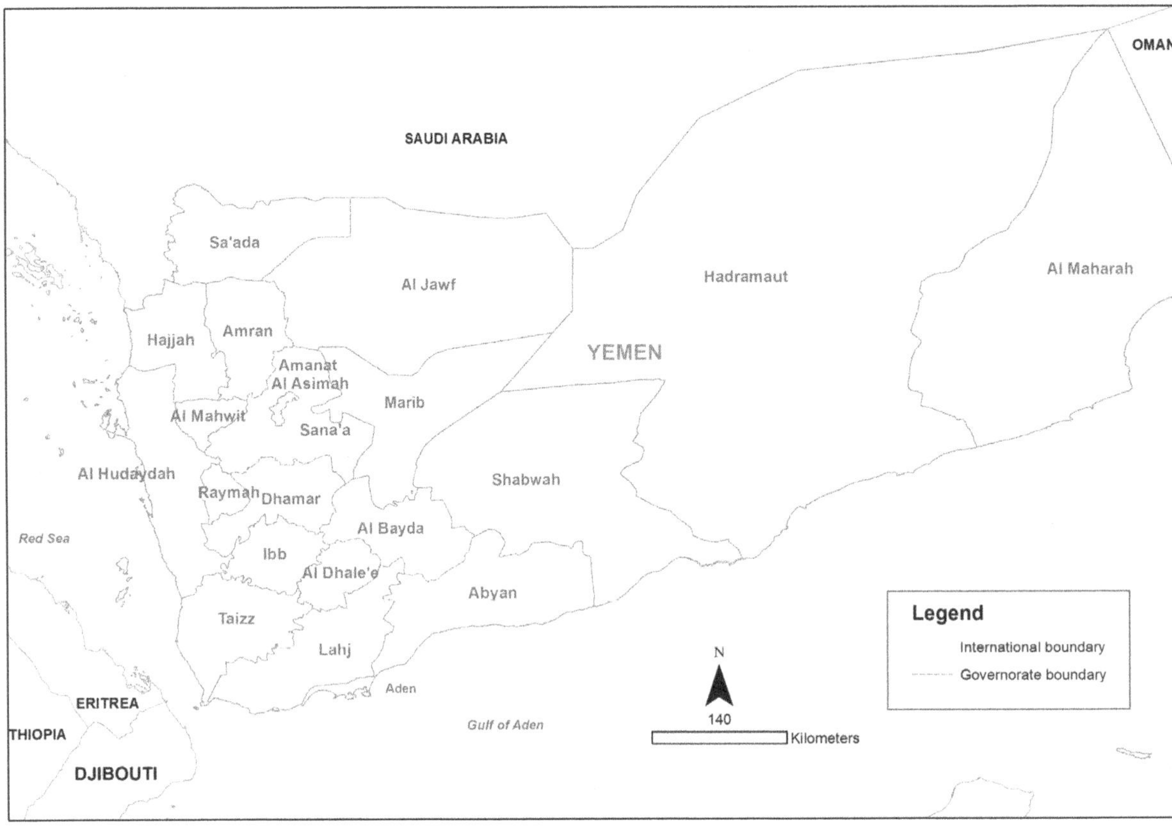

Figure 2. Map of Yemen showing administrative boundaries and locations of governorates.

In preparing the data for the social vulnerability analysis, raw data was converted either to percentages, averages or density functions and exported into statistical software called Statistical Package for Social Scientists (SPSS) version 18. The 80 variables were subjected to exploratory data analysis and tested for multicollinearity [6,45] to ensure that variables were not double counted. After testing for collinearity, 10 of the variables were removed as they showed a nearly perfect correlation with other variables. Considering that the variables were in different measurement units, the remaining 70 variables were standardized using the Z-score approach [6,46]. This was required in order to ensure that the variables could be aggregated [6].

Considering that social vulnerability is a multi-dimensional concept, we used a Principal Component Analysis (PCA) as the tool to quantify its multivariate nature and for data reduction [8,11,28]. With the 70 variables standardized, we ran the PCA several times using a varimax rotation option, eigenvalue greater than 1 and with Kaiser normalization applied with a pairwise exclusion option for missing values [17,21,47]. For optimal PCA solution, all the variables had correlation coefficients below 0.8, a Kaiser-Meyer-Olkin (KMO) minimum of 0.5, and we ensured that all variables in the communality table were also above 0.5 [48,49].

After several runs of PCA, the 70 variables were further reduced to 45. Table 1 shows the selected social variables used in constructing the index of social vulnerability to humanitarian emergencies. With optimal PCA solution, 12 factors were retained with no changes made. The 12 factors had eigenvalues greater than 1 and explained 92% of variance among the 22 governorates. Table 2 shows the 12 factors with the loaded variables. Factor scores were calculated for each of the 12 factors for each governorate using estimated scoring coefficients in SPSS (version 18.0; SPSS Inc.). Cutter, Boruff and Shirley [6] scaled all variables that increased social vulnerability to be positive and all those that reduced it to be negative. In this study, the extracted factors either have all the variables in same positive direction or have some variables in positive direction and others in the negative direction. Jolliffe [50] noted that the sign of any extracted PCA factor was completely arbitrary, and that reversing the sign of a factor does not change the variance. In this study, individual variables were not scaled.

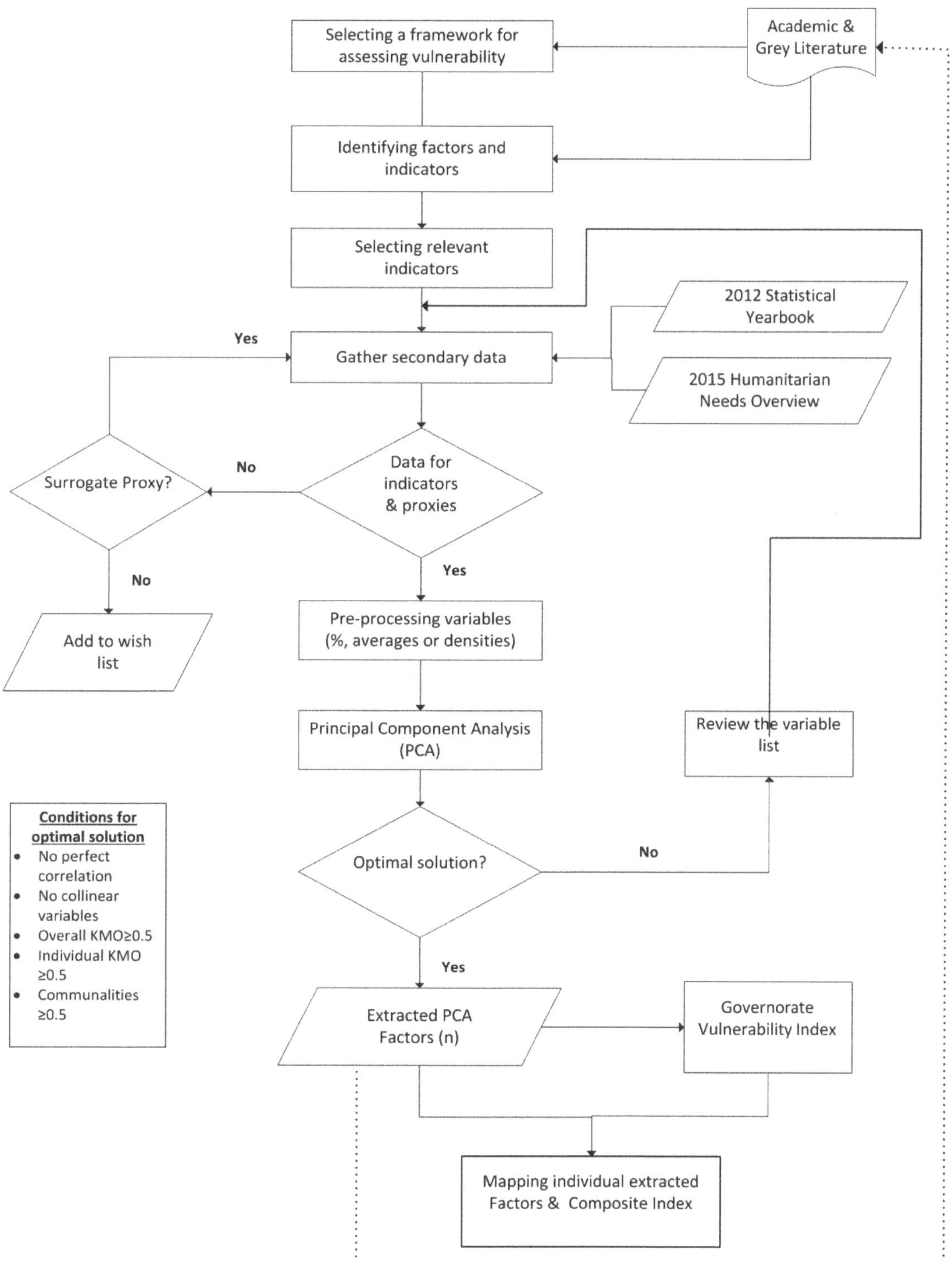

Figure 3. Methodology for assessing social vulnerability of governorates in Yemen.

Table 1. Social variables and the concept of social vulnerability each variable addresses.

Variables	Social Vulnerability Concept	Variables	Social Vulnerability Concept
Percent of population under five years old	Age	Percent of population under 5 years malnourished	Nutrition
Percent over 60 year old	Age	Percent of population under 5 years stunted	Nutrition
Total IDPs and Returnees	Displacement	Percent of female population on Social Welfare Fund	Poverty
Total Refugees and Migrants	Displacement	Percent of population on Social Welfare Fund	Poverty
Agricultural area under cultivation as a percent of total land	Economy	Percent households in poverty 2006	Poverty
Agricultural area cropped as a percent of total land	Economy	Number of gender based violence reported	Protection
Percent of population that cannot read and write	Education	Percent of households without safe drinking water	Sanitation
Percent of population with secondary education	Education	Percent of households without sanitary latrines	Sanitation
Percent of population with unknown education	Education	Percent of households with agricultural land	Socioeconomic status
Percent of population with masters education	Education	Percent of households living in huts	Socioeconomic status
Percent females population unemployed	Employment	Percent of households using Trucks for water sources	Wealth
Percent total population unemployed	Employment	Percent of households using 5 gallon water containers for hauling water	Wealth
Average number of people per household	Family structure	Percent of households using other water sources	Wealth
Percent females	Gender	Percent of households using public electricity	Wealth
Percent of population under 5 years without vaccination	Health	Percent of households using private electricity	Wealth
Percent of population under 5 years diarrhea infection	Health	Percent of households using cooperative electricity	Wealth
Child mortality rate	Health	Percent of households using private generator for electricity	Wealth
Under five mortality rate	Health	Percent of households using kerosene for lighting source	Wealth
Percent of asphalt roads	Infrastructure	Percent of households using gas for lighting source	Wealth
Hospital beds/10,000	Infrastructure	Percent of households using other light sources	Wealth
Hospitals/10,000	Infrastructure	Percent renter-occupied housing unit	Wealth
Health personnel/10,000	Infrastructure	Percent of households food insecure	Wealth
Percent of population under 5 years wasting	Nutrition	-	-

Due to the lack of a defensible weighting scheme and an established relationship between the variables and the extracted factors [51], two aggregation methods were used. In the first approach, the factors were given equal weighting and placed in an additive model [6,46] to produce the Yemen Social Vulnerability Index to Humanitarian Crisis (SoVIHC). In the second approach, using the PCA derived variances of the twelve factors, weights were derived and multiplied against the factor scores of each governorate and then placed also in an additive model. The 12 factor scores and the composite SoVIHC of both the unweighted and weighted factors were exported into ArcGIS 10.1 and mapped as standard deviations from the mean into five categories. Mapping vulnerability using this approach highlights the degree to which some governorates are more vulnerable than others whilst also highlighting their geographic variability.

Table 2. The twelve underlying factors used to construct the social vulnerability index of governorates of Yemen.

Factor	Name	Percent Variance Explained
1	Education	15.67
2	Lack of Basic Services: Poor Sanitation, electricity, housing, health	15.57
3	Social Welfare Fund Population	8.57
4	Poverty and Unemployment	8.04
5	Poor Housing and Illiteracy	6.63
6	Special Groups (Children and the Elderly)	6.57
7	Inadequate Water and Limited Education	6.34
8	Infrastructure; inadequate buildings, hospital facilities	6.11
9	Child Mortality and Health	5.30
10	Internal Displacement	5.22
11	Health, Water and Sanitation Infrastructure	4.63
12	Refugee and Migrant	3.35

3. Results and Discussion

Twelve composite factors were extracted that differentiated governorates of Yemen according to their relative level of social vulnerability to humanitarian emergencies. These factors explained 91.8% of the variance. Each of the twelve factors is described below.

Education: Educational attainment is linked to socioeconomic status with higher levels of education attained resulting in better jobs, higher earnings and savings. Lower education may affect people's ability to understand and access critical information during humanitarian emergency periods [34]. The first factor, explaining 15.7% of the variance, broadly relates to educational level of the population in the governorates. There is agreement that the educational level reduces vulnerability as the literate are able to participate in decision making as in camp environments or are able to access information relating to their entitlement. Years of conflict in Yemen has led to most of the schools to be dilapidated, children out of school and in some cases internally displaced populations housed in schools. Based on this factor, the most vulnerable governorate is Hadramaut, followed by Raymah and Dhamar (Figure 4a). The least vulnerable governorates are Sana'a, the capital city, Marib, Al Bayda, Al Dhale'e, Lahj and Aden.

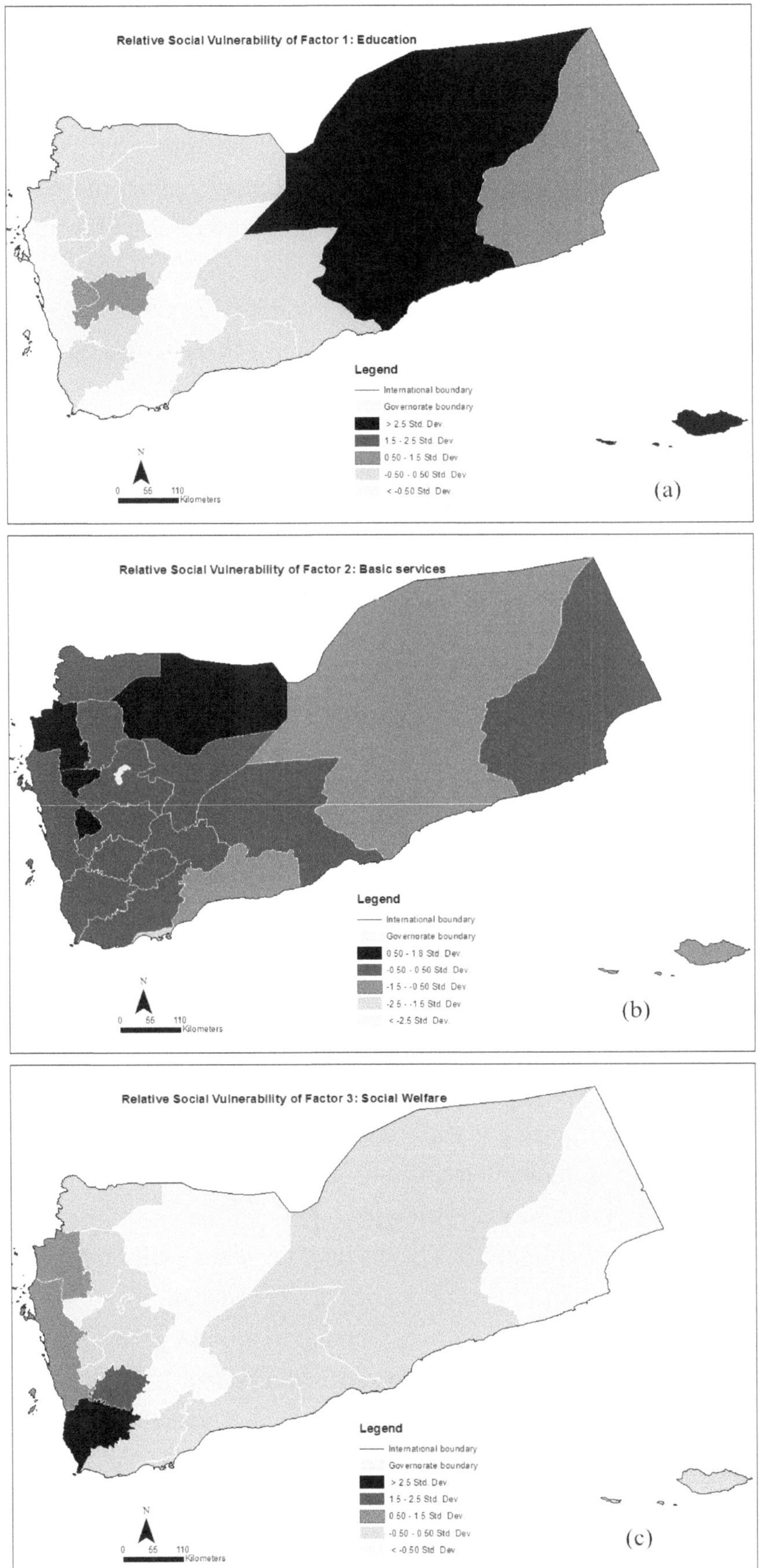

Figure 4. *Cont.*

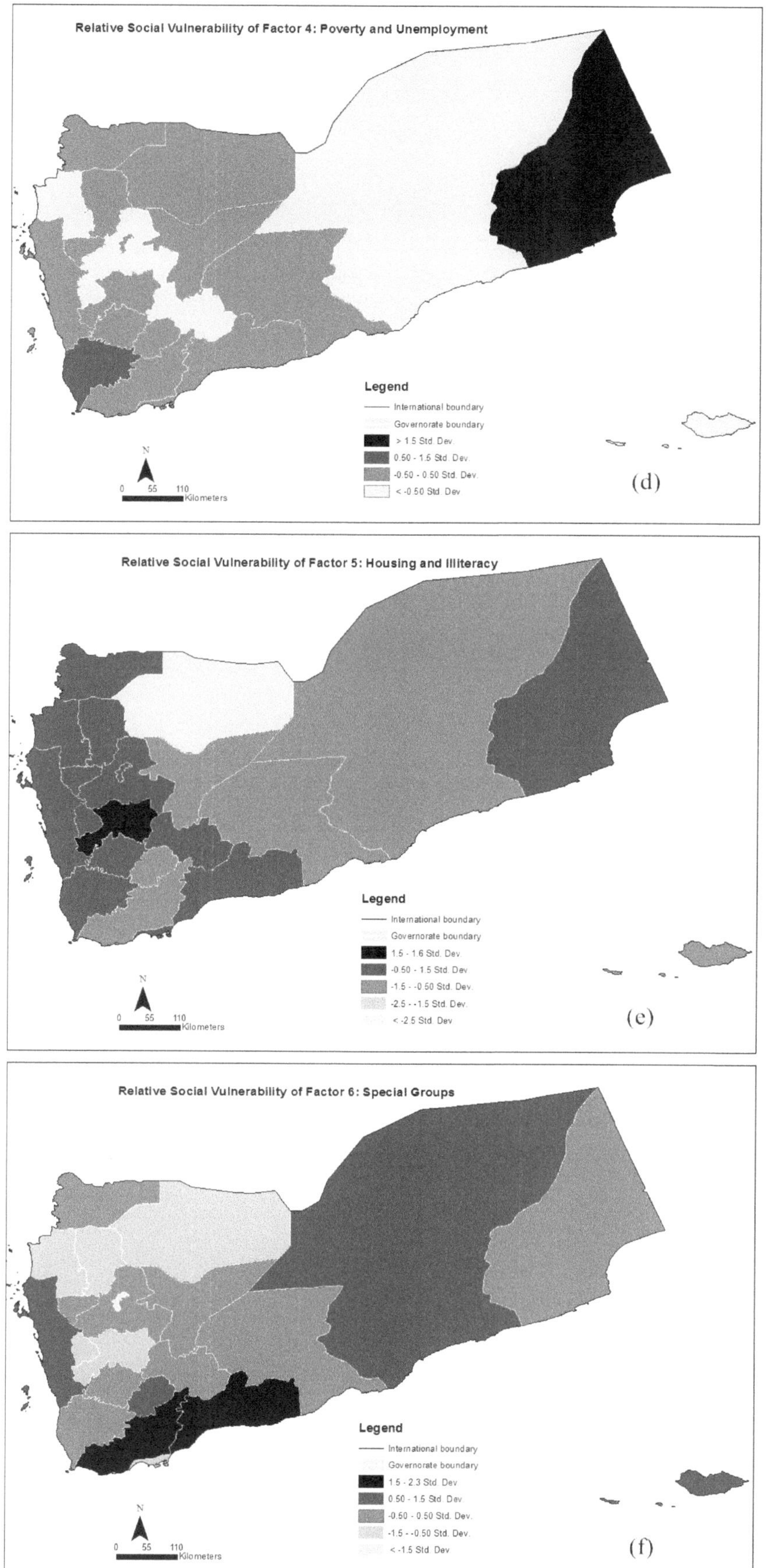

Figure 4. *Cont.*

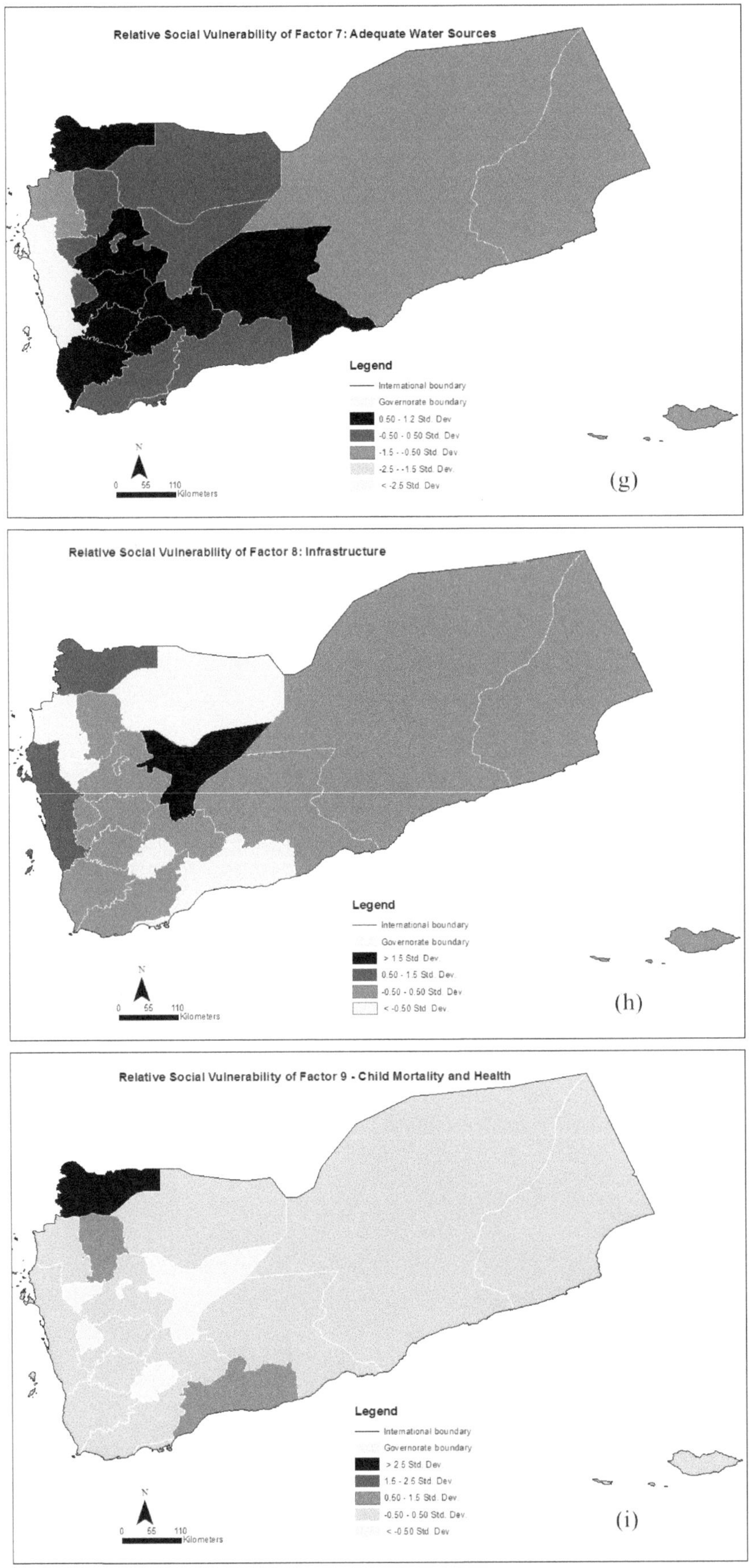

Figure 4. *Cont.*

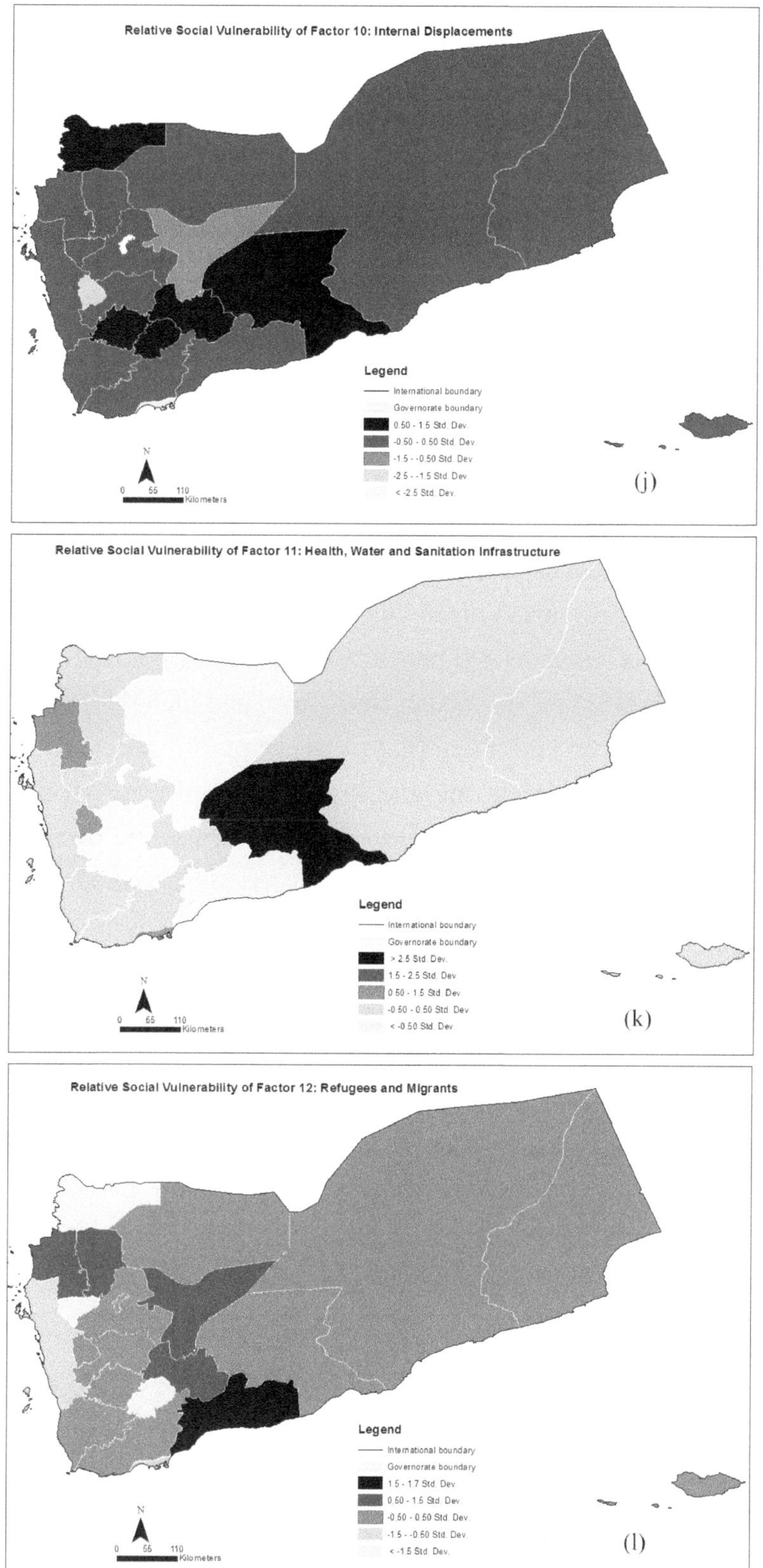

Figure 4. Spatial distribution and levels of social vulnerability for the different factors. (**a**) Education; (**b**) Basic Services; (**c**) Social Welfare; (**d**) Poverty and unemployment; (**e**) Housing and Illiteracy; (**f**) Special Groups; (**g**) Adequate Water Sources; (**h**) Infrastructure; (**i**) Child Mortality and Health; (**j**) Internal Displacement; (**k**) Health, Water and Sanitation Infrastructure (**l**) Refugees and Migrants. Governorate names are available from Figures 2 and 5.

Basic Services: The second underlying factor relates to a lack of basic services. The findings agree with present narrative of the underlying causes of vulnerability in Yemen in the humanitarian discourse. In humanitarian emergencies, such as the Yemen crisis following the airstrikes of 26 March 2015, basic services in the areas of electricity, clean water, rule of law and governance have collapsed. These have increased the vulnerability of the population. This factor explains 15.6% of the variance in the governorates. All governorates are generally affected, except Sana'a, the capital city. The most vulnerable includes Hajjah in the west, Al Jawf in the north and Raymah in the west (Figure 4b).

Social Welfare: The third underlying factor relates to governorates with higher populations on social welfare funding support. Recipients of the social welfare are categorized by the government as the poorest. In humanitarian emergencies, such as the 26 March Yemen war, the social welfare fund becomes suspended, leaving thousands of beneficiaries without the needed livelihood support. Households that benefitted from the fund may find it difficult to cope with the crisis. They may not be able to relocate to other safe governorates as they lack the means to do so. They may also find it difficult to recover even after the crisis. This result also agrees with the present narrative of the underlying causes of vulnerability in Yemen. This factor explains 8.6 percent of the variance in the governorates. The most vulnerable are coastal governorates of Hajjah, Hudaydah, and Taizz, and include Ibb in the west (Figure 4c).

Poverty: The fourth factor, poverty and unemployment, explains 8.0% of the variance in the governorates. Governorates with high poverty rate and higher unemployed population will be highly vulnerable in a humanitarian crisis. The World Bank describes Yemen as the poorest compared to other Middle Eastern countries, with 37.3% of Yemenis living in poverty (less than US$2 per day). Poverty is a determining factor of vulnerability during humanitarian emergencies as poor families are severely impacted. There are limited opportunities for Yemenis in improving their poverty level. Wealth enables communities to cope with humanitarian emergencies through social safety nets, insurance and other entitlement programs [6]. The most vulnerable governorates are Al Maharah in the east and Taizz in the south-west (Figure 4d).

Housing and Illiteracy: The quality of housing is an indicator of the state of the economy of a governorate, which may have consequences on its recovery state after an emergency. The fifth underlying factor relates to poor housing and a higher percentage of illiteracy. This factor explains 6.6% of the variance in the governorates. All governorates are greatly affected except Al Jawf (Figure 4e). There is a correlation between illiteracy population and the quality of houses they occupy. Their illiteracy may limit their chances of getting better jobs with most involved in manual jobs, which may be severely affected during the emergency.

Special Groups (Children and the Elderly): Two of the demographic groups that are mostly affected by humanitarian emergencies are the elderly (60 years and above) and children under five years, and these are represented in the sixth factor. The elderly may be sick or having mobility constraints, making them dependent on other members of the family who may be struggling themselves. The elderly may find themselves marginalized from decision making that may affect them [52]. Children, together with women, make up the bulk of refugees and internally displaced populations caused by war [53]. This factor explains 6.6% of the variance in the governorates and correlates with narratives on vulnerable groups in Yemen and in other disaster situations [6,26,46,54]. The most vulnerable are Lahj, Abyan, Hudaydah, AL Dhale'e, and Hadramaut (Figure 4f).

Water Sources: Access to adequate drinking water and sanitation are critical demands in humanitarian emergencies. Lack of adequate drinking water has been cited as a contributing factor for most of the mortality during humanitarian emergencies [55]. The current humanitarian emergency in Yemen due to the bombing by the coalition forces and internal armed clashes has seen water trucking collapse due to a lack of fuel, with hospitals and health centers without running water. The seventh factor of inadequate water sources explains 6.3 percent of the variance in the governorates. All governorates are generally affected except Hudaydah, Hajjah, Hadramaut and Al Mahara (Figure 4g). The Yemen humanitarian crisis has over 13 million people that require access to clean water and sanitation.

Infrastructure (buildings, hospitals): The eighth factor is inadequate infrastructure in terms of buildings, hospital facilities, sanitation and health. Lack of adequate buildings is indicative of a growing proportion of the Yemen population inhabiting poor quality buildings such as huts. Yemen is one of the poorest countries in the Middle East. It has two physicians per 10,000 people (2003–2012) compared to Jordan with 25.6 [56]. As a percentage of GDP, Yemen's health expenditure is 5.5% compared to Jordan with 8.4%. The 26 March 2015 war in Yemen has witnessed bombings leading to collapse of many of the buildings and near collapse of the hospitals and sanitation. This factor explains 6.1% of the variance in the governorates. The most vulnerable are Marib, Sa'ada and Hudaydah (Figure 4h).

Child Mortality and Health: The ninth factor (child mortality and health) explains 5.3% of the variance in the governorates. Compared to other Middle Eastern countries, such as Iraq, Jordan, Oman and Syrian Arab Republic, using 2003 to 2012 data, Yemen has the highest infant and under five mortality in the region [56]. Child mortality is highest in humanitarian emergencies due to poor care, lack of adequate nutrition, dehydration, diarrheal and water borne diseases [55,57]. The under-five mortality, child mortality, and diarrheal infections loaded positively on this factor. Governorates with high under five and child mortality compounded by diarrheal infections have relatively high vulnerability to humanitarian crisis. The most vulnerable are Sa'ada, Amran and Abyan (Figure 4i). These three governorates have also witnessed conflict in the past few years.

Internal Displacement: The United Nations Refugee agency (UNHCR) reports that the recent crisis has led to over 1 million Internally Displaced Persons (IDPs) moving from governorates and districts characterized by frequent bombings and violent armed clashes to other areas perceived to be calm. Internal displacement explains 5.2% of the variance in the governorates. Most of the IDPs stay in camps, public buildings and schools. However, many also stay with the host communities, most of whom are relatives. Over the years, the coping strategies of host communities harboring IDPs have diminished considerably, further adding to the vulnerability in these governorates. All governorates are generally affected except Raymah and Marib, with Sana'a City the most vulnerable (Figure 4j).

Health, Water and Sanitation Infrastructure: Health, Water and Sanitation Infrastructure explains 4.6 percent of the variance in the governorates. Health care providers, health facilities and access to water are important during a humanitarian crisis and even post-crisis. During humanitarian emergencies, these facilities are significantly affected, contributing to people's vulnerability. The most vulnerable are Shabwah, Hajjah, Raymah and Aden (Figure 4k).

Refugees and Migrants: The twelfth factor, relating to Refugees and Migrants, explains 3.4% of the variability in governorates. The most vulnerable governorates are Abyan, Hajjah, Amran, Marib and Al Bayda (Figure 4l). Yemen hosts about 243,000 refugees, most of them from Somalia. At the end of

December 2014, about 92,000 migrants, mostly from the Horn of Africa, crossed into Yemen. Hajjah hosts over 20,000 Ethiopian stranded migrants hoping to cross into the Kingdom of Saudi Arabia.

The overall calculated social vulnerability (SoVIHC) ranges from −0.7 to 0.42, with a mean vulnerability score of −0.00048 and standard deviation of 0.28 (Figure 5). There is a notable coastal and conflict dimension to social vulnerability of governorates to humanitarian crisis in Yemen. The most vulnerable governorates include Sa'ada and Amran in the north, Taizz in the southwest, Dhamar and Ibb in the central part, Abyan and Shabwah in the south and Al Mahara in the east of the country. Half of these are coastal (Taizz, Abyan, Shabwah and Al Maharah) and the others are mountainous (Sa'ada, Amran, Dhamar and Ibb), but also not far from the coast. With the exclusion of Taizz and Al Maharah, these vulnerable governorates have experienced conflicts, armed clashes and violence in the past few years. Between 2004 and 2010 the government of Yemen fought six wars with the Al Houthi tribes of Sa'ada governorate, leading to the displacement of over 300,000 persons [14]. Most of the IDPs moved to Hajjah and Amran. In Hajjah, thousands of IDPs have lived in camps since 2004 in deplorable conditions that lack most of the basic services.

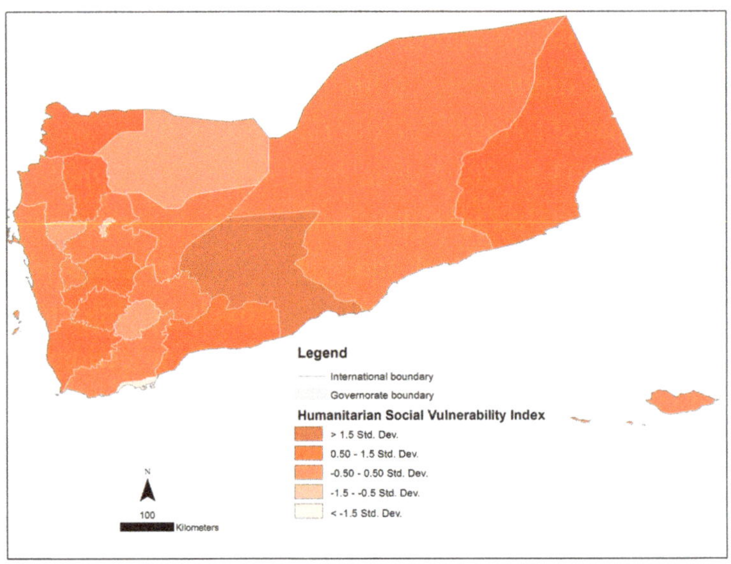

Figure 5. Social Vulnerability Index to Humanitarian Crisis of governorates in Yemen (unweighted).

For a very long time, Sa'ada governorate has been without the presence of a strong central governing authority, with the Houthi's as *de-facto* authorities. Limited employment opportunities and strict control by Houthies have left the governorate one of the most vulnerable in the country. Sa'ada is also one of the worst affected since the beginning of the war on 26 March 2015.

Most of the governorates of Yemen exhibit moderate levels of social vulnerability (Figure 5). These are mostly found in the west, central part and extend to the south, with the sole governorate of Hadramaut in the east. Sana'a City (Amanat Al Asimah), Al Mahwit, Al Dhale'e and Al Jawf exhibit the least social vulnerability (Figure 5).

In addition to conflict, the most vulnerable and moderately vulnerable governorates are characterized by weak governance, severe lack of basic services, high illiteracy, extreme poverty and limited livelihood opportunities to cope with the humanitarian crisis [2,3,16]. These underlying drivers of vulnerability are consistent with the derived 12 factors that influence social vulnerability to humanitarian crisis of governorates in Yemen.

To test the robustness and usefulness of the SoVIHC, we compared the mapped SoVIHC to the Humanitarian Needs map of the 2015 Yemen Humanitarian Programme Cycle (Figure 6). Visually there are correlations between some of the governorates such as Taizz, Ibb and Dhamar, which exhibit very high social vulnerability (Figure 5) and at the same time with a very high number of people in need (Figure 6). However, this is not the case for most of the other governorates. Al Mahara and Shabwah, two of the governorates with very high relative social vulnerability index have a low number of people in need. To further test this relationship, we conducted a simple correlation between the social vulnerability index and the 2015 humanitarian needs. There is a weak but positive relationship ($r = 0.19$, $S = 0.4$). This finding shows that social vulnerability does not necessarily correlate with current humanitarian needs.

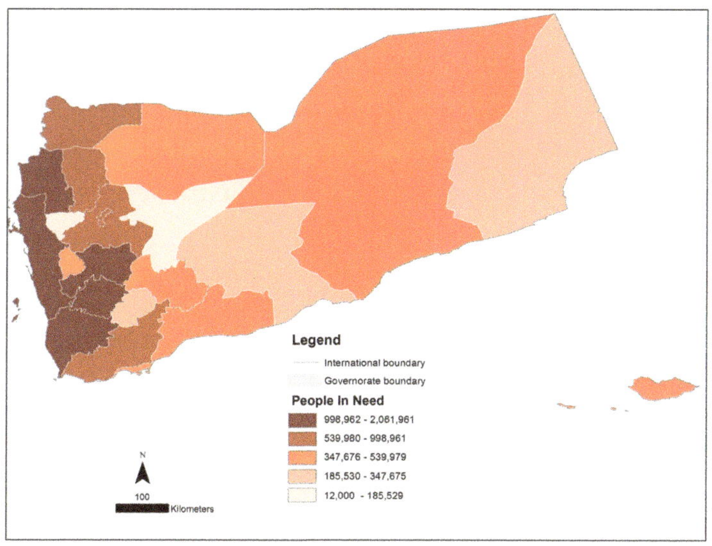

Figure 6. 2015 Population in needs in governorates in Yemen.

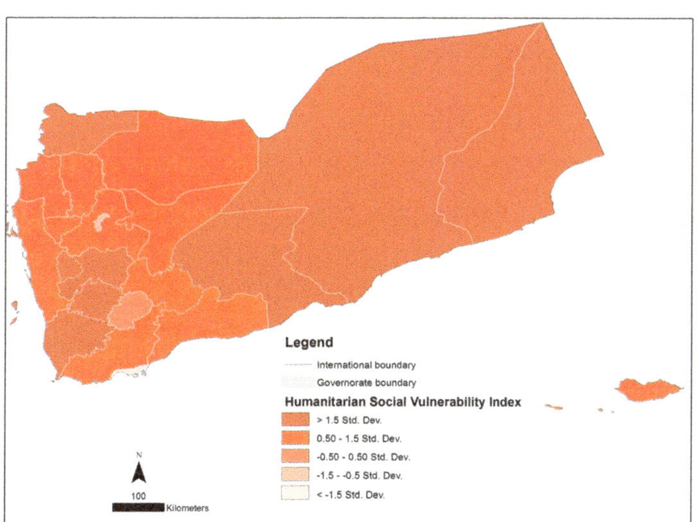

Figure 7. Social Vulnerability Index to Humanitarian Crisis of governorates (weighted).

Comparing the spatial variability between the unweighted and weighted SoVIHC (Figures 5 and 7), there is broad agreement between the two mapping methods except in the governorates of Amran, Abyan, Al Jawf, Al Mahwit, and Hadramaut.

4. Conclusions

A number of humanitarian crises have risen in the last decade [9] with consequences of increase in vulnerability of people and places. Although the link between social vulnerability and humanitarian crisis is understood, yet very little has been done, particularly in the area of its assessment to understanding the underlying factors that contribute to an increase or decrease of it in Yemen.

In this study, we used GIS to bring together socio-economic data that represent the underlying determinants of social vulnerability of governorates of Yemen to create a spatial index that measures social vulnerability towards humanitarian crisis. GIS was used to create a spatial index for each of the twelve underlying factors of social vulnerability. Using GIS we also mapped the social vulnerability index and the twelve factor indices to reveal the spatial distribution and variability of social vulnerability, thus contributing to our understanding of the geography of vulnerability in Yemen.

In this study, socio-economic and demographic data at governorate level for Yemen of mostly 2012 was used to construct a relative ranking of social vulnerability to humanitarian emergencies. Using data-reduction techniques and a GIS, a methodology to identify, rank and determine the underlying factors influencing social vulnerability to the humanitarian emergencies was developed. Using SoVIHC, the 22 governorates of Yemen were ranked based on their social vulnerability, putting Sa'ada and Amran in the north, Taizz in the southwest, Dhamar and Ibb in the central part, Abyan and Shabwah in the south and Al Mahrah in the east as the most vulnerable governorates. The SoVIHC index also ranked Al Jawf in the north, Al Mahwit in the west, Sana'a City in the center and Al Dhale in the south as the least vulnerable governorates. Our findings reveal that the differences in social vulnerability between the governorates are primarily driven by twelve factors, of which education, lack of basic services in health, water and sanitation, electricity, quality housing, poverty, limited livelihood opportunities, weak governance, and internal and external displacement are major determinants of social vulnerability to humanitarian crisis. This finding also shows that conflicts and armed clashes also contribute to social vulnerability of governorates in Yemen. These drivers of social vulnerability are consistent with the underlying drivers of vulnerability in the humanitarian crisis in Yemen [1,2,15]. The SoVIHC method used here provides a means of characterizing the multivariate nature of the social vulnerability of governorates of Yemen, identifying and mapping those governorates with greater susceptibility to humanitarian crisis. The study shows that social vulnerability of the governorates of Yemen to humanitarian crisis is not evenly distributed.

The results of the social vulnerability analysis and the maps provide empirical evidence in addressing the objectives of the study. Using the SoVIHC method, context relevant underlying factors influencing social vulnerability have been derived at governorate level. The derived factors are consistent with the literature for vulnerability studies conducted at local scales [6,36,58].

The SoVIHC method identifies the underlying dimensions of social vulnerability from a large set of socioeconomic data and reduces the original dataset to a smaller set of factors based on intercorrelated variables. We recognize that the derived social vulnerability indices are judged based on the input variables, the methodological choices made and the rigorous data reduction process. We also recognize that due to data limitation, indicators like remittances, female-headed households and population of minorities in each of the governorates were not captured. It is also important to point out that in naming the derived underlying factors and interpreting them, some element of subjectivity may have been introduced [49]. The study does not identify which Yemeni individuals will be most affected during the humanitarian crisis and their location in the governorates, however, it can throw light on which governorates with relatively higher vulnerability will be affected and pinpoint the underlying causes of their vulnerability. This information is particularly important for planning response-based needs of people bearing in mind their vulnerability. It is also important for vulnerability reduction to humanitarian crisis as it enables authorities to identify governorates and populations that require vulnerability and risk reduction, mitigation and adaptation. Knowledge of the underlying determinants of social vulnerability to humanitarian crisis and the geographies will enable development of policies and plans that are context driven. The conflict, armed clashes and war should be ended for any meaningful development to take place. Poverty as a determinant of social vulnerability can be addressed within the framework of poverty reduction, and increasing job and other livelihood opportunities. Education has been revealed as another determinant of social vulnerability. The government and the international community should undertake efforts to increase school enrolment. The government should embark on ensuring basic services are provided with access to affordable health, clean water and electricity to all to stimulate the economy. Embarking on these will contribute to reducing vulnerability to humanitarian crisis.

Author Contributions

Joseph Kandeh and Lalit Kumar conceived the idea and study; Joseph Kandeh analyzed the data; Joseph Kandeh and Lalit Kumar wrote the manuscript.

Conflicts of Interest

The authors declare no conflict of interest.

References

1. World Food Program. *Yemen Comprehensive Food Security Survey*; World Food Program: Washington, DC, USA, 2014.
2. United Nations Office for the Coordination of Humanitarian Affairs. *Yemen Humanitarian Needs Overview*; United Nations Office for the Coordination of Humanitarian Affairs: New York, NY, USA, 2014.
3. United Nations Office for the Coordination of Humanitarian Affairs. *Yemen Humanitarian Response Plan*; United Nations Office for the Coordination of Humanitarian Affairs: New York, NY, USA, 2015.
4. Reliefweb. Global Emergency Overview. Available Online: http://geo.acaps.org/ (accessed on 9 June 2015).
5. Kasperson, J.X.; Kasperson, R.E.; Turner, B. *Regions at Risk: Comparisons of Threatened Environments*; United Nations: New York, NY, USA, 1995.

6. Cutter, S.; Boruff, B.; Shirley, W. Social vulnerability to environmental hazards. *Soc. Sci. Q.* **2003**, *84*, 242–261.
7. Byrne, B.; Baden, S. *Gender, Emergencies and Humanitarian Assistance*; Institute of Development Studies: Brighton, UK, 1995.
8. Humanitarian Coalition. Available online: www.Humanitariancoalition.Ca (accessed on 9 June 2015).
9. Initiative, D. *Global Humanitarian Assistance Report 2014*; Development Initiative: Bristol, UK, 2014.
10. Alwang, J.; Siegel, P.B.; Jorgensen, S.L. *Vulnerability: A View from Different Disciplines*; The World Bank: Washington, DC, USA, 2001.
11. Birkmann, J. Measuring vulnerability to promote disaster-resilient societies: Conceptual frameworks and definitions. In *Measuring Vulnerability to Natural Hazards: Towards Disaster Resilient Societies*; United Nations University Press: Tokyo, Japan, 2006; pp. 9–54.
12. Cutter, S.L. Vulnerability to environmental hazards. *Prog. Hum. Geogr.* **1996**, *20*, 529–539.
13. Adger, W.N. Vulnerability. *Glob. Environ. Chang.* **2006**, *16*, 268–281.
14. United Nations Office for the Coordination of Humanitarian Affairs. *Yemen Humanitarian Needs Overview*; United Nations Office for the Coordination of Humanitarian Affairs: New York, NY, USA, 2012.
15. United Nations Office for the Coordination of Humanitarian Affairs. *Yemen Humanitarian Needs Overview*; United Nations Office for the Coordination of Humanitarian Affairs: New York, NY, USA, 2013.
16. United Nations Office for the Coordination of Humanitarian Affairs. *Yemen Humanitarian Needs Overview*; United Nations Office for the Coordination of Humanitarian Affairs: New York, NY, USA, 2015.
17. Boruff, B.; Cutter, S. The environmental vulnerability of caribbean island nations. *Geogr. Rev.* **2007**, *97*, 24–45.
18. Birkmann, J.; Fernando, N.; Hettige, S. Measuring vulnerability in Sri Lanka at the local level. In *Measuring Vulnerability to Natural Hazards: Towards Disaster Resilient Societies*; Birkmann, J., Ed.; United Nations University Press: New York, NY, USA, 2006; pp. 329–356.
19. Cutter, S.; Johnson, L.; Finch, C.; Berry, M. The US hurricane coasts: Increasingly vulnerable? *Environ. Sci. Policy Sustain. Dev.* **2007**, *49*, 8–21.
20. Cutter, S.L.; Barnes, L.; Berry, M.; Burton, C.; Evans, E.; Tate, E.; Webb, J. A place-based model for understanding community resilience to natural disasters. *Glob. Environ. Chang.* **2008**, *18*, 598–606.
21. Azar, D.; Rain, D. Identifying population vulnerable to hydrological hazards in San Juan, Puerto Rico. *GeoJournal* **2007**, *69*, 23–43.
22. Borden, K.A.; Schmidtlein, M.C.; Emrich, C.T.; Piegorsch, W.W.; Cutter, S.L. Vulnerability of US cities to environmental hazards. *J. Homel. Secur. Emerg. Manag.* **2007**, *4*, doi:10.2202/1547-7355.1279.
23. Füssel, H.-M. Vulnerability: A generally applicable conceptual framework for climate change research. *Glob. Environ. Chang.* **2007**, *17*, 155–167.
24. O'Brien, K.; Sygna, L.; Haugen, J.E. Vulnerable or resilient? A multi-scale assessment of climate impacts and vulnerability in Norway. *Clim. Chang.* **2004**, *64*, 193–225.

25. Vincent, K. *Creating an Index of Social Vulnerability to Climate Change for Africa*; Tyndall Centre for Climate Change Research: Norwich, UK, 2004.
26. Tapsell, S.; Tunstall, S.; Green, C.; Fernández-Bilbao, A. *Task 11 Social Indicator Set*; Flood Hazard Research Centre: London, UK, 2005.
27. López-Claros, A.; Hidalgo, S. *Humanitarian Response Index 2007: Measuring Commitment to Best Practice*; Palgrave Macmillan: London, UK, 2008.
28. Bennett, N.; Morris, B. IDP vulnerability assessment and profiling (IVAP) in Pakistan: A report and appraisal. *Humanit. Exch.* **2012**, *55*, 30–34.
29. Wiesmann, D. *2006 Global Hunger Index, A Basis for Cross-Country Comparisons*; International Food Policy Research Institute: Washington, DC, USA, 2006.
30. De Groeve, T.; Hachemer, P.; Vernaccini, L. *The Global Conflict Risk Index (GCRI): A Quantitative Model. Concept and Methodology*; Publications Office of the European Union: Rue Mercier, Luxembourg, 2014.
31. De Groeve, T.; Poljansek, K.; Vernaccini, L. *Index for Risk Management—Inform*; Publications Office of the European Union: Rue Mercier, Luxembourg, 2014.
32. Cutter, S.L.; Mitchell, J.T.; Scott, M.S. Revealing the vulnerability of people and places: A case study of Georgetown County, South Carolina. *Ann. Assoc. Am. Geogr.* **2000**, *90*, 713–737.
33. Walker, G.P. *Addressing Environmental Inequalities: Flood Risk*; Environment Agency: Bristol, UK, 2006.
34. Blaikie, P.; Cannon, T.; Davis, I.; Wisner, B. *At Risk: Natural Hazards, People's Vulnerability and Disasters*; Routledge: London, UK, 2014.
35. Buckle, P.; Mars, G.; Smale, S. New approaches to assessing vulnerability and resilience. *Aust. J. Emerg. Manag.* **2000**, *15*, 8–14.
36. Adger, W.; Brooks, N.; Bentham, G.; Agnew, M.; Eriksen, S. *New Indicators of Vulnerability and Adaptive Capacity*; Tyndall Centre for Climate Change Research: Norwich, UK, 2004.
37. Dwyer, A.; Zoppou, C.; Nielsen, O.; Day, S.; Roberts, S. *Quantifying Social Vulnerability: A Methodology for Identifying Those at Risk to Natural Hazards*; Geoscience Australia Canberra: Canberra, ACT, Australia, 2004.
38. King, D.; MacGregor, C. Using social indicators to measure community vulnerability to natural hazards. *Aust. J. Emerg. Manag.* **2010**, *15*, 52–57.
39. Thrush, D.; Burningham, K.; Fielding, J. *Flood Warning for vulnerable Groups: A Review of Literature*; Environment Agency: Bristol, UK, 2005.
40. Savage, K.; Harvey, P. *Remittances during Crises: Implications for Humanitarian Response*; Overseas Development Institute: London, UK, 2007.
41. Ruiz, I.; Vargas-Silva, C. Another consequence of the economic crisis: A decrease in migrants' remittances. *Appl. Financ. Econ.* **2010**, *20*, 171–182.
42. Central Statistical Organization. *Yemen Demographic And Maternal and Child Health Survey*; Central Statistical Organization: Sana'a, Yemen, 1998.
43. Aoláin, F.N. Women, vulnerability, and humanitarian emergencies. *Mich. J. Gend. Law* **2011**, *18*. Available online: http://repository.law.umich.edu/mjgl/vol18/iss1/1/ (accessed on 9 June 2015).
44. Central Statistical Organization. *Yemen Statistical Yearbook*; Central Statistical Organization: Sana'a, Yemen, 2012.

45. Field, A. *Discovering Statistics Using SPSS for Windows: Advanced Techniques for Beginners (Introducing Statistical Methods Series)*; Sage Publications Ltd.: London, UK, 2005.
46. Boruff, B.J.; Emrich, C.; Cutter, S.L. Erosion hazard vulnerability of US coastal counties. *J. Coast. Res.* **2005**, *21*, 932–942.
47. Schneiderbauer, S. Risk and Vulnerability to Natural Disasters from Broad View to Focused Perspective. Ph.D. Thesis, Freie Universität Berlin, Berlin, Germany, 2007.
48. Burton, C.; Cutter, S.L. Levee failures and social vulnerability in the Sacramento-San Joaquin Delta Area, California. *Nat. Hazards Rev.* **2008**, *9*, 136–149.
49. Wood, N.J.; Burton, C.G.; Cutter, S.L. Community variations in social vulnerability to cascadia-related tsunamis in the US Pacific Northwest. *Nat. Hazards* **2010**, *52*, 369–389.
50. Jolliffe, I.T. *Principal Component Analysis and Factor Analysis*, 2nd ed.; Springer: New York, NY, USA, 2002.
51. Jones, B.; Andrey, J. Vulnerability index construction: Methodological choices and their influence on identifying vulnerable neighbourhoods. *Int. J. Emerg. Manag.* **2007**, *4*, 269–295.
52. Wisner, B. *Who? What? Where? When? In an Emergency: Notes on Possible Indicators of Vulnerability and Resilience: By Phase of the Disaster Management Cycle and Social Actor*; Institute for Environment and Human Security: Bonn, Germany, 2002.
53. Women's Refugee Commission. *Halfway Home: Unaccompanied Children in Immigration Custody*; Women's Refugee Commission: New York, NY, USA, 2009.
54. Morrow, B.H. Identifying and mapping community vulnerability. *Disasters* **1999**, *23*, 1–18.
55. Brennan, R.J.; Nandy, R. Complex humanitarian emergencies: A major global health challenge. *Emerg. Med.* **2001**, *13*, 147–156.
56. United nations Development Programme. *Sustaining Human Progress: Reducing Vulnerabilities and Building Resilience*; United nations Development Programme: New York, NY, USA, 2014.
57. Toole, M.J.; Waldman, R.J. The public health aspects of complex emergencies and refugee situations. *Annu. Rev. Public Health* **1997**, *18*, 283–312.
58. Gall, M. Indices of Social Vulnerability to Natural Hazards: A Comparative Evaluation. Ph.D. Thesis, University of South Carolina, Columbia, SC, USA, 2007.

13

Local Edge Matching for Seamless Adjacent Spatial Datasets with Sequence Alignment

Yong Huh

Spatial Information Research Institute, Korea Cadastral Survey Corp., Seoul 150-911, Korea;
E-Mail: spinsgis@gmail.com

Academic Editor: Wolfgang Kainz

Abstract: This study proposes a local edge matching method with a sequence alignment technique for adjacent spatial datasets. By assuming that the common boundary edges of the datasets are point strings, the proposed method obtains the sequence for point edit operations to align the edges by using the string matching algorithm with the following operations: (1) snapping two points from each string to their average position, (2) removing a point from one string and (3) removing a point from the other string. The costs for these operations are derived from the deformation of the involved line segments in terms of the angle and length changes. The corresponding point pairs are then considered point pairs for which the snapping operation is chosen in a sequence. Based on these pairs, a border area of adjacent spatial datasets can be partitioned into sub-border areas where distinctive matching and alignment processes can be performed.

Keywords: edge matching; adjacent spatial dataset; corresponding point pair; sequence alignment; string matching

1. Introduction

Edge matching between adjacent spatial datasets finds and removes the geometric differences of common boundary edges between adjacent spatial datasets [1–4]. There are two components in this task: feature matching to identify the corresponding point or edge pairs and map alignment to remove the differences in the identified pairs.

Because map alignment is performed with the corresponding point pairs, point feature matching is generally used in the field of geospatial information. For this reason, diverse point matching methods have been proposed. Beard and Chrisman proposed the Zipper algorithm, which finds the closest point pairs within a distance threshold and determines that these pairs are the corresponding point pairs [1]. This algorithm is simple and efficient; thus, it has been adopted in many studies [2–5]. However, one problem with this algorithm lies in estimating an appropriate threshold because even though the threshold is related to the positional accuracies of the spatial datasets involved, it is not clear how the accuracies are translated into a threshold value [6]. To address this problem, a statistical analysis of the sampled corresponding point pairs in the training data was applied. Lolonis [4] proposed a chi-square analysis on the positional differences of candidate corresponding point pairs and then discriminated mismatched pairs according to the level of statistical confidence. Masuyama [5] determined the threshold with a Monte-Carlo simulation-based analysis of the sampled pairs. Huh et al. [7] used a box-plot analysis on the sampled pairs because the analysis can discriminate proper and outlier values in the sampled pairs without any assumption of an underlying statistical distribution. Meanwhile, map alignment methods without the above feature matching were recently proposed [6,8]. They represent the geometric differences of adjacent spatial datasets as gaps and overlaps. The gaps and overlaps stand for erroneous areas where no spatial data coverage exists and where coverage from different datasets co-exists, respectively. These areas are decomposed into triangles by using computational geometry techniques such as a constrained Delaunay triangulation method. Then, each of the triangles [6] or divided parts of each of the triangles [8] is assigned to one of the spatial datasets and seamless common boundary edges without gaps and overlaps are obtained.

Although the above studies have presented successful results, additional improvements are necessary in feature matching to address locally auto-correlated positional differences. In the field of map conflation, which finds corresponding point pairs or corresponding objects between super-imposed spatial datasets, a transformation model such as an affine or rigid transformation model is applied to explain the auto-correlated positional differences. These types of differences also arise in the edge matching shown as sub-border areas in Figure 1. By considering the horizontal direction differences of the corresponding point pairs in sub-border area 1, ($a_{(1)}$, $b_{(1)}$), ($a_{(2)}$, $b_{(2)}$), ($a_{(3)}$, $b_{(3)}$) and ($a_{(4)}$, $b_{(4)}$) are proper corresponding point pairs. However, ($a_{(3)}$, $b_{(2)}$) and ($a_{(5)}$, $b_{(4)}$) are closest pairs and thus would be chosen for the corresponding point pairs when a general matching method is applied. Meanwhile, there are vertical direction differences for the corresponding point pairs in sub-border area 3. When there are many sub-border areas whose positional difference directions are distinctive, these differences cannot be explained by a single transformation model.

Moreover, this problem can be more severe when there are many road or river objects that are divided by boundaries of adjacent spatial datasets and are partially represented in each dataset. When the two datasets are constructed by the same mapping agency with the same mapping rule, divided spatial objects have the same attributes, including the name or identification code. In this case, these objects are found first and then the corresponding point pairs of each corresponding object pair can be easily found. When the two datasets are not constructed by the same mapping agency with the same mapping rule, only geometric properties such as distance or shape are available to find corresponding point pairs. However, feature matching with these geometric properties is vulnerable to the aforementioned problem in Figure 1.

Moreover, this problem could not be addressed by a transformation model because the border area's length is often more than several kilometres and a simple model cannot explain locally distinctive differences.

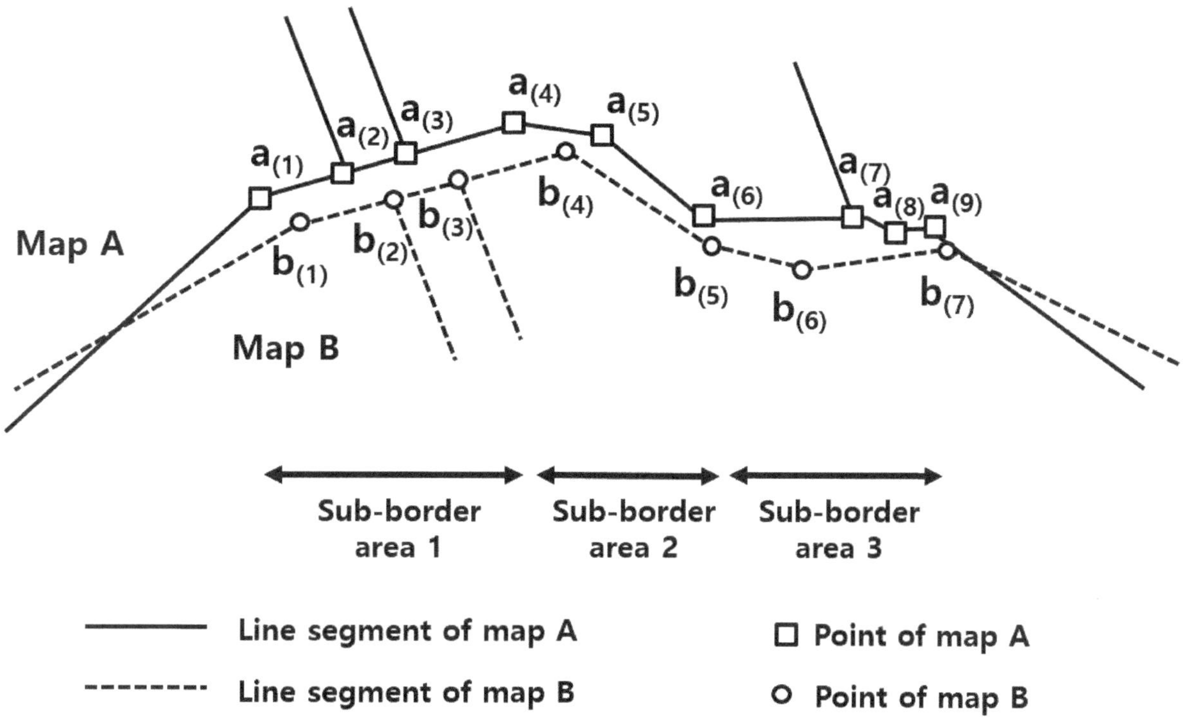

Figure 1. Positional differences between adjacent spatial datasets.

To address this problem, this study proposes a local edge matching method with sequence alignment by using a string matching algorithm. By assuming that the border boundary edges of two adjacent maps are point strings, the proposed method obtains the sequence for point edit operations to align the edges by using the following operations: (1) snapping two points from each string to their average position, (2) removing a point from one string and (3) removing a point from the other string. The costs for these operations are derived from the deformation of the involved line segments in terms of the angle and length changes. The optimal operation sequence with the minimum total cost is obtained by means of a dynamic programming optimization technique [9,10]. The corresponding point pairs are then considered point pairs for which the snapping operation is chosen in a sequence. Based on these pairs, a border area of adjacent spatial datasets can be partitioned into sub-border areas where distinctive matching and alignment processes can be performed.

The remainder of the paper is structured as follows. In the next section, the details of the proposed method are presented. In Section 3, the results of the proposed method are evaluated and discussed. The proposed method is applied to two cadastral maps of neighbouring local authorities. The result of the proposed method is compared with manually chosen corresponding point pairs and with a conventional distance threshold method. Finally, the conclusion is given in Section 4.

2. Proposed Method

The proposed method consists of four steps: Step 1. Extraction of point strings from adjacent spatial datasets, Step 2. String matching for extracted point strings, Step 3. Local map transformation and matching for disconnected line segments and Step 4. Local map alignment to remove gaps and overlaps as shown in Figure 2. The details for this method are as follows.

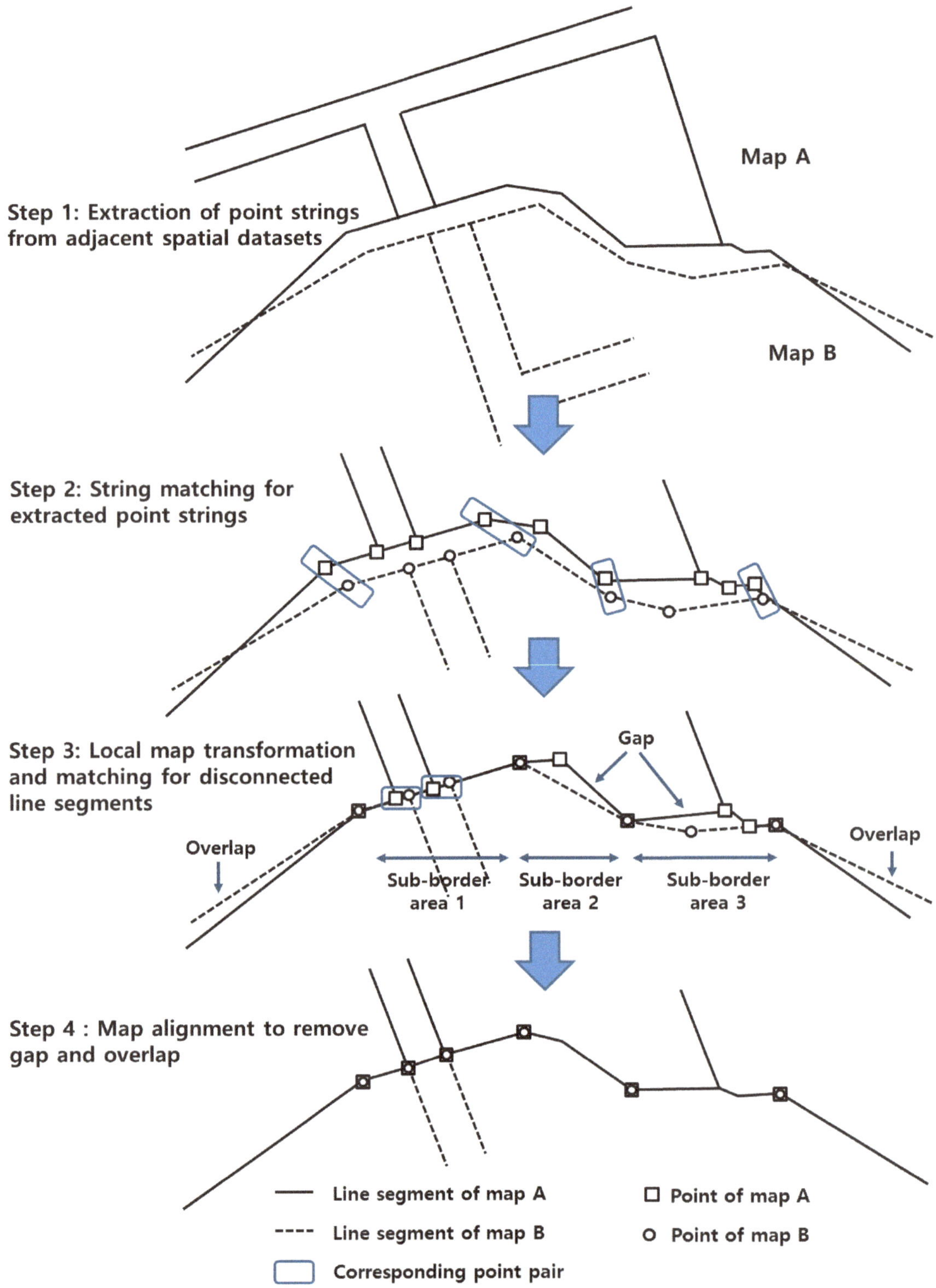

Figure 2. Workflow of the proposed method.

2.1. Extraction of Point Strings from Adjacent Spatial Datasets

This step is a preprocessing step to extract boundary point strings from spatial datasets. Technically, different data manipulations are necessary according to the data type of involved spatial datasets. In this study, there are two assumptions regarding spatial datasets. The first assumption is that these datasets are built and maintained by mapping agencies only for their own catchment areas, so that the boundaries of the datasets present a seamless spatial dataset after edge matching. The other assumption is that the datasets are composed of polygon objects that are mutually exclusive and collectively exhaustive, such as a cadastral map.

Currently, it is necessary to define a border area where point strings to be matched and aligned are extracted. This area is obtained as a union area of buffered boundaries for two spatial datasets (Figure 3c). With a buffer distance Th, each spatial dataset's boundaries are extended (Figure 3a,b) and all line segments and points for these datasets within the union area are extracted. Then, line segments that are not for each dataset's boundary edges are obtained as duplicate pairs because of the aforementioned assumptions (Figure 3d). These duplicate segments are removed and the remaining line segments are concatenated into polylines, which are represented as point strings (Figure 3e). Later, these removed line segments connected to these point strings are used for matching the disconnected line segments in Step 3.

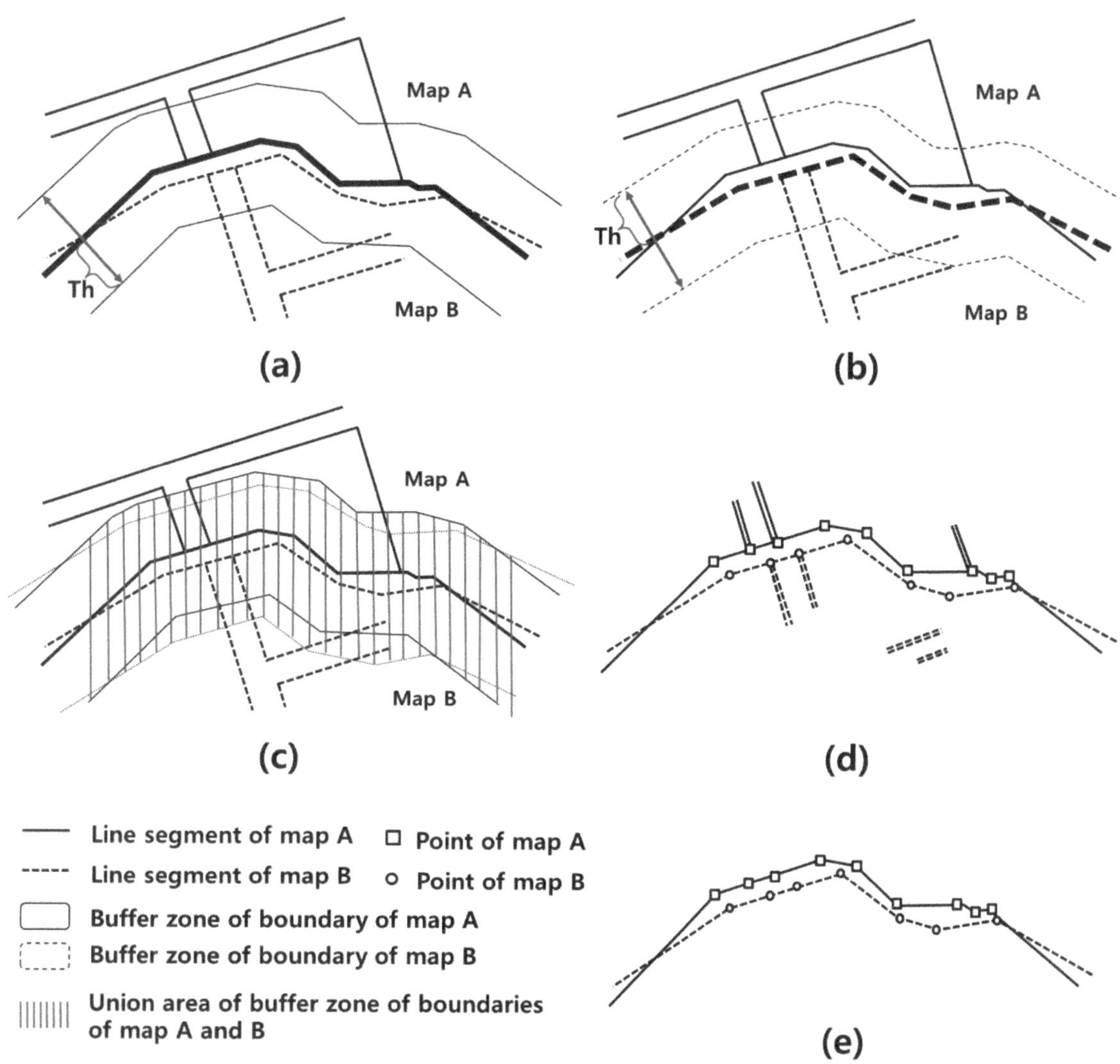

Figure 3. Extraction of point strings from adjacent spatial datasets; (**a**) buffer zone of boundary of map A, (**b**) buffer zone of boundary of map B, (**c**) union area of buffer zones of boundaries of map A and B, (**d**) line segments extracted by union area, (**e**) point strings to be aligned.

It is notable to determine the order of the point strings. In general, the point order of a simple polygon is counter-clockwise. Similarly, end points for each point string are connected temporarily, and then the orders are determined to be counter-clockwise.

2.2. String Matching for Extracted Point Strings

2.2.1. Cost Function

The proposed method uses three types of point edit operations. The snapping operation $o(a_{(i)}, b_{(j)} \rightarrow v)$ moves two points, one from each string, to their average position v, as shown in Figure 4a. The removing operation $o(a_{(i)} \rightarrow \phi)$ or $o(b_{(j)} \rightarrow \phi)$ removes a point of one string, as shown in Figure 4b. Because the removing operations remove the points of the boundaries, only point pairs for snapping operations remain and are used for subsequent edit operations. Thus, p is the point of the average position of two points for the latest snapping operation.

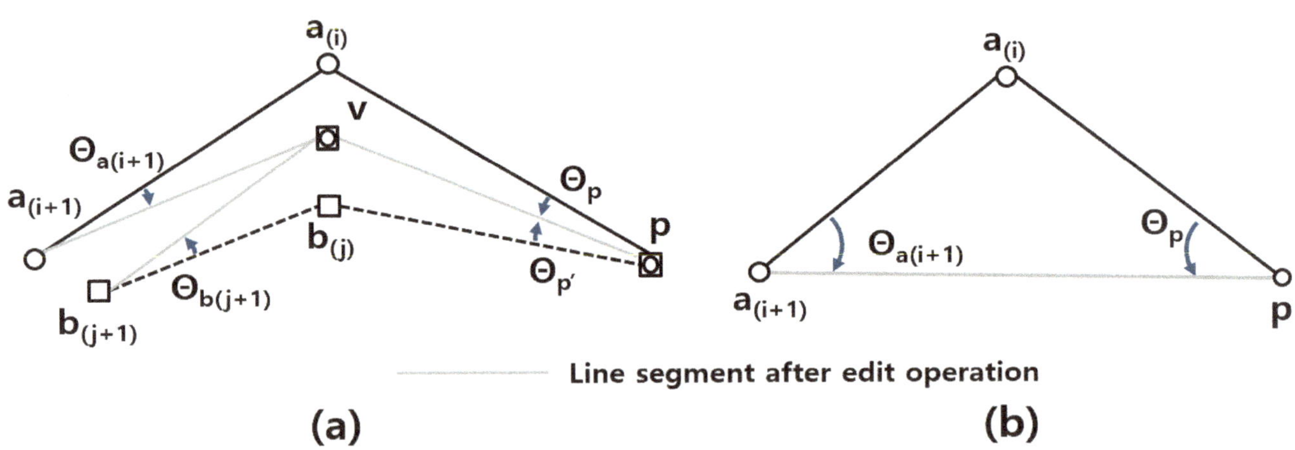

Figure 4. Cost functions of string matching in this study; (**a**) snapping operation, (**b**) removing operation.

These operations change the shapes of the boundary edges by moving or removing involved points. There have been studies to align boundaries for the corresponding object pairs with point edit operations [7,11–13]. These studies proposed various cost functions quantitatively to measure the change in the shape by finding the optimal point edit operation sequence with the minimum total cost. Among the cost functions, we choose the deformation energy model [7]. It measures the change in terms of the stretching and bending energy, as expressed by Equation (1).

$$E = E_{stretching} + E_{bending} = \lambda(\Delta l)^2 + (1-\lambda)(\Delta\theta)^2 \tag{1}$$

Here, Δl and $\Delta\theta$ are the length and angle differences caused by an operation, respectively, and $\lambda(0 \leq \lambda \leq 1)$ is a coefficient for the weighted summarization of the stretching and bending energy.

In Figure 4a, the snapping operation $o(a_{(i)}, b_{(j)} \rightarrow v)$ changes the lengths and angles of four boundary line segments. Thus, according to Equation 1, the cost of the snapping operation is determined by Equation (2).

$$c(a_{(i)}, b_{(j)} \rightarrow v) = \lambda\{(l_{p,a(i)} - l_{p,v})^2 + (l_{a(i),a(i+1)} - l_{v,a(i+1)})^2 + (l_{p,b(j)} - l_{p,v})^2 + (l_{b(j),b(j+1)} - l_{v,b(j+1)})^2\} + (1-\lambda)\{\theta_p{}^2 + \theta_{a(i+1)}{}^2 + \theta_{p'}{}^2 + \theta_{b(j+1)}{}^2\} \quad (2)$$

In this equation, $l_{x,y}$ is the length (meter) of a boundary line segment between points x and y. θ_x is the angle change (degree) at point x after an operation.

In Figure 4b, the removing operation $o(a_{(i)} \rightarrow \phi)$ changes the lengths and angles of two line segments of one boundary. Thus, the cost of the removing operation is formulated as Equation (3).

$$c(a_{(i)} \rightarrow \phi) = 2\lambda\left\{\left(\frac{l_{p,a(i)} + l_{a(i),a(i+1)} - l_{p,a(i+1)}}{2}\right)^2\right\} + (1-\lambda)\{\theta_p{}^2 + \theta_{a(i+1)}{}^2\} \quad (3)$$

The other removing operation $o(b_{(j)} \rightarrow \phi)$ is similarly formulated.

2.2.2. String Matching to Find Corresponding Point Pairs

Given two point strings $A = [a_{(1)}, a_{(2)}, \cdots, a_{(n)}]$ and $B = [b_{(1)}, b_{(2)}, \cdots, b_{(m)}]$ whose sizes are n and m, respectively, a string matching method calculates the total cost between A and B, as described by Equation (4).

$$C(A,B) = \min\left\{\sum_O c_{(i)} \,|\, O = [o_{(1)}, o_{(2)}, \cdots, o_{(N)}]\right\} \quad (4)$$

Here, $o_{(i)}$ is one type of edit operation and $c_{(i)}$ is the cost of $o_{(i)}$. N is the total number of edit operations. Given the two point strings and the cost functions, the optimal sequence O is obtained by the dynamic programming technique shown in Figure 5 based on the following property of Equation (5) [14].

$$C(A_{<1,i>}, B_{<1,j>}) = \min\begin{cases} C(A_{<1,i-1>}, B_{<1,j-1>}) + c(a_{(i)}, b_{(j)} \rightarrow v \,|\, P(i-1, j-1)) \\ C(A_{<1,i-1>}, B_{<1,j>}) + c(a_{(i)} \rightarrow \phi \,|\, P(i-1, j)) \\ C(A_{<1,i>}, B_{<1,j-1>}) + c(b_{(j)} \rightarrow \phi \,|\, P(i, j-1)) \end{cases} \quad (5)$$

Here, $A_{<1,i>} = [a_{(1)}, a_{(2)}, \cdots, a_{(i)}]$ and $B_{<1,j>} = [b_{(1)}, b_{(2)}, \cdots, b_{(j)}]$ denote two partial point strings from $a_{(1)}$ to $a_{(i)}$, and from $b_{(1)}$ to $b_{(j)}$, respectively. $c(\cdot\,|p)$ indicates the cost for a point edit operation whose point pair of the latest snapping operation is p. As explained earlier, the above cost functions need the point pairs for the latest snapping operation. To support this information when calculating $C(A_{<1,i>}, B_{<1,j>})$, matrix P is constructed in Figure 5. A distance constraint for the snapping operation to improve the matching accuracy is added; the distance between the corresponding points is less than Th. This parameter is the same as the buffer distance in the previous step.

However, the point pairs for the snapping edit operations cannot be directly obtained from the total cost. They are obtained from an edit operation matrix T, where $T(i,j)$ denotes the latest chosen edit operation of $C(A_{<1,i>}, B_{<1,j>})$. From this matrix, the optimal edit sequence for $C(A_{<1,i>}, B_{<1,j>})$ can be obtained by using a backtracking analysis from $T(i,j)$. This finds the previously chosen edit operation among $(a_{(i)} \rightarrow \phi)$ from $T(i-1,j)$, $o(b_{(j)} \rightarrow \phi)$ from $T(i,j-1)$ and $o(a_{(i)}, b_{(j)} \rightarrow v)$ from $T(i-1,j-1)$ [7]. Thus, starting from $T(n,m)$, the point edit operation sequence to $T(0,0)$ is traced and the corresponding point pairs are obtained as the pairs for which the snapping operations are chosen.

```
C(0,0) = 0;   T(0,0) = null;   P(0,0) = (0,0)        //initialization
for i = 1 to n
  C(i,0) = C(i-1,0)+ c(a(i) →ϕ|P(i-1,0))
  T(i,0) = o(a(i) →ϕ);  P(i,0) = P(i-1,0)
end
for j = 1 to m
  C(0,j) = C(0,j-1)+ c(b(j) →ϕ|P(0,j-1))
  T(0,j) = o(b(j) →ϕ);  P(0,j) = P(0,j-1)
end
for i = 1 to n
  for j = 1 to m

    if dist(a(i),b(j)) ≤ Th                       //distance constraint
      C1 = C(i-1,j-1)+ c(a(i),b(j) →v|P(i-1,j-1))
    else
      C1 = ∞
    end
    T1 = o(a(i),bj →v);  P1(i,j) = ½(a(i)+b(j))

    C2 = C(i-1,j)+ c(a(i) →ϕ|P(i-1,j))
    T2 = o(a(i) →ϕ);    P2(i,j) = P(i-1,j)

    C3 = C(i,j-1)+ c(b(j) →ϕ|P(i,j-1))
    T3 = o(b(j) →ϕ);    P3(i,j) = P(i,j-1)

    k = argmin(C1, C2, C3)              //cost minimization
    C(i,j) = Ck;  T(i,j) = Tk;  P(i,j) = Pk
  end
end
```

Figure 5. Pseudo code for string matching in this study.

2.3. Local Map Transformation and Matching for Disconnected Line Segments

The point strings extracted in Step 1 are divided into sub-point strings and are locally aligned by a rigid transformation according to the average position of corresponding points as shown in Figure 6. Although the string matching in this study finds most of the corresponding point pairs between common boundaries of adjacent spatial datasets, it does not work well for points of disconnected lines segments as shown in Figure 6a. This is because the string matching in this study searches the edit sequence to align two point strings with the minimum cost. Generally, points on the disconnected line segments are incident to almost straight segments and their removing operation often causes a smaller amount of deformation compared with their snapping operation. Therefore, it often misses the corresponding point pairs of the disconnect line segments. To solve this problem, the mutually closest point pairs within a tolerance distance tol are also chosen in post-processing as shown in Figure 6a. Meanwhile, there are no point pairs within tol in Figure 6b,c.

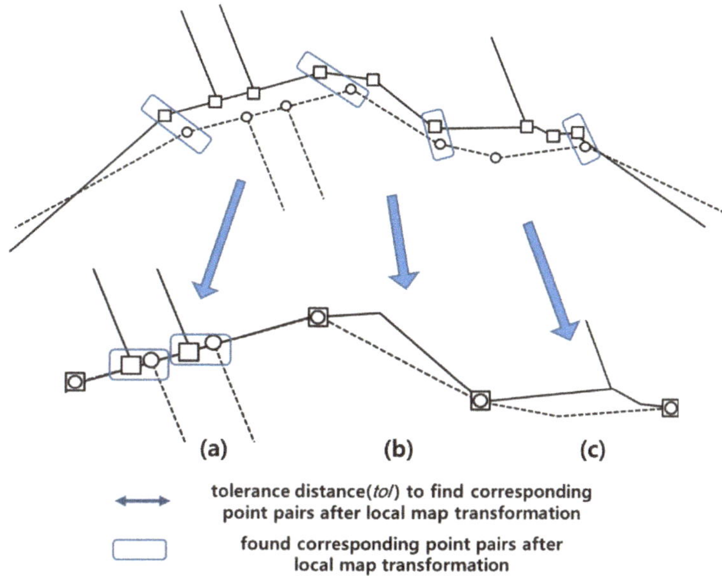

Figure 6. Local map transformation of corresponding sub-point string pair; (**a**) two corresponding point pairs are found after transformation, (**b,c**) no additional corresponding point pairs are found and there remain spatial inconsistences after transformation.

2.4. Map Alignment to Remove Gap and Overlap

After the above map transformation, there are three types of geometric differences: linear difference (Figure 7a), triangle difference (Figure 7b) and polygonal difference (Figure 7c). The linear difference is a case when every point for the sub-point strings has its own corresponding points so that snapping these point pairs to their average positions removes their geometric differences. The triangle difference is a case when one sub-point string is a line segment and the other is a polyline with two line segments. In this case, an aligned boundary edge is obtained by connecting the snapped corresponding points via the centroid of the triangle.

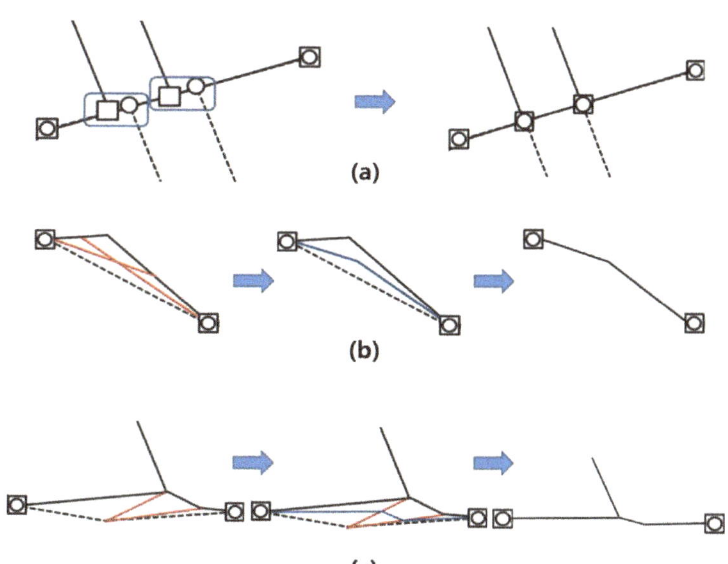

Figure 7. Map alignment method for geometric difference case in this study; (**a**) linear difference, (**b**) triangle difference and (**c**) polygon difference.

The polygon difference is a complicated problem compared with the above two cases. In shape analysis, this difference is studied as a skeleton problem that finds a thin version of a shape that is equidistant to its boundaries. Among skeleton methods, a constrained Delaunay triangulation (CDT) method is applied to obtain an aligned boundary edge. CDT is a generalization of the Delaunay triangulation that forces certain line segments into the edges of triangulation. In this study, the aforementioned constraint is imposed on the line segments of the transformed boundary edge. Then, an aligned boundary edge is obtained by connecting snapped corresponding points via the mid-points of two internal edges of each triangle. The details for this CDT are explained in [15].

Sometimes, transformed corresponding sub-point strings intersect as shown in Figure 8. In this case, these strings are divided according to the intersection points and the above map alignment method is applied for each section by assuming that these intersection points are snapped corresponding points.

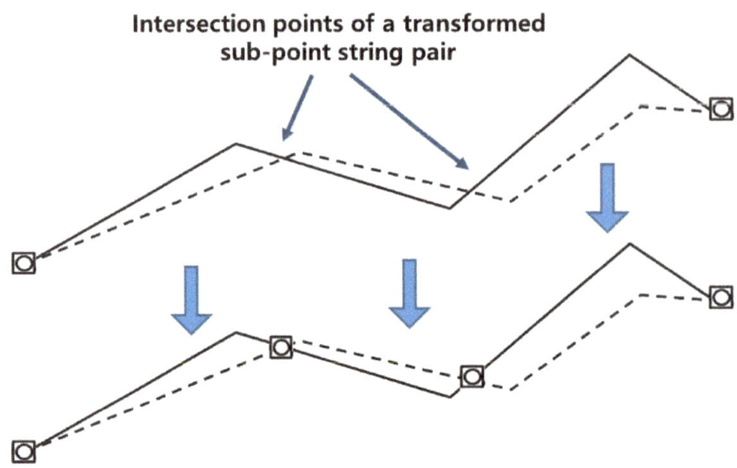

Figure 8. Polygon difference with intersection points of a transformed sub-point string pair.

3. Experiment and Discussion

3.1. Dataset and Parameter Determination

We applied the proposed method to two adjacent cadastral maps, as shown in Figure 9. Map A is a cadastral map of the Geum-Cheon district in Seoul city, and map B is a cadastral map of the Gwang-Myeong city in the Gyeong-gi province. The length of the border area is approximately 10 km. Because these maps are created by joining the respective legacy parcel maps and they are maintained independently by local authorities, irregular positional discrepancies arise between the boundary edges of the maps.

The proposed method has three parameters: Th, λ and tol. Among them, Th is determined by a statistical analysis of 351 corresponding point pairs in the training area of Figure 9. We apply the boxplot method [16] to determine Th because the threshold should be a feasible upper limit of the lengths for the corresponding point pairs. The method begins by finding the median of the training data and then doing the same for each of the halves. These upper and lower quartiles define the centre box and are often referred to as the upper hinge (UH) and lower hinge (LH), respectively. The upper inner fence (UIF) is defined as an upper fence of the box that is extended by 1.5 times the length of the box towards the maximum, and the upper whisker (UW) is defined as the farthest observation inside the UIF, as expressed by Equations (6) and (7), respectively, and as used for Th. UW is used for Th, and it has a value of 8.89 m as calculated from the pairs in the training data, as shown in Figure 10.

$$UIF = UH + 1.5 \cdot (UH - LH) \quad (6)$$

$$UW = \max(x \mid x < UIF) \quad (7)$$

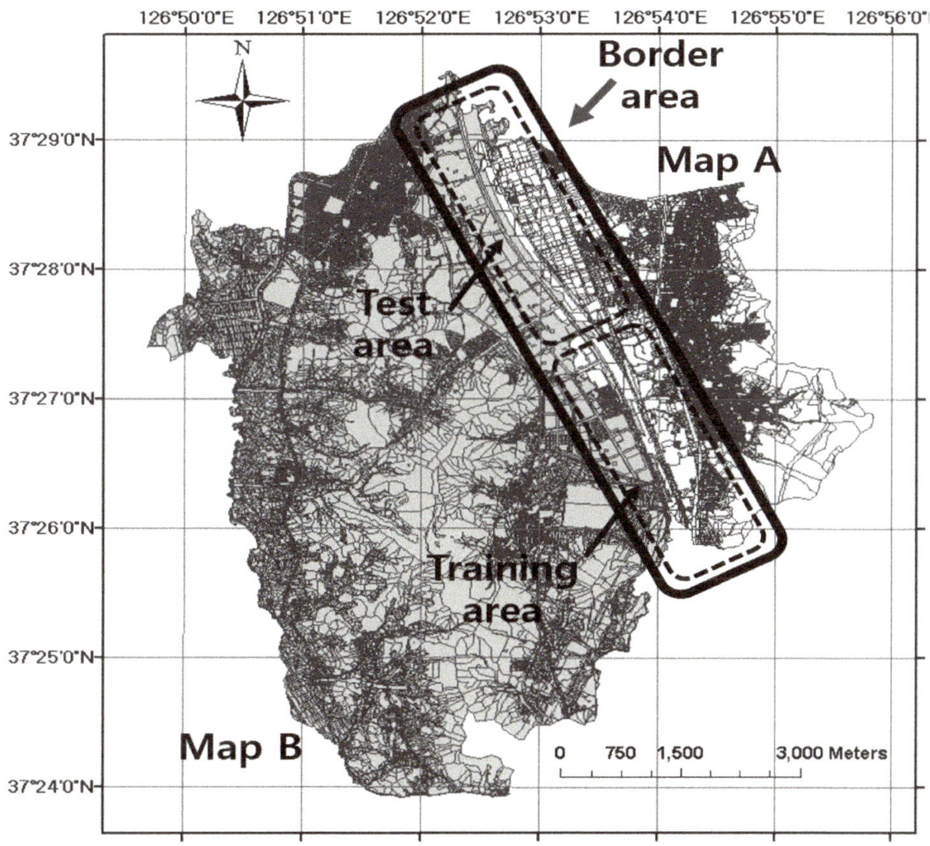

Figure 9. Two adjacent cadastral maps (map A and map B) for the experiments.

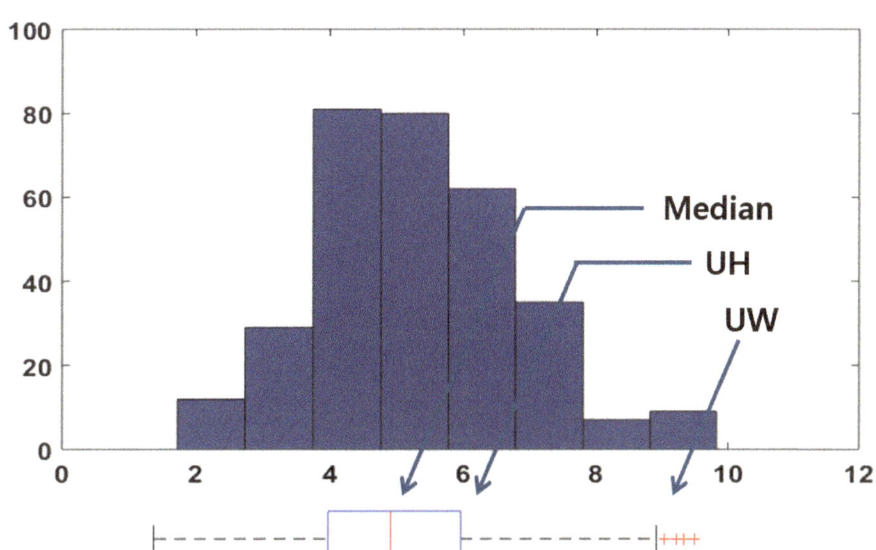

Figure 10. A histogram and boxplot analysis of manually chosen corresponding point pairs.

Meanwhile, the remaining parameters λ and tol cannot be directly trained by this analysis. Thus, various values of the parameters are evaluated and then the optimal parameters with the highest level of matching accuracy are obtained. We used three types of measures for the accuracy: precision, recall and the F-measure. Precision refers to the ratio of correctly found pairs over the total number of found pairs and recall denotes the ratio of correctly found pairs to the total number of correct pairs. The F-measure is defined by Equation (8). In this equation, P and R represent precision and recall, respectively.

$$F - Measure = \frac{2 \cdot P \cdot R}{P + R} \tag{8}$$

We applied 21 candidate λs from 0 to 1 with 0.05 intervals and five candidate tols from 1 m to 9 m with 2 m intervals. The highest precision (0.924) was obtained when λ and tol were 0.95 and 5 m, respectively, as shown in Figure 11a. In addition, the highest recall (0.887) was obtained when λ and tol were 0.25 and 7 m, respectively, as shown in Figure 11b. According to λ, precision and recall presented a trade-off relationship. When λ increased and became closer to one, the degree of precision increased. Meanwhile, when λ decreased and became closer to zero, the degree of recall increased. Thus, λ serves as a matching threshold for the proposed method. In general, a tighter threshold presents a smaller number of matching pairs with higher precision and lower recall, whereas a looser threshold presents a larger number of matching pairs with lower precision and higher recall. In this study, the proposed method uses the snapping and removing edit operations. In addition, one of the operations that presents the minimum total cost is chosen for a given point. Accordingly, a matching result with high precision and low recall according to a large λ indicates an increase in the cost of the snapping operation. Meanwhile, a matching result with low precision and high recall according to a small λ indicates a decrease in the cost of the snapping operation. Compared with λ, tol does not have a meaningful effect on the matching performance. As a result, the highest F-measure (0.902) was obtained when λ and tol were 0.25 and 5 m, respectively, as shown in Figure 11c.

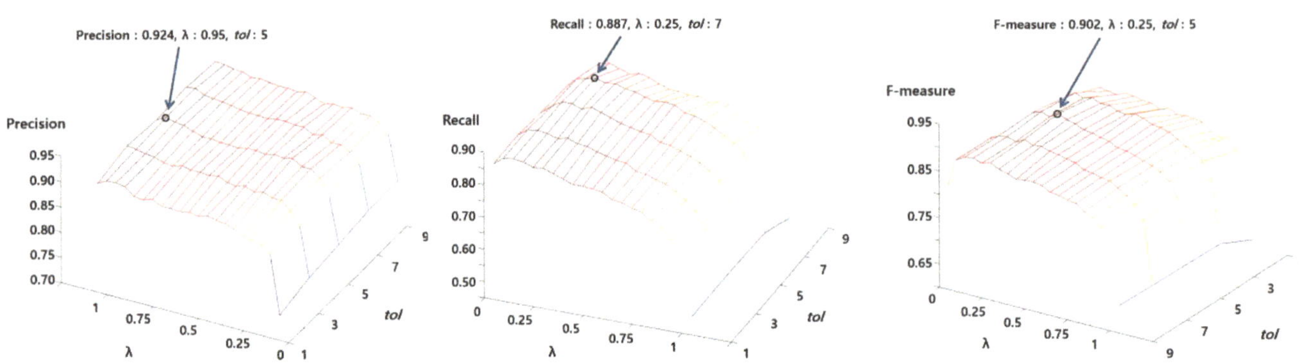

(a) (b) (c)

Figure 11. Accuracy evaluations of the proposed method according to the two parameters λ and tol: (**a**) precision, (**b**) recall and (**c**) the F-measure.

3.2. Result and Discussion

To compare the performance and find the corresponding point pairs for edge matching, a statistical evaluation of the proposed method and a distance threshold method is performed in the test area. As shown in Table 1, precision, recall and F-measure values of the distance threshold method with *Th* of 8.89 m were 0.761, 0.927 and 0.835, respectively. Meanwhile, the proposed method with the parameters determined in the previous section showed higher precision and F-measure values; however, they also showed a lower recall value. These findings indicate that the proposed method has a tighter matching criterion and that the overall matching result of the proposed method was better than that of the conventional distance threshold method in terms of the F-measures.

Table 1. The statistical evaluation of the proposed method and a distance threshold method in the test area in Figure 9.

	The Proposed Method with *Th*: 8.89 m, λ:0.25 and *tol*: 5 m	**A Distance Threshold Method with *Th*: 8.89 m**
Precision	0.912	0.761
Recall	0.895	0.927
F-measure	0.903	0.835

A higher precision for the proposed method was obtained especially from points on disconnected line segments, as shown in Figure 12b,c. When the distances between these points are sufficient compared with the positional discrepancies between adjacent maps, as shown in Figure 12a, the two methods presented nearly the same results. However, the distances for the erroneous point pairs are coincidently shorter than those of true point pairs, as shown in Figure 12b, which shows that the distance threshold method was vulnerable to this problem and presented an inaccurate result with low precision. This problem became even more severe for the points on parallel disconnected line segments, which describe the roads as shown in Figure 12c. Occasionally, the widths of these roads are not sufficient compared with the positional discrepancies between the maps. Therefore, many erroneous point pairs can be obtained by the distance threshold method because the closest points are simply chosen for the pairs regardless of the neighbouring auto-correlated positional discrepancies. Meanwhile, the proposed method performs a matching process by considering these discrepancies; hence, it presents an improved matching performance in terms of precision. Figure 13 is the result of map alignment with these corresponding point pairs.

Compared with the higher degree of precision, the proposed method presented lower recall, as shown in Table 1. This occurred because it tends to choose corresponding point pairs between salient corner points rather than those on nearly straight line segments, as shown in Figure 14a. The proposed method determines corresponding point pairs as the point pairs for which the snapping operation is chosen in the optimal sequence with which two point strings are aligned at the minimum total cost. In general, the removing operation of a salient corner point causes a considerable amount of shape deformation on the line segments involved, which results in a larger edit operation cost than that of the snapping operation. This relationship is opposite for the points connected by nearly straight line segments. Thus, as shown in the previous figures, the proposed method accurately found the corresponding point pairs between salient corner areas with a higher degree of recall accuracy. However, except for those disconnected line segments, the corresponding point pairs between points connected by nearly straight line segments were not sufficiently found because the choices in removing operations for them present a lower cost. This property prevents irregular stretching or shrinking of the neighbouring border areas, as shown in Figure 14b. The proposed method found the corresponding point pairs between salient corner areas by allowing gradual map alignments along the border area to be obtained. However, when all of the mutually closest points are used, the map alignment result by itself could be abrupt, which decreases the overall performance. Based on the above comparisons to the manually chosen corresponding point pairs and the matching property, the proposed method is superior to previous distance threshold methods.

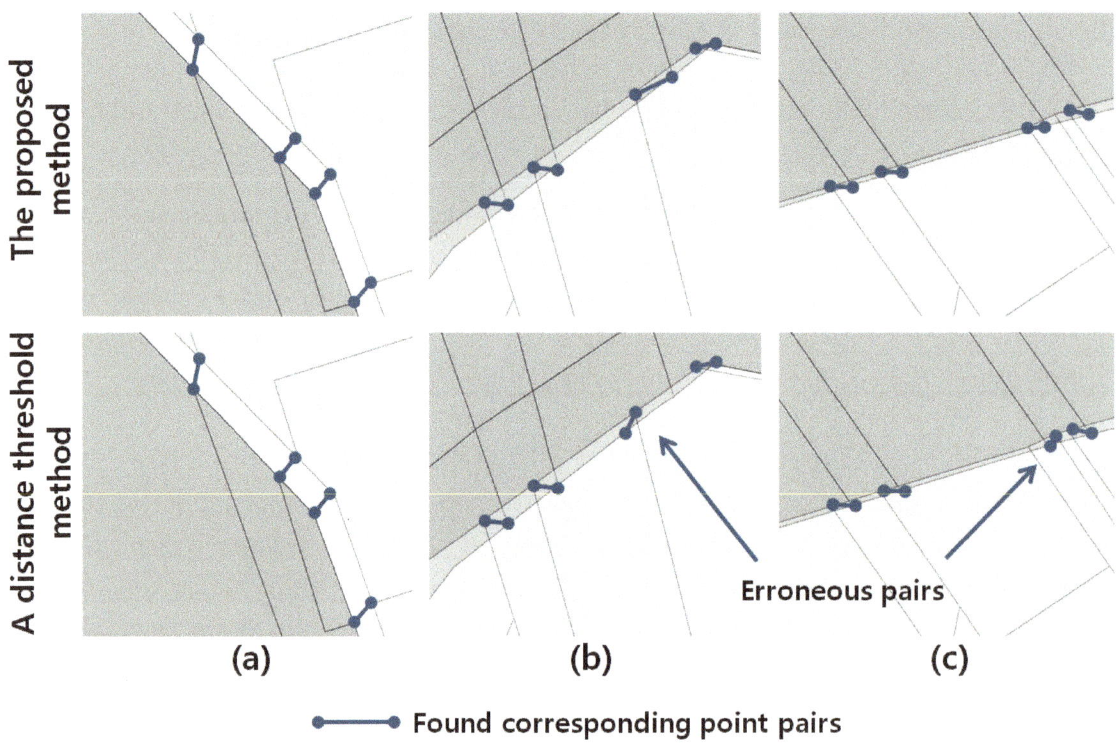

Figure 12. Comparison of corresponding point pairs found by the proposed method and the distance threshold method; (**a**) both methods find correct pairs, (**b,c**) the proposed method finds correct pairs of disconnected line segments between two maps meanwhile the distance threshold methods finds erroneous pairs.

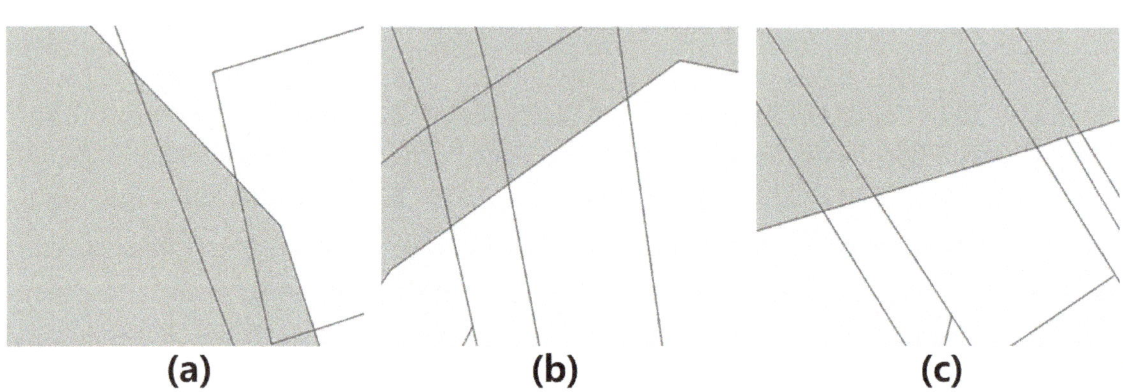

Figure 13. Map alignment to remove gap and overlap of Figure 12; (**a**) alignment result of Figure 12a, (**b**) alignment result of Figure 12b, (**c**) alignment result of Figure 12c.

Additionally, it is necessary to discuss the effect of point string order. In Section 2.1, the orders were determined to be counter-clockwise. However, according to the order of whether it is counter-clockwise or clockwise, matching results for the corresponding point pairs can be different. Although both results are nearly the same, this cannot be explained at this stage and further research is necessary.

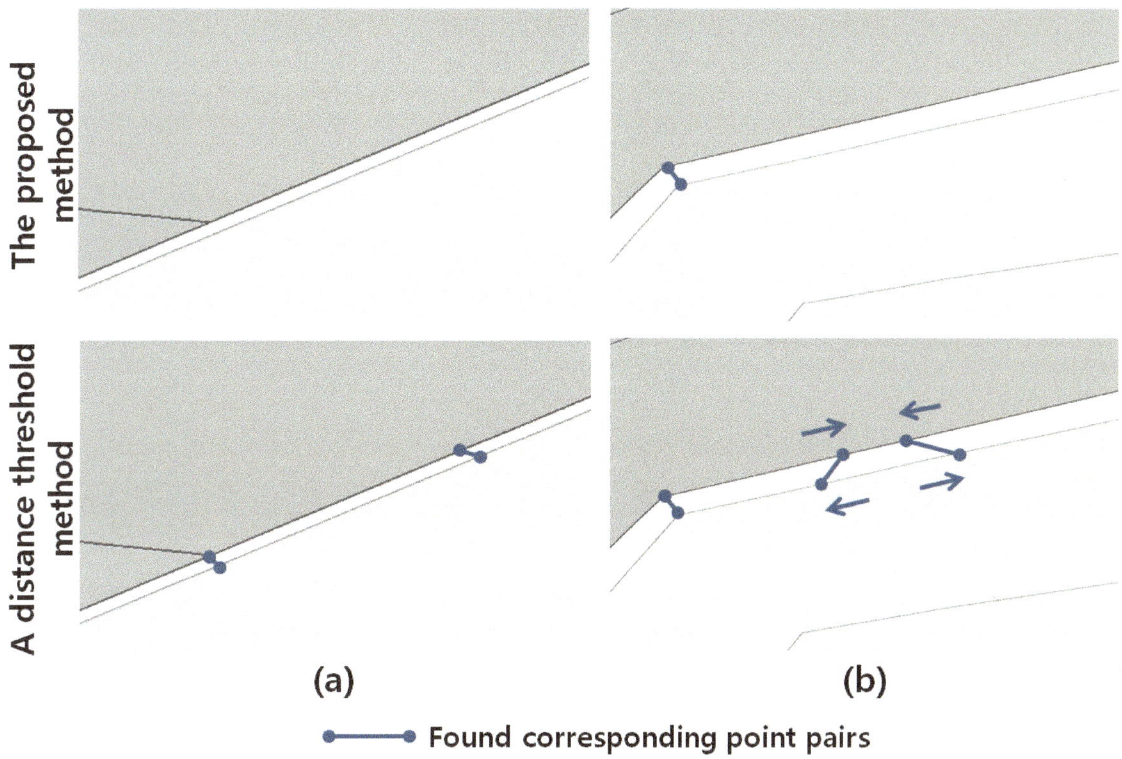

Figure 14. Comparison of corresponding point pairs found by the proposed method and the distance threshold method for straight line segments; (**a**) the proposed method does not find corresponding point pairs between points on nearly straight line segments meanwhile the distance threshold method finds correct pairs, (**b**) the distance threshold method finds corresponding point pairs which result irregular stretching or shrinking after map alignment meanwhile the proposed method does not cause such problem.

4. Conclusions

This paper proposed a new method to find corresponding point pairs for common boundary edge matching between adjacent spatial datasets by using a string matching technique with three types of point edit operations. These three operations snapped two points of the strings to their average position, removed a point of one string, and removed a point of the other string. Unlike previous distance threshold methods, the proposed method can consider local auto-correlated positional discrepancies because the choice in edit operation changes the geometries of the boundary line segments involved, after which the changed geometries affect the remaining points' edit costs. Thus, the proposed method presented an improved matching performance compared with a previous method especially for the precision measure. This demonstrates that the proposed method finds more robust corresponding point pairs. However, it found fewer pairs from true corresponding point pairs than the previous method because according to the proposed cost functions, the costs of removing points connected by nearly straight line segments are generally lower than the costs of snapping the points. For this reason, the proposed method tends to find corresponding pairs between salient corner points and not between points connected by nearly straight line segments. However, this property results in gradual map alignments along the border area by preventing the irregular stretching or shrinking of neighbouring borders. As a result, the proposed local edge matching method showed an improved ability to find corresponding point pairs for seamless adjacent spatial datasets.

Acknowledgments

This research was supported by the Ministry of Land, Infrastructure, and Transport in Korea under the Urban Planning and Architecture (UPA) research support program and was supervised by the Korea Agency for Infrastructure Technology Advancement (KAIA) (13 Urban Planning & Architecture 02).

Conflicts of Interest

The author declares no conflict of interest.

References

1. Beard, M.; Chrisman, N. Zipper: A localized approach to edge matching. *Am. Cartogr.* **1988**, *15*, 163–172.
2. Coren, I.; Doytsher, Y. Cartographic edge matching: Mapping the problem towards a computer based solution. In Proceedings of the 1998 ACSM annual convention, Baltimore, MD, USA, 28 February–5 March 1998.
3. Gregory, I.; Ell, P. Error sensitive historical GIS: Identifying areal interpolation errors in timeseries data. *Int. J. Geogr. Inf. Sci.* **2006**, *20*, 135–152.
4. Lolonis, P. A Chi-square statistic for checking satisfactory edge matching of maps and diagrams that depict adjacent areas. In Proceedings of FIG XXII International Congress, Washington, DC, USA, 19–26 April 2002.
5. Masuyama, A. Methods for detecting apparent differences between spatial tessellations at different time points. *Int. J. Geogr. Inf. Sci.* **2006**, *20*, 633–648.
6. Ledoux, H.; Ohori, K. Edge-matching polygons with a constrained triangulation. In Proceedings of Symposium GIS Ostrava 2011, Ostrava, Czech, 24–26 January 2011; pp. 377–390.
7. Huh, Y.; Yu, K.; Heo, J. Detecting conjugate-point pairs for map alignment between two polygon datasets. *Environ. Urban Syst.* **2011**, *35*, 250–262.
8. Cho,S; Punithan, M.; Gim, J.; Huh, Y. Tagging-the-triangle algorithm for partitioning features with inconsistent boundaries. *Int. J. Geogr. Inf. Sci.* **2014**, *28*, 2533–2550.
9. Kruskal, J.B. An overview of sequence comparison: Time warps, string edits, and macromolecules. *SIAM Rev.* **1983**, *25*, 201–237.
10. Kaygin, S.; Bulut, M. Shape recognition using attributed string matching with polygon vertices as the primitives. *Pattern Recognit. Lett.* **2002**, *23*, 287–294.
11. Maître, H.; Wu, Y. Improving dynamic programming to solve image registration. *Pattern Recognit.* **1987**, *20*, 443–462.

12. Wang, C.; Stefanidis, A.; Croitoru, A.; Agouris, P. Map registration of image sequences using linear feature. *Photogramm. Engin. Remote Sens.* **2008**, *74*, 25–38.
13. Huh, Y.; Yang, S.; Ga, C.; Yu, K.; Shi, W. Line segment confidence region-based string matching method for map conflation. *ISPRS J. Photogramm. Remote Sens.* **2013**, *78*, 69–84.
14. Bunke, H.; Buhler, U. Applications of approximate string matching to 2D shape recognition. *Pattern Recognit.* **1993**, *26*, 1797–1812.
15. Zou, J.J.; Chang, H.H.; Yan, H. Shape skeletonization by identifying discrete local symmetries, *Pattern Recognit.* **2001**, *34*, 1895–1905.
16. Reimann, C.; Filzmoser, P.; Garrett, R.G. Background and threshold: Critical comparison of methods of determination. *Sci. Total Environ.* **2005**, *345*, 1–16.

Permissions

All chapters in this book were first published in ISPRSIJGI, by MDPI; hereby published with permission under the Creative Commons Attribution License or equivalent. Every chapter published in this book has been scrutinized by our experts. Their significance has been extensively debated. The topics covered herein carry significant findings which will fuel the growth of the discipline. They may even be implemented as practical applications or may be referred to as a beginning point for another development.

The contributors of this book come from diverse backgrounds, making this book a truly international effort. This book will bring forth new frontiers with its revolutionizing research information and detailed analysis of the nascent developments around the world.

We would like to thank all the contributing authors for lending their expertise to make the book truly unique. They have played a crucial role in the development of this book. Without their invaluable contributions this book wouldn't have been possible. They have made vital efforts to compile up to date information on the varied aspects of this subject to make this book a valuable addition to the collection of many professionals and students.

This book was conceptualized with the vision of imparting up-to-date information and advanced data in this field. To ensure the same, a matchless editorial board was set up. Every individual on the board went through rigorous rounds of assessment to prove their worth. After which they invested a large part of their time researching and compiling the most relevant data for our readers.

The editorial board has been involved in producing this book since its inception. They have spent rigorous hours researching and exploring the diverse topics which have resulted in the successful publishing of this book. They have passed on their knowledge of decades through this book. To expedite this challenging task, the publisher supported the team at every step. A small team of assistant editors was also appointed to further simplify the editing procedure and attain best results for the readers.

Apart from the editorial board, the designing team has also invested a significant amount of their time in understanding the subject and creating the most relevant covers. They scrutinized every image to scout for the most suitable representation of the subject and create an appropriate cover for the book.

The publishing team has been an ardent support to the editorial, designing and production team. Their endless efforts to recruit the best for this project, has resulted in the accomplishment of this book. They are a veteran in the field of academics and their pool of knowledge is as vast as their experience in printing. Their expertise and guidance has proved useful at every step. Their uncompromising quality standards have made this book an exceptional effort. Their encouragement from time to time has been an inspiration for everyone.

The publisher and the editorial board hope that this book will prove to be a valuable piece of knowledge for researchers, students, practitioners and scholars across the globe.

List of Contributors

Ryan Hile and Thomas J. Cova
Center for Natural and Technological Hazards, University of Utah, 260 S. Central Campus Dr. Rm. 270, Salt Lake City, UT 84112-9155, USA
Department of Geography, University of Utah, 260 S. Central Campus Dr., Rm. 270, Salt Lake City, UT 84112-9155, USA

Yongjun Yang, Yaqin Sun, Kuoyin Wang, Huping Hou and Shishuo Xu
School of Environment Science and Spatial Informatics, China University of Mining and Technology, Daxue Road No.1, Xuzhou 221116, China

Songnian Li
School of Environment Science and Spatial Informatics, China University of Mining and Technology, Daxue Road No.1, Xuzhou 221116, China
Department of Civil Engineering, Ryerson University, 350 Victoria St., Toronto, ON M5B 2K3, Canada

Shaoliang Zhang
Low Carbon Energy Institute, China University of Mining and Technology, Daxue Road No.1, Xuzhou 221116, China

Michal Lichter, Asher Yair Grinberger and Daniel Felsenstein
Department of Geography, Hebrew University of Jerusalem, Mount Scopus 91900, Israel

Jon D. Blower, Charles J. Roberts, Guy H. Griffiths and Jane P. Lewis
Department of Meteorology, Harry Pitt Building, University of Reading, Reading RG6 6BB, UK

Joan Masó and Daniel Díaz
Grumets Research Group, CREAF, Edifici C, Universitat Autònoma de Barcelona, 08193 Bellaterra, Catalonia, Spain

Xiaoyu Yang
Computer Network Information Center, Chinese Academy of Sciences, Beijing 100190, China

Xavier Pons
Grumets Research Group, Dep Geografia, Edifici B, Universitat Autònoma de Barcelona, 08193 Bellaterra, Catalonia

Jihong Dong, Wenting Dai, Guoquan Shao and Jiren Xu
School of Environment and Spatial Informatics, China University of Mining and Technology, Xuzhou 221116, China

Wei Tu
College of Information Engineering, Shenzhen University, Shenzhen 518060, China
Shenzhen Key Laboratory of Spatial Information Smart Sensing and Services, College of Civil Engineering, Shenzhen University, Shenzhen 518060, China
Key Laboratory for Geo-Environmental Monitoring of Coastal Zone of the National Administration of Surveying, Mapping and GeoInformation, Shenzhen University, Shenzhen 518060, China

Qingquan Li
Shenzhen Key Laboratory of Spatial Information Smart Sensing and Services, College of Civil Engineering, Shenzhen University, Shenzhen 518060, China
Key Laboratory for Geo-Environmental Monitoring of Coastal Zone of the National Administration of Surveying, Mapping and GeoInformation, Shenzhen University, Shenzhen 518060, China
State Key Laboratory of Information Engineering in Surveying, Mapping, and Remote Sensing, Wuhan University, Wuhan 430079, China

Zhixiang Fang
State Key Laboratory of Information Engineering in Surveying, Mapping, and Remote Sensing, Wuhan University, Wuhan 430079, China

Baoding Zhou
Shenzhen Key Laboratory of Spatial Information Smart Sensing and Services, College of Civil Engineering, Shenzhen University, Shenzhen 518060, China
Key Laboratory for Geo-Environmental Monitoring of Coastal Zone of the National Administration of Surveying, Mapping and GeoInformation, Shenzhen University, Shenzhen 518060, China

Alexander Fekete and Katerina Tzavella
Institute of Rescue Engineering and Civil Protection, Cologne University of Applied Sciences, Betzdorfer Str. 2, Köln 50679, Germany

Iuliana Armas
Faculty of Geography, Department of Geomorphology-Pedology-Geomatics, University of Bucharest, NicolaeBalcescu 1, Sector 1, 010041 Bucharest, Romania

Jane Binner
Department of Finance, Business School, University House, University of Birmingham, Birmingham B15 2TT, UK

Matthias Garschagen
Institute for Environment and Human Security (UNU-EHS), United Nations University, Bonn D-53113, Germany

Carlo Giupponi and Vahid Mojtahed
Venice Centre for Climate Studies, Department of Economics, Ca' Foscari University, Venice 30121, Italy

Marcello Pettita
Amigo S.R.L., via flaminia 48, Roma 00196, Italy
European Academy-EURAC, Institute for Applied Remote Sensing, European Academy of Bozen/Bolzano, Viale Druso 1, Bolzano 39100, Italy

Stefan Schneiderbauer
European Academy-EURAC, Institute for Applied Remote Sensing, European Academy of Bozen/Bolzano, Viale Druso 1, Bolzano 39100, Italy

Damien Serre
UMR ESPACE 7300 CNRS, Université d'Avignon et des Pays du Vaucluse, Case 19, 74 rue Louis Pasteur, 84029 Avignon Cedex, France

Shasha Lu
School of Economics and Management, Beijing Forestry University, No. 35, Tsinghua East Road, Beijing 100083, China

Xingliang Guan and Linjie Zhou
National Academy for Mayors of China, Huixin West Street, Beijing 100029, China

Danlin Yu
Department of Earth and Environmental Studies, Montclair State University, Montclair, NJ 07043, USA

Yu Deng
Institute of Geographic Sciences and Natural Resources Research, Chinese Academy of Sciences, Datun Road, Beijing 100101, China

Thorsten Förster and Agnes Mainka
Department of Information Science, Heinrich-Heine-University Düsseldorf, Universitätsstraße 1, 40225 Düsseldorf, Germany

Wen Luo and Sheng-Jun Xiao
Key Lab of Virtual Geographic Environment, Ministry of Education, Nanjing Normal University, Nanjing 210023, China

Zhao-Yuan Yu, A-Xing Zhu and Lin-Wang Yuan
Key Lab of Virtual Geographic Environment, Ministry of Education, Nanjing Normal University, Nanjing 210023, China
State Key Laboratory Cultivation Base of Geographical Environment Evolution (Jiangsu Province), Nanjing 210023, China
Jiangsu Center for Collaborative Innovation in Geographical Information Resource Development and Application, Nanjing 210023, China

Youhei Kawamura, Hajime Nobuhara, Takeshi Shibuya and Itaru Kitahara
Faculty of Engineering, Science and Information Systems, University of Tsukuba, Ibaraki Prefecture 305-8573, Japan

Markus Wagner
School of Computer Science, University of Adelaide, Adelaide, SA 5005, Australia

Hyongdoo Jang
Department of Mining Engineering and Metallurgical Engineering, Curtin University, Perth, WA 6433, Australia

Ashraf M Dewan and Bert Veenendaal
Department of Spatial Sciences, Curtin University, Perth, WA 6433, Australia

Joseph Kandeh
United Nations Office for the Coordination of Humanitarian Affairs (UN OCHA), Sana'a, Yemen

Lalit Kumar
School of Environmental and Rural Science, University of New England, Armidale NSW 2351, Australia

Yong Huh
Spatial Information Research Institute, Korea Cadastral Survey Corp., Seoul 150-911, Korea